D0386384

Climatic Change and Variability

Climatic Change and Variability

A Southern Perspective

Editors: A. B. Pittock   L. A. Frakes
D. Jenssen   J. A. Peterson   J. W. Zillman
on behalf of the
Australian Branch, Royal Meteorological Society

Based on a conference at Monash University, Melbourne, Australia,
7–12 December 1975 which was co-sponsored by the Australian Academy
of Science

CAMBRIDGE UNIVERSITY PRESS
CAMBRIDGE
LONDON·NEW YORK·MELBOURNE

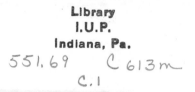

Library
I.U.P.
Indiana, Pa.
551.69   C 613 m
C. 1

Published by the Syndics of the Cambridge University Press
The Pitt Building, Trumpington Street, Cambridge CB2 1RP
Bentley House, 200 Euston Road, London NW1 2DB
32 East 57th Street, New York, NY 10022, USA
296 Beaconsfield Parade, Middle Park, Melbourne 3206, Australia

© Cambridge University Press 1978

First published 1978

Printed in Great Britain at the
University Press, Cambridge

Library of Congress Cataloguing in Publication Data
Main entry under title:
Climatic change and variability.
Based on edited papers from the Australasian Con-
ference on Climate and Climatic Change.
Bibliography: p.
Includes indexes.
1. Climatic changes. 2. Climatic changes –
Southern Hemisphere. I. Pittock, A. Barrie,
1938– II. Royal Meteorological Society, London.
Australian Branch. III. Australasian Conference
on Climate and Climatic Change, Monash University,
1975.
QC981.8.C5C56   551.6'9   76-53521

ISBN 0 521 21562 5

Contents

Contents

Contents

Contents

Contents

Contributors

J. L. Anderson: Department of Economics, Latrobe University, Victoria 3083, Australia

G. W. Arnold: CSIRO Division of Land Resources Management, Western Australia 6014, Australia

R. G. Barry: Institute of Arctic and Alpine Research, University of Colorado, Boulder, Colorado 80302, USA

R. J. Bray: PO Box 494, Nelson, New Zealand

R. A. Bryson: Institute for Environmental Studies, University of Wisconsin, Madison, Wisconsin 53706, USA

W. Budd: Antarctic Division, Melbourne 3004, Australia

J. Chappell: Department of Geography, Australian National University, ACT 2600, Australia

M. Charlesworth: Philosophy Department, Deakin University, Belmont, Victoria 3216, Australia

D. M. Churchill: Royal Botanic Gardens, Melbourne, South Yarra, Victoria 3141, Australia

M. J. Coughlan: Bureau of Meteorology, Melbourne, Victoria 3000, Australia

R. W. Fairbridge: Columbia University, New York 10027, USA

H. Flohn: Meteorological Institute, University of Bonn, Federal Republic of Germany

L. A. Frakes: Department of Earth Sciences, Monash University, Victoria 3168, Australia

J. R. Freney: CSIRO Division of Plant Industry, Canberra, ACT 2601, Australia

I. E. Galbally: CSIRO Division of Atmospheric Physics, Aspendale, Victoria 3195, Australia

K. A. Galbraith: CSIRO Division of Land Resources Management, Western Australia 6014, Australia

R. W. Galloway: Division of Land Use Research, CSIRO, Canberra, ACT 2601, Australia

W. J. Gibbs: Bureau of Meteorology, Melbourne, Victoria 3000, Australia

J. S. Godfrey: CSIRO Division of Fisheries and Oceanography, Cronulla, NSW 2230, Australia

J. S. A. Green: Atmospheric Physics, Imperial College, London SW7 2RH, UK

B. V. Hamon: CSIRO Division of Fisheries and Oceanography, Cronulla, NSW 2230, Australia

S. Hastenrath: Meteorology Department, University of Wisconsin, Madison, Wisconsin 53706, USA

B. G. Hunt: Australian Numerical Meteorology Research Centre, Melbourne, Victoria 3000, Australia

D. Jenssen: Meteorology Department, University of Melbourne, Parkville, Victoria 3052, Australia

B. McInnes: Meteorology Department, University of Melbourne, Parkville, Victoria 3052, Australia

J. V. Maher: Bureau of Meteorology, Melbourne, Victoria 3000, Australia

G. R. Manley: Department of Environmental Sciences, University of Lancaster, Lancaster, UK

H. A. Martin: Department of Botany, University of New South Wales, NSW 2033, Australia

W. J. Maunder: New Zealand Meteorological Service, Wellington, New Zealand

B. R. Morton: Department of Mathematics, Monash University, Victoria 3168, Australia

J. Oliver: Geography Department, James Cook University, Townsville, Queensland 4811, Australia

G. W. Paltridge: CSIRO Division of Atmospheric Physics, Aspendale, Victoria 3195, Australia

G. I. Pearman: CSIRO Division of Atmospheric Physics, Aspendale, Victoria 3195, Australia

J. A. Peterson: Department of Geography, Monash University, Victoria 3168, Australia

A. B. Pittock: CSIRO Division of Atmospheric Physics, Aspendale, Victoria 3195, Australia

R. A. S. Ratcliffe: British Meteorological Office, Bracknell RG12 2SZ, UK

N. J. Shackleton: Subdepartment of Quaternary Research, University of Cambridge, Cambridge CB2 1ER, UK

G. Singh: Research School of Pacific Studies, Australian National University, ACT 2600, Australia

G. B. Tucker: CSIRO Division of Atmospheric Physics, Aspendale, Victoria 3195, Australia

H. van Loon: National Center for Atmospheric Research,[*] Boulder, Colorado 80303, USA

[*] The National Center for Atmospheric Research is sponsored by the National Science Foundation.

List of contributors

D. Walker: Research School of Pacific Studies, Australian National University, ACT 2600, Australia

P. J. Walsh: CSIRO Division of Building Research, Victoria 3190, Australia

J. Williams: International Institute for Applied Systems Analysis, 2361 Laxenburg, Austria

P. B. Wright: Department of Oceanography, University of Hawaii, Honolulu, Hawaii 96822, USA

J. W. Zillman: Bureau of Meteorology, Melbourne, Victoria 3000, Australia

Foreword*

It is a very remarkable thing that climatology faces such an important, possibly a dramatic future, when it is the oldest science we have. Though astronomy rivals this latter claim, it was climate that decided early man where and how to live and when to migrate. He came to study the heavens *because* he had learnt that the passage of the sun across them determined the day, and that variations in its trajectory determined the seasons. Since the big light composed the theme of the seasons, perhaps the little lights would be found to compose the variations. Probably this was man's first scientific speculation.

Over such a long period, from then till now, effort has waxed and waned. The most recent stagnant period was that following the development of meteorological instruments, when the thought reigned that one had merely to amass enough years of simple figures, and the averages would describe the climate.

However, about 100 years ago, a dynamic and synoptic climatology began to emerge, based on the tracks and centres of action of the synoptic systems. For the Southern Hemisphere, names like Russell, Kidson and Griffith Taylor provide the landmarks. There was a classification climatology too, of which the architect was Köppen. But it was only within the working span of old-timers like myself, that we have seen the beginnings of a physical climatology.

Compared with its sedate history, the present scene in climatology is one of bustle. We recognise that climate is a many-splendoured thing. And yet we have no accepted strict definition of what the word climate means. No wonder then that some people's concept of what can be called climatic change differs widely from others.

Unfortunate climatic episodes, sequences of years with drought or flood, are an inevitable consequence of the year-to-year variability, and need not by any means imply that the climate is changing. The upward trend in world population and communications, with increasing use of marginal land, means that a climatic episode today tends to cause much more suffering and attract wider notice than a similar episode of the past. While there are great uncertainties about climate changes being on the way,

* Adaptation of Opening Address at Australasian Conference on Climate and Climatic Change, Monash University, 7 December 1975

there is no uncertainty about one thing. The world population must learn to live more successfully with the year-to-year variability. One would like to see much more work, by economists, social scientists, agriculturalists, and subsequently educationists, on how man is best to live with this variability. If they could quantify what they have to say, this could do much to avoid and alleviate suffering.

These social and economic questions lead us into the future, where the other big questions, and I stress that they are two different questions, are the predictability of climatic change and the predictability of the year-to-year variation. It is very tempting for we meteorologists, who tend to get hauled over the coals if a weather forecast is proved wrong, as happens very occasionally, to take up climatic predictions. Then we will only be proved wrong after we are dead. But this is a real challenge to our sense of responsibility as a profession. There have been rather too many claims of recent years, made on too little evidence.

Those who have read Shakespeare's plays will know that he often starts a scene with what he calls 'alarums and excursions', and those who have seen these plays know that this means trumpets blowing, bells ringing and people rushing here, there and everywhere. Environmental science, of which climatology is an important part, has a weakness for writing some of its scenarios in the same way, and this is an image which the mass media love to project, because it is good theatre. But it is not good science, and others of us may breathe more freely about climatic change because we hope that we are now through the alarums and excursions, and settling in to Act 2, Scene 1.

There is a consensus that a vital key to the future lies in the numerical model, but here again we must be careful not to overstate the case in order to attract funds. Going round the world during 1975, I found few who felt hopeful of real breakthroughs in climate predictability in less than ten years. On the other hand, ten years' delay of investment now will mean ten years' postponement of ultimate understanding. This is the sort of story which we should be telling to the public. They pay the piper, and have the right to know what tune they are calling.

A second key to the future is also widely considered to lie in the Global Atmospheric Research Programme, the first real test of blending the numerical models with the fuller potential of the new satellite-oriented observing techniques. By supplementing and re-shaping the tools of the future, the ultimate benefits of GARP to the understanding of climate may be immense, particularly in the Southern Hemisphere where the present gaps are so much wider. It is hoped that the scientists of all countries, most of all our own, will succeed in persuading their governments to give the necessary support to this imaginative and vital programme.

Whilst I have pleaded for balance in public utterances, I know that within our own discipline there is plenty of room for controversy in this complex the vitally important subject, and a little extremism here can serve a very useful purpose. It remains for me to congratulate our Branch of the Royal Meteorological Society on its initiative in calling the original conference, which the Academy was delighted to co-sponsor and which led to this publication. The great mathematician, John von Neumann, described the general circulation of the atmosphere, that is to say the quantitative machinery of the climate, as the most difficult unsolved problem still to confront the scientific intellect of man, and his words strike a suitable note on which to introduce this book.

C. H. B. PRIESTLEY
Australian Academy of Science

Preface

The study of climatic change and variability involves a vast range of disciplines, interests and techniques. It is unfortunate, therefore, that it has suffered in the past from a lack of effective communication across disciplinary barriers. Thus, for example, the geological literature contains many examples of sound geology of palaeoclimatic interest which is nevertheless coupled with inadequate meteorological interpretation. Similarly, too much of the meteorological literature, including the new breed of numerical modelling, is lacking in perspective as to the reality of what has happened in the past (and *ipso facto* is possible). Rigorous interdisciplinary collaboration is lacking in most areas of study, although the CLIMAP programme (CLIMAP Project Members, 1976) is a notable exception.

We aim therefore, in this book as in the conference on which it is based, to bridge these barriers. We have sought to exclude jargon, and to avoid technical details better found in the specialised literature to which there is ample reference. In so doing, we hope to have produced a book intelligible to the educated layman and particularly to tertiary students and specialists in related fields. We hope it will open some windows, and even a few doors, to stimulating ideas and interactions across disciplinary and institutional barriers.

The book arises out of, but is not a straight proceedings of, a conference on the theme 'Climatic change and variability, with particular reference to the Australian or Southern Hemisphere region', which was held at Monash University in December 1975. The aim of the conference was 'to bring together people particularly from the Australasian region interested in questions of climate and climatic change so that they may familiarise themselves with the broad scope of the field, its relevance to their individual needs, interests and disciplines and interest others in the contribution they can make to the broader fields'.

The conference largely succeeded in its aim, with 94 papers and 224 participants including a number from overseas. We believe the original aim will be further achieved by the publication of this book based on edited versions of the invited review papers, supplemented by a small selection of heavily edited versions or amalgams of contributed papers. These latter contributions have been selected because they usefully illustrate or round off themes developed in the invited papers, and describe either some major new development from a Southern Hemisphere worker or some unique insights to be found from a study of Southern Hemisphere climate.

The Southern Hemisphere theme is further expounded in Chapter 1. Suffice it to say here that the study of Southern Hemisphere climate is different from that of the Northern Hemisphere in several important aspects. Not only are there significant climatic differences due to different land–sea distributions, but the smaller land base and different historic/ cultural development of the Southern Hemisphere populations have led to much greater ignorance and neglect of Southern Hemisphere climate. Much has been written about Northern Hemisphere climate, sometimes as if it were the global story, but relatively little is available which focuses on the south.

The atmosphere and oceans are of course not bounded at the equator, and it is both impossible and undesirable to exclude rigorously Northern Hemisphere contributions, data, or results from this volume. Nevertheless we have borne the Southern Hemisphere emphasis in mind while editing this book in the belief that it provides a useful and stimulating focus.

Differing views and inconsistencies exist in the climatological literature, and in the available contributions, on numerous questions, particularly on matters of definition and usage. The editors believe that these should be pointed out and discussed, but not editorially eliminated. To eliminate them would be to present one particular view rather than a balanced over-view of the state of the art. Such differing views will be found most obviously in Chapters 7 and 8, but also in Chapter 4 where the term 'climatic variability' is used by some authors to mean variability *of* climate, and by others to mean variability *within* climate, depending on the time scale used implicitly or explicitly to define 'climate'. This inconsistency within the book has led the editors to accept consciously the ambiguity implicit in its very title: *Climatic Change and Variability*. This may appear to be unduly confusing, but it accurately reflects a confusion which exists far beyond the confines of this book.

One other question that has concerned the editors is whether our theoretical understanding of climatology as a science justifies a logical development in this book from theory to example and practice. The more pessimistic view is that climatology is still essentially an observational science in which the data must be presented first, as a basis from which to examine critically the many competing and largely qualitative hypotheses about the nature and causes of climatic change.

We have chosen a compromise in which the physical framework which we believe underlies climate is first described (Chapter 2), the climatic record is next presented and analysed (Chapters 3 and 4), and then various models of climatic change are discussed (Chapter 5). These are followed by sections on the inadvertent modification of climate (Chapter 6), the economic and social impact of climatic change (Chapter 7), and a final review of progress and prospects.

Much controversy has been generated recently by various hypotheses and statements as to the likely onset of rapid climatic change (e.g. Alexander, 1974; Bryson, 1973 a; Calder, 1974; International Federation of Institutes for Advanced Study, 1974; Lamb, 1974 a). Given the present state of knowledge about the causes and mechanisms of climatic change and variability, many of these must be regarded as somewhat speculative, but they have tended to divide climatologists into those who might be termed climatic optimists on the one hand, and pessimists on the other. Issues of social responsibility and scientific objectivity are raised when positions are taken on such matters which are of obvious importance to humanity.

These questions are touched upon particularly in Chapters 7 and 8. Whatever the long-term trends and prospects, however, even a continuation of the normal year-to-year climatic variability described in the instrumental record of the last 100 years or so (see Chapter 4) will become increasingly critical to human survival as the world population increases, world food reserves are used up, more and more marginal land is brought into agricultural production, and water resources are more fully utilised.

Climatic variability is part of Australian folklore, and it is perhaps instructive as well as amusing to ponder one of the classics of popular Australian literature, the ballad 'Said Hanrahan', by P. J. Hartigan ('John O'Brien', 1879–1952), a Catholic priest who for many years was stationed at Narrandera in south-central New South Wales. Hanrahan is perhaps the archetypal pessimist.

Said Hanrahan*

'We'll all be rooned,' said Hanrahan,
In accents most forlorn,
Outside the church, ere Mass began,
One frosty Sunday morn.

'It's lookin' crook,' said Daniel Croke;
'Bedad, it's cruke, me lad,
For never since the banks went broke,
Has seasons been so bad.'

'If rain don't come this month,' said Dan,
And cleared his throat to speak –
'We'll all be rooned,' said Hanrahan,
'If rain don't come this week.'

---

* Verses from 'Said Hanrahan' from *Around the boree log*, reprinted by permission of Angus and Robertson Publishers. © John O'Brien 1921 and Fr F. A. Mecham 1952.

Preface

'We want a inch of rain, we do,'
O'Neil observed at last;
But Croke 'mantained' we wanted two
To put the danger past.

'If we don't get three inches, man,
Or four to break this drought,
We'll all be rooned,' said Hanrahan,
'Before the year is out.'

In God's good time down came the rain;
And all the afternoon
On iron roof and window-pane
It drummed a homely tune.

And every creek a banker ran,
And dams filled overtop;
'We'll all be rooned,' said Hanrahan,
'If this rain doesn't stop.'

And stop it did, in God's good time:
And spring came in to fold
A mantle o'er the hills sublime
Of green and pink and gold.

And, oh, the smiles on every face,
As happy lad and lass,
Through grass knee-deep on Casey's place
Went riding down to Mass.

While round the church in clothes genteel
Discoursed the men of mark,
And each man squatted on his heel,
And chewed his piece of bark.

'There'll be bush-fires for sure, me man,
There will, without a doubt;
We'll all be rooned,' said Hanrahan,
'Before the year is out.'

Increasing scientific knowledge should bring optimists and pessimists closer together and eventually enable climatologists to discharge their social responsibilities with somewhat fewer confusing 'alarums and excursions'.

Preface

The conference on which this book is based was organised by the Australian Branch of the Royal Meteorological Society, with the co-sponsorship of the Australian Academy of Science. Major financial contributions to the success of the conference came from:

Government of Victoria
Australian Government
Australian Academy of Science
Jennings Industries Ltd
CSR Co. Ltd
Broken Hill Proprietary Co. Ltd.
Commonwealth Foundation (London)
Rockefeller Foundation

with additional support from Monash University, Trans Australia Airlines and numerous individuals and their supporting institutions.

Thanks are due to all these, plus of course, individual contributors, and Cambridge University Press, for making this book possible.

In addition, special thanks are due to Dr U. Radok (University of Melbourne), Dr W. Budd (Antarctic Division), and Mr M. Coughlan (Bureau of Meteorology) for scientific help, and to Mrs M. Leicester, Mrs H. Macdonald, Mrs L. Masson, Mrs V. Murphy, and Mrs R. Hampson for typing and re-typing many mangled manuscripts.

Melbourne
July 1976

A. B. PITTOCK
L. A. FRAKES
D. JENSSEN
J. A. PETERSON
J. W. ZILLMAN

1. An overview

A. B. PITTOCK

If we would discover the seat of those forces which produce...[the] difference in the dynamical status of the two great aerial oceans that envelop our planet, we should search for them in the unequal distribution of land and water over the two hemispheres. In one the wind is interrupted in its circuits by the continental masses, with their wooded plains, their snowy mantles in winter, their sandy deserts in summer, and their mountain ranges always. In the other there is but little land and less snow.
Matthew Fontaine Maury, *The Physical Geography of the Sea* (1855)

1.1 Concepts and perspectives

The atmosphere and oceans respond to the unevenly distributed driving force of energy from the sun's radiation, storing, redistributing and re-emitting it in various ways. The dynamic and thermodynamic manifestations of these responses are seen, instantaneously, as 'weather'. Viewed over longer time spans these responses are described as 'climate', although there is no clear consensus as to where weather ends and climate begins.

Every quantifiable element of the weather, such as temperature, pressure and rainfall, varies continuously or discontinuously with time on all time scales, so that an essential aspect of climate is variability. The concept of climate thus includes the totality of variations in weather with time, yet there is no time span over which the moving averages and variabilities of the weather elements remain constant. Some meteorologists have defined climate in terms of the weather over 30-year periods, and the years 1931–60 have been regarded as the standard. This particular period, however, is not necessarily representative even of this century, and an examination of the highly variable records of countries such as Australia indicates that longer intervals are often necessary before estimated means and standard deviations of such elements as rainfall or temperature even approximate constancy.

Climatic change is likewise a difficult concept to define. Clearly it must include an element of continuing trend, or a discontinuity, in the average value or variability of some climatic element. Numerous statistical tests may be applied to determine the significance of trends or changes in a

1

strictly statistical sense, but these tests usually make assumptions as to the statistical distributions and behaviour of the climatic elements being tested which are often only approximately true. It is also quite possible for certain properties, such as average annual rainfall, to remain constant while related properties, such as the seasonal distribution of rainfall, or its variability about the mean, vary significantly. The question as to whether a given variation in climate is a statistically significant change as contrasted with a manifestation of the normal 'random' variability of climate is in any case largely academic to the people affected, and the statistical answer will often depend on the time and space scales considered.

It should not be surprising therefore if various concepts such as climatic change and variability are sometimes used in different senses, nor that controversy should develop as to the 'reality' of a hypothesised climatic change. Such controversy assumes different aspects according to one's view-point and perspective. Meteorologists or climatologists working with the instrumental record of a mere 100 years or so, as they must in most of the Southern Hemisphere, are working on a time scale dominated by short-term variability in which longer-term trends are difficult to establish with any statistical certainty.

The time scale of most interest to agriculturalists, economists, planners and politicians is perhaps even shorter, of the order of decades. On this time scale a run of several bad years constitutes a major event of increasingly disastrous proportions as world populations increase and the margins between food production and consumption, or water supply and water use, decrease. To people involved at this practical level, the perceived consequences of a supposed climatic change may be seen to be so great as to justify shifting the onus of proof from those who hypothesise change to those who hypothesise no change. From this perspective 'social responsibility' may well run counter to a conservative interpretation of 'scientific responsibility', and 'caution' may demand that decision-makers allow for the possibility of climatic change rather than dismiss its likelihood. Similar differences in perspective, with their implicit value judgements, run through many debates on scientific issues of practical significance (e.g. see Weinberg, 1972).

Palaeoclimatologists consider time scales ranging from centuries through millennia to hundreds of millions of years. On these time scales the evidence of major climatic changes is irrefutable, although difficult to quantify both as to precise timing and the range and exact nature of the changes. Here controversy often centres on the precise climatic interpretation and scaling of secondary evidence derived from such sources as tree rings, pollen deposits, geomorphological features, or isotope ratios which are the result of many complex physical or biological processes.

2

For those unfamiliar with palaeoclimatological terms, it is perhaps as well to explain the meanings attached to terms such as 'ice age', 'glacial' and 'stadial', and to set these in some time perspective.

Geological evidence shows that during the 4,700 million years of Earth's existence it has gone through several 'ice ages' each of which lasted a few million years. Between these events there were no ice caps on Earth. The most recent ice age, the Quaternary Period, began about 1.8 million years ago and continues today. During this ice-age period the quantity of ice at high latitudes and in mountainous regions has waxed and waned in a complex manner at least partly related to global-scale climatic variations. There have been a number of long periods (of the order of 10^5 years each) during which the ice volume and its geographical extent were greater than at present. During several of these there was at least three times the present amount of ice. These colder periods should more properly be termed 'glacial stages' (although they are sometimes referred to as 'ice ages'), and the warmer times between them 'interglacial stages'. Over the last half-million years or so, the interglacials have occupied much less time than the glacials.

Shorter cold periods with smaller ice volumes than the full glacial stages are called 'stadials', and the warmer intervals between them 'interstadials'. These variations are superimposed on the longer-term fluctuations of glacials and interglacials.

Typical palaeoclimatological curves exhibiting these features will be found in Sections 3.1, 3.4, 5.1 and 5.2. Time spans for various named geological Eras, Periods and Epochs will be found in standard texts such as Flint (1971), Schwarzbach (1963) and Frenzel (1973). With the exception of Fairbridge (Section 5.1), the present book deals with the Cenozoic Era (~ 65 million years ago to the present), and particularly the Quaternary Period which encompasses the Pleistocene Epoch and the post-glacial Holocene (or 'Recent') Epoch. The latter includes the most recent historical times and the years of instrumental records.

1.2 The Southern Hemisphere Aspect

Despite the excellence of much recent work on climate and climatic change emanating from the Northern Hemisphere (e.g. Schwarzbach, 1963; Royal Meteorological Society, 1966; Flohn, 1969; Lamb, 1972a; Frenzel, 1973; US Committee for GARP, 1975; World Meteorological Organization, 1975a), coverage of Southern Hemisphere climate has in general been rather small. The geographical, historical and cultural reasons for this bias are understandable, but the consequences in terms of a true global understanding of climate are serious. This is of global rather than merely regional concern, because the nature of the global ocean–atmosphere system is such that a complete understanding of climatic behaviour in one hemisphere is not possible without an understanding of that in the other, and the two hemispheres are different in a number of important respects.

The reasons for Northern Hemisphere bias in the literature are not merely northern parochialism and the remoteness of the Southern Hemisphere from the major centres of modern scientific culture, but also more fundamental limitations determined until the advent of meteorological satellites by a much smaller and more recent ground-based network of meteorological stations, and a much smaller area and latitudinal range over which to establish a land-based palaeoclimatic record in the south.

These limitations are now rapidly being overcome by the use of satellites (Barrett, 1974), isotopic analysis of ice and ocean-bottom cores, more refined techniques of tree-ring and pollen analysis and other methods which extract the maximum information from scanty and mainly non-glacial land records. The paucity of Southern Hemisphere data and palaeoclimatic sites has necessitated different approaches, for example, a growing emphasis on pattern analysis to determine the representativeness of particular sites and to help select sites for the most profitable deployment of scarce research personnel and resources.

Major differences between the Northern and Southern Hemispheres which have climatological significance stem essentially from the shapes and distribution of land, sea and ice as to latitude, longitude and altitude.

Only about 20% of the Southern Hemisphere is covered by land or ice caps as against nearly 40% of the Northern Hemisphere. Yet the Antarctic ice sheet, including the floating ice shelves, is about 14 million km² in area and contains more than 90% of all the world's ice. The additional area of Antarctic sea-ice varies seasonally from around 4 million km² in March to some 20 million km² in September. The total Arctic sea-ice cover, on the other hand, varies from a minimum of around 7 million km² in September to about 12 million km² in March and April.

Differences in the heat budget of the Arctic Ocean with its thin and partly seasonal ice cover compared to that of the permanent elevated Antarctic ice cap leads to winter air temperatures over the Arctic Ocean of about -35 °C as against temperatures down to -70 °C or lower over Antarctica. Since the temperature of the tropics varies little between hemispheres, the temperature gradient from the equator to pole, which is closely related to the intensity of the westerlies and the general circulation, is nearly 40% larger over the Southern Hemisphere than over the Northern Hemisphere. This is one major factor leading to the subtropical high-pressure belt in the Southern Hemisphere being on average some 7° of latitude nearer the equator than in the north and largely accounts for the asymmetry of the intertropical convergence zone or 'meteorological equator' (Flohn, 1969). Seasonal displacements of wind and pressure belts depend largely on direct heating of the atmosphere (i.e. on the 'sensible heat') as opposed to heating by evaporation and condensation of water (i.e. on the 'latent heat') and, as this is greater over continental land masses, seasonal displacements are on average greater in the Northern Hemisphere.

The so-called 'polar monsoon' regime of the subarctic, where polar easterly winds blow in summer and westerlies in winter, owes its existence to the pattern of land–sea distribution and has no counterpart in the south.

Other major circulation differences arising from the different land–sea distributions and topography of the two hemispheres include the greater relative importance in the Southern Hemisphere of mean zonal (east–west) motion as opposed to eddy motion (large-scale turbulence), and of transient eddies as compared to standing eddies (moving eddies or wave-motions as opposed to those more or less fixed in longitude) (Adler, 1975). Standing eddies consisting of two, or more often three, waves around each circle of latitude, with ridge–trough positions tied to the land–sea distribution appear to play a key role in the climate of the Northern Hemisphere. Variations in their intensity and precise location at middle and high latitudes may play a major role in climatic change. Lower latitudes of the Southern Hemisphere are also strongly influenced by three standing waves associated with Africa, the Indonesia–New Guinea region, and South America, but at higher southern latitudes one and two waves appear to dominate, variously said to be due to the influence of the Andes and the Antarctic Peninsula, the eccentricity of the Antarctic ice cap, or to the dominance of the southeast Asian archipelago region in exporting energy from the tropics (van Loon & Jenne, 1972; Vonder Haar, 1968).

These and other significant interhemispheric differences could provide vital clues to the mechanisms underlying global climatic change. A key element in unravelling this question may well turn out, therefore, to be the degree of synchronisation and the time lags between climatic variations in the two hemispheres. In the absence of positive feedback processes and major storage of heat, atmospheric and oceanic transports of heat will tend to smooth out temperature differences between similar latitude zones at a rate depending on the magnitude of the transports between them, and probably of the order of a few years at most. Oceanic and ice-cap storage terms, which differ greatly between hemispheres, will introduce major lags, and feedback processes due most likely to albedo differences between land, sea, and ice, snow, or cloud cover may maintain even longer-term differences.

Field evidence as to time lags between hemispheres both on the time scale of the major glacial–interglacial sequences and on that of the instrumental record are thus of great diagnostic significance, so that the Southern Hemisphere record is indeed vital to the global picture. Recent evidence (Sanchez & Kutzbach, 1974; Tucker, 1975; Salinger & Gunn,

1975; Burrows, 1975) throws considerable doubt on the supposed parallelism of the Southern and Northern Hemisphere instrumental records (Mitchell, 1961, 1963), although broad parallelism in the late Quaternary Record does exist, albeit with time lags which could be as large as a couple of thousand years. Accurate absolute dating and fine time resolution is essential to answering this question on the palaeoclimatic time scale.

1.3 Intentional climate modification

A notable omission from the conference proceedings (apart from Flohn's brief reference in Section 3.7) was in consideration of *intentional* climate modification. In view of Australia's pioneering and significant efforts in cloud-seeding research, and the obvious benefits to be gained from increasing rainfall in a generally dry country, this omission is worthy of some discussion.

Research by CSIRO in Australia (Smith, 1974; Warner, 1974) and by other groups around the world has demonstrated that in the case of individual clouds having suitable characteristics, cloud seeding can increase precipitation. The same can be said for a small number of experiments aimed at increasing precipitation over areas ranging from 1,000 to about 10,000 km². However, the attitude of at least some of the workers in this field is well summed up by Hosler (1974) who in introducing a review on 'Overt Weather Modification' stated: 'Global weather modification and climatic change are not discussed, since the techniques employed thus far in overt weather modification present us with no physical reason for supposing that they affect anything on a grid larger than a few hundred kilometres.'

On the other hand, it could be argued that a significant increase in precipitation over a catchment area of 10,000 km² would significantly increase the runoff available for irrigation and thus be tantamount to a climatic change over the area served by such irrigation. Even so, the affected areas would remain small on a global scale.

Views of other workers as to the state of the art may be found in Hess (1974) and Sax *et al.* (1975). Amongst the most optimistic are Huff & Semonin (1975) who claim somewhat cautiously that in some circumstances 'successful cloud seeding operations could contribute occasionally to temporary alleviation of water shortages over portions of an extensive drought region'.

There has been considerable expenditure of money and effort in attempts to suppress hail and to modify hurricanes or tropical cyclones. However, the comparative rarity of these phenomena and the absence of adequate theoretical models of their behaviour have so far prevented any wide agreement being reached as to the effectiveness of the techniques used.

Grandiose engineering schemes aimed at climate modification, such as diverting rivers which flow into the Arctic Ocean, damming Bering Strait, or creating huge artificial lakes in the Saharan region of Africa, have been suggested. Southern Hemisphere examples include damming Drake Passage, creating an inland sea in Australia, and coating parts of the Australian desert with bitumen. Several of these proposals are hardly practicable, and the probable climatic consequences of each is not yet understood. Until suitably complex and realistic theoretical/numerical model experiments have been performed on these proposals they remain completely unproven and too uncertain in their potential effects to be undertaken with any degree of social responsibility (Kellogg & Schneider, 1974).

It is interesting, however, to note that preliminary studies and primitive attempts at such model experiments for certain Northern Hemisphere cases are beginning to be made (e.g. Vowinckel & Orvig, 1974; Gray *et al.* 1974; Rapp and Warshaw, 1974), and that the conclusions, while highly tentative, are not entirely negative. Perhaps this is an area worthy of more serious thought in a Southern Hemisphere context.

1.4 A problem for all humanity

Having stressed the Southern Hemisphere perspective and contribution to the problem of climatic change and variability, it is worth while to pause briefly to consider the unifying aspects of our subject, which crosses boundaries of nations, regions, disciplines and cultures.

The study of climatic change and variability requires the contributions of and interactions between not only meteorologists, geographers, geologists, oceanographers, glaciologists, mathematicians and statisticians, but also chemists, botanists, biologists, and even historians, archaeologists and economists. The subject is so fraught with complex physical, biological and cultural interactions and feedbacks that any worker from a single discipline must inevitably trespass outside his own area of expertise. If such a worker does so in isolation from other relevant expertise the result can too often be oversimplification and error. The very nature of climatic interactions therefore demands interdisciplinary approaches and exchanges at a scientific level.

Geographically speaking, climatic interactions are truly global in character, although subject to important local and regional modifications. Significant climatic variations other than those induced by topographic changes are never confined to single regions or locations, but occur as part of global patterns of variation (see Chapter 4). In a world increasingly interconnected physically and culturally in terms of communications, trade, economics, food supplies, and political and ideological movements and conflicts, the influence of climatic variations in any particular region will have ever-wider implications elsewhere (see Chapter 7).

As world population edges closer to the limits of world food production and to the full utilisation of existing fresh-water supplies, profound cultural and philosophical questions are being raised as to the limits of growth (Mesarovic & Pestal, 1974), the ability of science and technology to provide viable 'solutions' to problems raised by growth (Crowe, 1969; Moncrief, 1970), and as to the proper relationship between humankind and nature. Should we accept the limitations imposed by the natural climate on food production and water supply, or should we go all out for the technological 'fix' of global-scale climatic engineering? If we do not choose to adapt to the natural environment and live 'in harmony with nature', as so-called 'primitive Man' is supposed to have done for millennia, do we not expose ourselves to great risks of global disasters brought on by global pollution and inadvertent, unanticipated, climatic modification? (See Chapter 6.) Could so-called 'Civilisation' generate a climatic disaster such as a melting of the ice caps or a new ice age? Might our survival, on the other hand, depend on human intervention to prevent just such events from occurring naturally?

These are profound human questions which go far beyond science as such to deep matters of philosophy and human values. The global nature of the questions demands global answers. Are we, as members of an intelligent species of beings who are still divided by nationality, ideology, race and religion, ready to provide the answers which are required of us at this stage in our cultural and historical development? I for one have my doubts, but for this very reason I hope the problems raised by climatic change and variability will be considered not only by scientists but by religious and secular philosophers, politicians, administrators and members of the public at large.

2. The physical basis of climate

2.1 The climatic system*

Subsequent chapters will summarise what is known of the climates of the past and the variability of the earth's climate through geological and historical time. If we are to appreciate the significance of these changes and begin to unravel their causes, we must first understand something of the processes that determine the nature of climate itself. The complete problem is enormously complex and is still far from solved. It is possible in this chapter to introduce only the broad framework. This must include at least the sun, the earth's atmosphere, the oceans, the cryosphere (ice masses and snow deposits), the land surface and the biomass. The sun is the ultimate energy source. Its influence on climate depends not only on the sun–earth distance but on other orbital characteristics such as the rotation rate of the earth and its angle of tilt to the sun. Differential solar heating of the earth–atmosphere system drives atmospheric and oceanic circulation systems which constitute the essential physical basis of climate.

The energy source

The sun is known to emit energy in various forms as electromagnetic radiation (radiowaves, thermal radiation, X-rays) and streams of charged particles including cosmic rays. The precise state of climate on earth may depend directly or indirectly on several of these but by far the largest part of the solar power output is in the form of thermal electromagnetic radiation in the wavelength range 0.2–5 micrometres. The total radiant power output of the sun is about 3.88×10^{26} watts which is equivalent to the thermal emission from a black body the size of the solar photosphere at a temperature of 5,790 K.

The radiant energy crossing a unit of area normal to the solar beam per unit time at mean sun–earth distance is known as the solar constant. Its value lies in the range $1,360 \pm 20$ watts per square metre (W m^{-2}).

* Editorial contribution.

9

2.1 Physical basis of climate

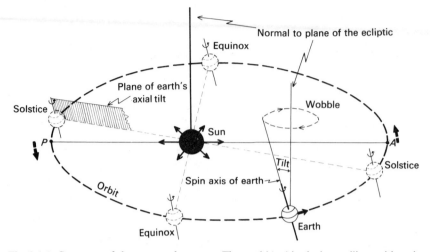

Fig. 2.1.1. Geometry of the sun–earth system. The earth's orbit, the large ellipse with major axis *AP* and the sun at one focus, defines the plane of the ecliptic. The plane of the earth's axial tilt (shaded) is shown passing through the sun corresponding to the time of the southern summer solstice. The earth moves around its orbit in the direction of the solid arrow (period one year) while spinning about its axis in the direction shown by the thin curved arrows (period one day). The broken arrows shown opposite the points of aphelion (*A*) and perihelion (*P*) indicate the direction of the very slow rotation of the orbit.

Sun–earth geometry

Figure 2.1.1 shows the main features of the sun–earth geometry that are of importance to climate. The earth moves on a slightly elliptical orbit around the sun, at present distant about 147.1 million km at perihelion and 152.1 million km at aphelion. This causes the solar radiation incident on the earth to wax and wane about 3½% either side of the solar constant in the course of a year. At present the incident radiation is strongest in January during the Southern Hemisphere summer.

The earth's axis of rotation is not upright with respect to the plane of its orbit (known as the ecliptic). This tilt, coupled with the earth's annual journey around the sun, leads to the normal march of the seasons. The earth is at one of the solstices (December, June) when the sun–earth vector lies in the plane of its axial tilt; at the equinoxes (March, September) when it is normal to this plane. The various orbital parameters change slowly with time as follows. Firstly, the elliptical orbit itself revolves slowly in space in the same direction as the earth's travel in its orbit, completing one revolution in space after an irregular interval that averages 96,600 years. Secondly, the eccentricity of the orbit (the difference between the sun–earth distances at aphelion and perihelion divided by their sum) varies as the orbit revolves in space with the same period of about 96,600 years.

10

Thirdly, the earth wobbles in its orbit like a spinning top, its spin axis describing a cone in space completing one revolution in about 26,000 years, a phenomenon known as precession of the equinoxes. Because the precession is in the opposite sense to the rotation of the orbit, the perihelion recurs at the same time of year after a period of less than 26,000 years, about 20,600. Fourthly, the tilt of the earth's axis from the normal to the ecliptic (known as the obliquity of the ecliptic) itself undergoes a very slow oscillation, with a period of about 40,000 years, between 24° 36' and 21° 59'. At present the angle is 23° 27' and decreasing.

Planetary energy balance

It might be thought that a knowledge of the radiant energy output from the sun and the location and attitude of the earth in its orbit would be sufficient to determine uniquely at least the broad characteristics of the earth's global mean climate. If, for example, we make the plausible assumption that the solar (short-wave) radiant energy absorbed by the earth–atmosphere system is just balanced by terrestrial (long-wave) radiation emitted by space by the earth–atmosphere system, we may write:

$$\begin{pmatrix} \text{Solar} \\ \text{constant} \end{pmatrix}\begin{pmatrix} 1- \dfrac{\text{planetary}}{\text{albedo}} \end{pmatrix}\begin{pmatrix} \text{cross-section} \\ \text{area of earth} \end{pmatrix} = \begin{pmatrix} \text{surface area of} \\ \text{the earth} \end{pmatrix}\sigma T^4, \quad (2.1.1)$$

so that, with the Stefan–Boltzmann constant σ known, it would seem a straightforward matter to solve for what we might refer to as the 'planetary temperature' T. Unfortunately the problem is not so simple. The main difficulty is that the planetary albedo, i.e. the reflectivity of the complete earth–atmosphere system to solar radiation, cannot be determined *a priori*. It depends, for example, on the type and distribution of clouds in the atmosphere, the extent of snow and ice fields, and the state of the sea; and thus on complex processes occurring *within* the atmosphere and oceans. We can, however, estimate or measure the planetary albedo existing with the present climate. It is about 0.3. A simple evaluation of Equation (2.1.1) gives a planetary temperature of about 250 K, some 35 degrees lower than the mean temperature of the earth's surface. This is consistent with our understanding that the atmosphere is only partially transparent to the long-wave radiation from the earth's surface so that a part of the planetary emission originates in the colder upper atmosphere.

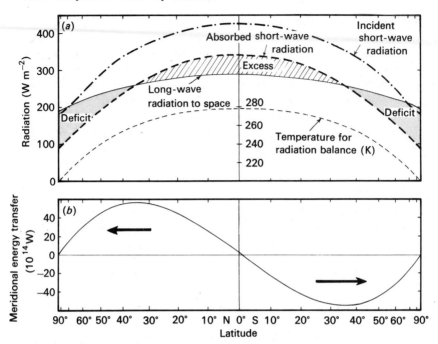

Fig. 2.1.2. (a) Radiation balance of the earth–atmosphere system as a function of latitude. The thin broken curve indicates an equivalent radiative temperature (centre scale) for each latitude band to be individually in radiative balance with the absorbed short-wave radiation (heavy broken curve). The thin solid curve shows the actual observed long-wave radiation emitted to space as a function of latitude. (b) Poleward energy transfer within the earth–atmosphere system needed to balance the resulting meridional distribution of net radiative gains (hatched) and losses (stippled) as indicated in (a). The data are from Kondratyev (1972) and Newton (1972). Note that the horizontal scale is compressed towards the poles so as to be proportional to the surface area of the earth between latitude circles.

Effects of latitude

Consider now the effects of the spherical shape of the earth in causing the incident solar radiation per unit of surface area to decrease from the equator to the poles. As before, we cannot *a priori* determine the meridional distribution of climate but, if the meridional distribution of albedo is known and we imagine the atmosphere and ocean somehow constrained to prevent internal energy flow across latitude circles, an effective radiative temperature can be determined for each latitude band (Fig. 2.1.2(a)). It turns out, however, that the various latitude bands are not individually in radiative balance with space, the long-wave radiation emitted by the planet exceeding the absorbed solar radiation in high latitudes and being correspondingly less in the equatorial belt as shown in Fig. 2.1.2(a).

12

It is evident that, if a long-term state of balance is to be maintained, the equatorial energy excess must be transferred meridionally within the earth–atmosphere system to make up the deficit in polar regions. The poleward energy transfer across latitude circles required to effect this transfer is shown in Fig. 2.1.2(*b*). If we look further at the distribution of net radiation income, we find that it varies also around latitude circles so that a pattern of large-scale energy sources and sinks is distributed over the planet. The needed energy transfer from (predominantly equatorial) energy sources to (predominantly polar) sink regions is achieved through the action of the general circulation of the atmosphere and oceans.

Components of the climatic system

In the sense that 'climate' is to be identified with the long-term mean state of the atmosphere, it is primarily a product of the atmospheric general circulation. To understand the physical basis of climate, we must therefore explore the nature of the general circulation in terms of the response of a physical system to external forcing. We have already hinted at an interdependance between the behaviour of the atmosphere and the oceans. In fact, it is now widely accepted (e.g. Joint Organizing Committee for GARP, 1975) that if we are to develop a satisfactory understanding

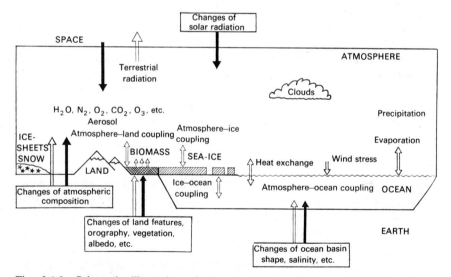

Fig. 2.1.3. Schematic illustration of the components of the coupled atmosphere–ocean–cryosphere–land surface-biomass climatic system. The full arrows indicate processes external to the climatic system and the open arrows indicate internal processes in climate formation and climatic change (from Joint Organizing Committee for GARP, 1975). Reproduced with permission of the National Academy of Sciences.

of climate and climatic variation, we must take account of a wide range of interactions between the atmosphere, oceans, cryosphere, land surface and biomass. These five can be regarded as the essential components of the earth's climatic system. The climatic system and the principal interactions between its components are shown schematically in Fig. 2.1.3.

The atmosphere

The atmosphere is the most variable part of the system. It is the gaseous envelope of the earth composed mainly of nitrogen (about 78%) and oxygen (about 21%) with much smaller quantities of other gases including carbon dioxide, and a variable admixture of water vapour. The density decreases with height, with half the atmospheric mass in the layer below about 5–6 km. Temperature also decreases with height through the troposphere, the well-mixed region below about 10–15 km in which most of the clouds and weather occur. Above the troposphere is the stratosphere (temperature constant or gradually increasing with height to about 50 km) and mesosphere (temperature decreasing to about 80 km). Higher still, above about 80 km, where the temperature rises again, is the ionised region known as the ionosphere. The lower atmosphere is set in motion by differential heating and the motion transports heat in both its sensible and latent forms, moisture (vapour, cloud, rain and snow), solid material carried up from the surface (dust, salt and smoke particles) and various liquid and gaseous forms of pollution. The lower atmosphere interacts in various ways with the underlying land, ocean and ice surfaces and with the ionised regions above. In particular the lower layers of the atmosphere are heated by contact with the underlying land and ocean. Also water vapour, evaporated from the oceans by the incident solar radiation, condenses in the atmosphere and the released latent heat provides one of the main energy sources for driving the atmospheric circulation.

The oceans

The oceans comprise the salt water of the world oceans and adjacent seas. Compared with the atmosphere, the oceans appear as a sluggish component of the climatic system. The ocean circulation is primarily driven by the winds. The resulting currents carry large amounts of heat from equatorial regions towards the poles and thus contribute to maintenance of the global energy balance. The upper layers of the ocean interact with the overlying atmosphere on time scales of months to years while the deeper ocean waters have thermal adjustment times of the order of centuries. The oceans also exchange carbon dioxide with the atmosphere.

14

The cryosphere

The cryosphere comprises the world's ice masses and snow deposits including the continental ice sheets, mountain glaciers, sea-ice and snow cover. Snow-cover and sea-ice extent show large seasonal variations while the glaciers and ice sheets vary much more slowly. Nearly 80% of the earth's fresh water supply resides in the ice sheets of Antarctica and Greenland. These are nowadays seen, not merely as passive stores of water which would otherwise be found in the oceans and which are modulated by changes of atmospheric origin, but rather as active participants in the climate-forming process involving, possibly, glacial surges which set in train changes in sea level, sea-surface temperatures and surface albedo.

The land surface

The land surface comprises the land masses of the continents including the mountains, surface rock, sediments and soil and may be taken to include also lakes, rivers and ground water. These are variable parts of the climate system on all time scales. Over very long periods, uplift, subsidence and continental drift lead to changes in the elevations and positions of land masses. The earth's surface is an important source of airborne particles which may have an influence on climate. The soil, in turn, evolves in response to climate and vegetation.

The biomass

The biomass includes the plants on the land and in the ocean, and the animals of the air, sea and land, including man. Although their response characteristics differ widely, these biological elements are sensitive to climate and, in turn, may influence climate. The biomass plays an important role in the carbon dioxide budget of the atmosphere and ocean and in the production of aerosols and in the related chemical balances of other constituent gases and salts. Natural changes in plants occur over periods ranging from seasons to thousands of years in response to changes in temperature, radiation and precipitation, and, in turn, alter the surface albedo and roughness, evaporation and ground hydrology.

Underlying physics

The behaviour of the individual components of the climatic system and the interactions between them are governed by a number of basic physical laws. Fluid motion, for example, is constrained by conservation principles for mass, momentum and energy. These may be expressed, for the atmosphere and oceans, by a coupled set of mathematical equations which prescribe the time rate of change of the various elements in terms of their initial configuration and certain external influences. Mass conservation is expressable by continuity equations for the basic fluid (air, water) and for the various active admixtures (e.g. water vapour in the atmosphere, salt in the ocean). Momentum conservation is expressed through the equations of motion (Newton's Law), and energy conservation through the so-called thermodynamic energy equation. It is likewise possible to produce a mathematical representation of processes governing the dynamics of ice sheets or to set up equations describing moisture and heat flow in the soil.

One approach to understanding the physical basis of climate is to use these mathematical expressions of the governing laws to build mathematical simulation models of the various parts of the climatic system. This approach is discussed in Chapter 5. But they provide also the theoretical and conceptual framework within which to examine the climate machine itself at work. The following two sections examine in turn the workings of the atmosphere and oceans – the two major components of the climatic system.

2.2 The general circulation of the atmosphere
G. B. TUCKER

If the earth's surface were uniform with no mountains or land–sea contrasts and if the dynamical effects of the earth's rotation could be suppressed, the atmospheric response to differences in heating between equator and poles would be rather simple. A large toroidal Hadley* circulation cell would develop with air rising in the region of low-latitude solar heating and sinking in the region of cooling near the poles. The circulation would produce a net poleward transfer of heat to balance the sources and sinks. If the circulation ran too fast, more heat would be carried poleward than is available at the tropical source, the tropics would cool and the circulation would be slowed down. If it ran too slowly, the tropics would overheat and the circulation would accelerate. Provided the complicating effects of the moisture cycle could also be ignored, the climate on such a planet would be fairly easy to describe. There would be a steady temperature decrease from equator to poles and a steady equatorward surface wind of a few metres per second.

The real situation is much more complex. A simple thermally-driven Hadley cell carrying heat from equatorial source to polar sink cannot be maintained on the rotating earth. Strong westerly winds develop aloft as

---

* So named after George Hadley who, in 1735, proposed such a circulation as an explanation for the Trade Winds.

a result of the poleward-flowing air in the upper branch of the circulation conserving the absolute angular momentum it has gained from the earth's surface in the tropics. It turns out that this westerly current is unstable and, even without the effects of land–sea contrasts, the upper winds develop a wave structure. The direct Hadley cell is confined to the tropics. Clearly the global heat balance must be achieved by some mechanism other than a direct circulation from heat source to heat sink. To gain some understanding of the mechanisms at work, it is useful to consider the main features of the observed circulation.

The form of the circulation

In both hemispheres and in all seasons the dominant feature of the atmospheric circulation is a large tropospheric circumpolar vortex of westerly winds, strongest and most extensive in the upper troposphere. On any single day the vortex appears full of embedded waves and eddies – the familiar pattern of cyclones and anticyclones that control our daily weather. Figure 2.2.1(*a*) shows a typical Southern Hemisphere summer circulation pattern at the surface and about 5½ km above the surface.

If the individual daily patterns are averaged over a month, a season or many seasons to produce the type of representation of the circumpolar vortex that we might relate to climate in the same way that the daily patterns relate to weather, we find that most of the eddy structure disappears (Fig. 2.2.1(*b*)). Because of influences resulting from the irregular distribution of land and ocean, the averaged vortex is not completely symmetrical about the poles although the Southern Hemisphere vortex is more nearly circular than its northern counterpart. If, instead of the single mid-tropospheric chart shown in Fig. 2.2.1(*b*), the winds at a series of higher levels (10–15 km) are examined, a distinct core of maximum westerlies will be seen to occur approximately above the dividing line (the subtropical ridge) between westerly and easterly winds at the surface and where the poleward decrease of temperature in the troposphere steepens. This is evident from Fig. 2.2.2, in which the zonal winds and temperatures have been further averaged around latitude circles. The cores of strong westerlies are known as the subtropical jet streams.

Although, as in Fig. 2.2.1, quite strong poleward and equatorward winds may be found both at the surface and aloft in any particular longitude, when these are averaged around the hemisphere the resulting mean meridional flow is weak – a few metres per second at most. This is because the large-scale flow in the atmosphere is very nearly geostrophic (i.e. the wind blows parallel to the isobars with a speed that depends only on their spacing and the latitude) and thus southward flow at one longitude is just balanced by northward flow at another longitude. Only where the

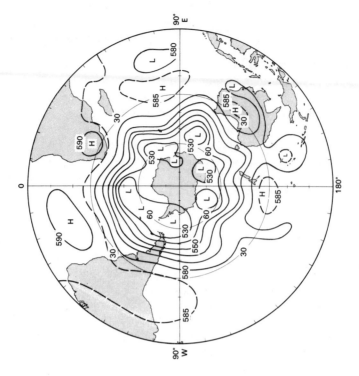

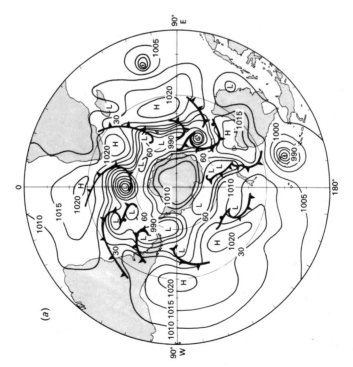

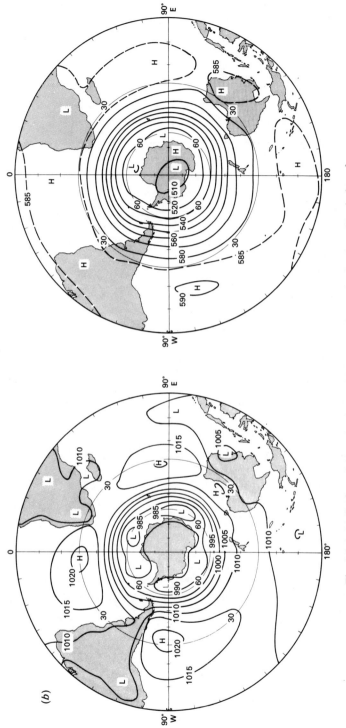

Fig. 2.2.1. (a) Surface (left) and 500 mb (right) circulation in the Southern Hemisphere for 00 GMT 16 January 1976. (b) Corresponding long-term means for January (lower). Isobars of the mean sea-level pressure field are labelled in millibars and contours of the 500 mb height in geopotential dekametres.

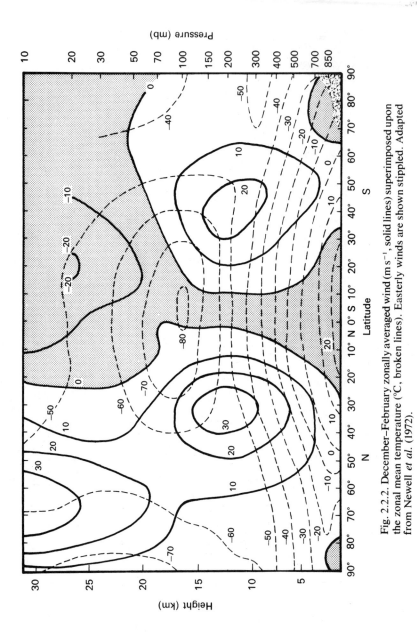

Fig. 2.2.2. December–February zonally averaged wind (m s⁻¹, solid lines) superimposed upon the zonal mean temperature (°C, broken lines). Easterly winds are shown stippled. Adapted from Newell *et al.* (1972).

20

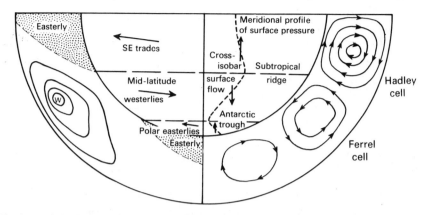

Fig. 2.2.3. Schematic representation of the principal features of the general circulation for the Southern Hemisphere showing the meridional circulation (right) and zonal circulation (left).

air is accelerating or decelerating does it blow slightly across the isobars. One such region is the thin layer near the earth's surface where frictional effects cause it to cross the isobars at a slight angle from high to low pressure. From the meridional profile of atmospheric pressure at the surface evident from Fig. 2.2.1 we can thus infer the broad nature of the mean meridional circulation and this is borne out by careful analysis of actual wind data. It is shown schematically in Fig. 2.2.3 from which we can now conveniently summarise the basic features of the general circulation as follows:

(i) a shallow belt of polar easterlies extending equatorward to about 65° latitude at the surface but varying with the seasons;

(ii) a broad band of mid-latitude westerlies increasing with height to a maximum which is strongest (greater than 40 m s^{-1}) and at its lowest latitude (around 30°) in winter;

(iii) a zone of Trade-Wind easterlies separated from the westerlies by subtropical ridges which slope equatorward with height from around 30° at the surface to about 10° at 200 mb;

(iv) a three-cell mean meridional circulation with ascending motion in the equatorial belt and around latitude 60° and with descent over the poles and in the vicinity of latitude 30°.

Mean and eddy states

We know that there is an overall global balance between absorbed solar radiation and emitted terrestrial radiation – though the fluctuations in this balance on extended time scales have yet to be determined. We know also that the net radiative heating of the earth–atmosphere system is

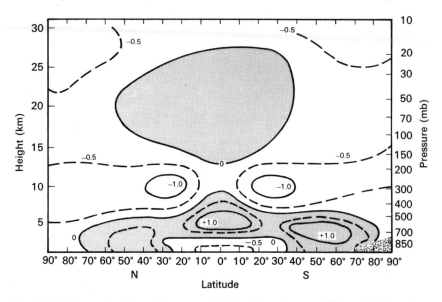

Fig. 2.2.4. Mean annual zonally averaged diabatic heating rate (°C per day) based on estimates summarised by Newell *et al.* (1974). Regions of net diabatic heating are stippled.

concentrated in low latitudes and that there is net radiative cooling in high latitudes. As the combined result of (i) direct atmospheric radiative heating and cooling, (ii) heating of the lower layers of the atmosphere by contact with the radiatively warmed underlying land or ocean, and (iii) release within the atmosphere of the latent heat of water vapour evaporated from the oceans by solar radiation, the net atmospheric heating is distributed in a quite complex way with height, latitude and longitude. Figure 2.2.4 shows a recent estimate of the zonally averaged distribution of net atmospheric heating. The circulation of the atmosphere is the means by which balance is achieved, transferring energy from source regions to sink regions. The circulation must fulfil other balance requirements also. The atmosphere experiences a source of westerly angular momentum due to 'surface stress' in the zone of low-latitude easterlies and a sink of momentum in the zone of surface westerlies. These sources and sinks must balance. Also, the pattern of horizontal and vertical fluxes of momentum in the atmosphere must be such as to ensure a zero net accumulation of momentum everywhere in the long-period mean.

In 1926, Sir Harold Jeffreys had shown that the frictional meridional drift of axially symmetric zonal circulations could not account for the required momentum exchange and proposed that the necessary meridional flux of angular momentum was accomplished mainly by horizontal eddies with trough axes that tilt westwards towards the equator in both hemispheres (Jeffreys, 1926). This led to the concept of mean and eddy modes of transfer of other properties, particularly energy, by the general circulation. In their simplest form, the time-average meridional and zonal components of the flux of a quantity s per unit volume at a particular location can be written as:

meridional flux $\qquad \overline{vs} = \bar{v}\,\bar{s} + \overline{v's'},$ $\qquad\qquad$ (2.2.1)

zonal flux $\qquad\qquad \overline{us} = \bar{u}\,\bar{s} + \overline{u's'},$ $\qquad\qquad$ (2.2.2)

where v and u are the northward and eastward wind components and the overbar and prime represent respectively a time average and the deviation from that average. Each thus takes the form of the sum of a term coresponding to flux by the time-averaged velocity component and a term which results from the covariance between instantaneous values of the velocity component and the quantity s.

The concept was further developed by Priestley (1949) who separated the time and zonally averaged large-scale fluxes into three components as follows:

meridional flux $\qquad (\overline{vs}) = (\bar{v})\,(\bar{s}) + (\bar{v}^*\bar{s}^*) + (\overline{v's'}),$ $\qquad$ (2.2.3)

zonal flux $\qquad\qquad (\overline{us}) = (\bar{u})\,(\bar{s}) + (\bar{u}^*\bar{s}^*) + (\overline{u's'}),$ $\qquad$ (2.2.4)

where, in addition to the terminology introduced above, the brackets and asterisk represent a zonal average and the deviation from that average respectively. The first terms in Equations (2.2.3) and (2.2.4) represent respectively the effects of mean meridional (toroidal) and mean zonal circulations shown in Fig. 2.2.3. The second terms represent the effects of standing eddies, that is, of time-averaged patterns (Fig. 2.2.1(*b*)) with the mean meridional and mean zonal components removed. The third terms represent the effects of transient (local) eddies.

Zonally averaged budgets

To a certain extent, the large-scale behaviour of the atmosphere in the form of zonally averaged budgets of momentum, heat and moisture has been determined from observations with a reasonable degree of accuracy, although there are still major deficiencies in the evaluation of some important terms because of observational deficiencies and the need to determine accurately vertical velocities.

Figure 2.2.5 illustrates the relative contribution of the mean meridional circulation and the eddies to heat and angular momentum balance for the global atmosphere in the annual mean. The numerical values are based on the work of Newell *et al.* (1972, 1974) and Newton (1972).

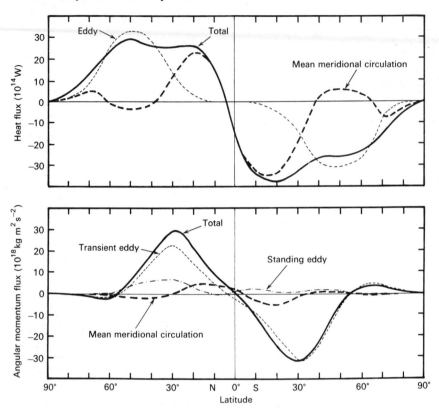

Fig. 2.2.5. Annual mean heat (upper) and relative angular momentum (lower) budgets based on data given by Newton (1972) and Newell *et al.* (1972) respectively. Note that although the contributions of the standing and transient eddies to the momentum flux are shown separately, only their combined effect is shown for the heat flux.

The global energy cycle

Turning now to the important question of how the energy provided initially by solar heating maintains the observed global wind systems against frictional dissipation, Margules (1903) was the first to point out that only a small part of what he referred to as the 'total potential energy' (potential plus internal energy) of the atmosphere is 'available' for conversion to kinetic energy. Consider some natural state of the atmosphere with cold and warm air masses side by side such that isentropic* surfaces slope upwards from the warm to cold air. It has a certain store of total potential energy. Now suppose that the cold air sinks and the warm air

---

* Isentropic surfaces are surfaces of constant potential temperature. Potential temperature is the temperature a sample of air would attain if transferred without loss or addition of heat from its actual level in the atmosphere to that where the pressure is 1,000 mb.

24

Library
I.U.P.
Indiana, Pa.

551.69 C 613m
C.1

rises until the isentropic surfaces are flat and horizontal. In the process some of the potential energy is released into kinetic energy. The resulting flat 'reference' state still contains a large store of potential energy but there is no more available for conversion to kinetic energy. To generate a new store of potential energy available for conversion to kinetic energy it is necessary to heat or cool the atmosphere differentially so that the isentropes again incline to the horizontal.

The two-component energy cycle

Lorenz (1955) applied Margules' ideas to the global atmosphere. The difference in total potential energy between the natural state and the hypothetical flat reference state he called the 'available potential energy' (A). The global energy cycle for the atmosphere could thus be depicted

(*a*)

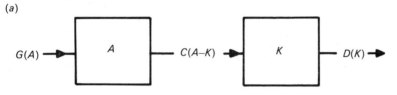

(*b*)

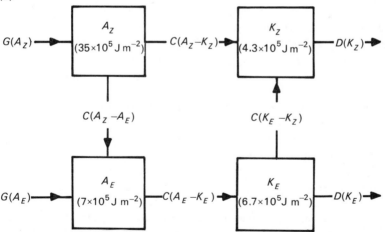

Fig. 2.2.6. The global mean energy cycle of the atmosphere in terms of available potential (A) and kinetic energy (K). When the available potential and kinetic energies are partitioned into zonal and eddy components the simple two-component representation (*a*) is replaced by the four-component model (*b*).

25

2.2   Physical basis of climate

in terms of the available potential energy A and the kinetic energy K as in Fig. 2.2.6(a).

The generation ($G(A)$) of available potential energy was shown to result from diabetic heating at high pressure and cooling at low pressure on isentropic surfaces. The conversion of available potential to kinetic energy ($C(A-K)$) in the global cycle results from the predominance of ascent of warm air and descent of cold air, or, equivalently, the predominance of horizontal flow from high to low pressure. The energy cycle is completed by the frictional dissipation of the kinetic energy ($D(K)$).

The four-component energy cycle

Lorenz (1955) was the first to show that if the available potential and kinetic energies depicted in Fig. 2.2.6(a) are each partitioned into portions associated with the zonally averaged and eddy structure of the atmosphere, then the conversion from one of the four resulting energy forms to another depends largely on the large-scale flux:gradient relation of relevant quantities.

The four-component energy cycle as devised by Lorenz is shown in Fig. 2.2.6(b) with some recent estimates of the various terms shown in brackets. The four energy forms can be interpreted broadly as follows:

(i) the zonal available potential energy (A_Z) is a measure of the meridional temperature variation;

(ii) the eddy available potential energy (A_E) is a measure of the temperature variation around latitude circles;

(iii) the zonal kinetic energy (K_Z) is a measure of the kinetic energy associated with the zonally averaged motion;

(iv) the eddy kinetic energy (K_E) is a measure of the kinetic energy of the departures from the zonally averaged motion.

The generation and conversion terms correspond to physical processes in the atmosphere as follows:

(i) $G(A_Z)$ is a measure of differential heating at low latitudes relative to cooling at high latitudes;

(ii) $G(A_E)$ is a measure of differential heating of warm air relative to cooling of cold air at the same latitude;

(iii) $C(A_Z-A_E)$ is a measure of the horizontal eddy flux of heat down the temperature gradient;

(iv) $C(K_Z-K_E)$ is roughly a measure of the eddy flux of relative angular momentum down the gradient of relative angular velocity;

(v) $C(A_Z-K_Z)$ is a measure of the rate at which warm air is rising relative to cold air sinking in meridional planes;

(vi) $C(A_E-K_E)$ is a measure of the extent to which warm air rises and cold air sinks at the same latitudes.

26

Of particular interest is the conversion term $C(K_Z-K_E)$ which is shown as negative. This means that the eddy flux of relative angular momentum is up-gradient (negative eddy viscosity (Starr, 1968)) and the eddies may be regarded as driving the circumpolar vortex.

Physically, the global atmospheric energy cycle depicted may be interpreted as follows:

(i) the zonal mean diabatic heating (Fig. 2.2.4) generates zonal available potential energy by net heating of the tropics and cooling in the polar regions;

(ii) baroclinic eddies (i.e. eddies in a current characterised by horizontal temperature gradients) transport warm air poleward and cold air equatorward, thus transforming the zonal available potential energy to eddy available potential energy;

(iii) eddy available potential energy in converted into eddy kinetic energy by sinking of cold air and rising of warm air in the eddies;

(iv) the zonal kinetic energy is maintained primarily by the conversion from eddy kinetic energy;

(v) the energy is dissipated by surface and internal friction in the eddies and the mean flow.

Since Lorenz's (1955) study, further contributions to available energy theory have come in papers by van Mieghem (1956), Dutton & Johnson (1967) and others, and a large number of authors have evaluated various terms in the energy budgets from observational data. A recent comprehensive survey of the global energy cycle is given by van Mieghem (1973) and a number of estimates of the various terms from atmospheric data are included in the book by Newell *et al.* (1974). It is not yet possible, however, to determine definitively the magnitudes of the various energy contents, generation rates and conversions.

The globally integrated energy budget approach has been one of the main ways of assessing the verisimilitude between numerical models of the general circulation and the real atmosphere. However, because this approach has focused on combined time and zonally averaged values of eddy fluxes, it has to some extent diverted attention from their equally important spatial distribution.

The longitude-dependent circulation

When considering how the time-average features of the atmosphere behave over longer time periods, perhaps the most noticeable feature is the way in which the phase speed of patterns decreases as the interval over which the time average is taken is extended. Thus instantaneous synoptic systems in middle and higher latitudes generally propagate eastwards with speeds typically about 10 ms^{-1}, whereas monthly-average flow

27

patterns show no progressive movement but rather occur in positions within only 10 or 20 degrees of longitude from their very long-term 'normal' positions.

It is probable that the geographically fixed lower boundary anchors the pattern and is mainly responsible for the waxing and waning of amplitude with season. The question arises, however, as to the relative roles of the time-averaged flow and the transient eddies, and their response to the geographically fixed heat sources in controlling the sequential behaviour of monthly-averaged patterns of the general circulation. While it is not yet possible to answer such questions definitively, a simple manipulation of the time-averaged vorticity* equation indicates a dependence of the phase speed of the long waves on the constant k in an observationally plausible parameterisation of the meridional transient eddy angular momentum flux

$$\overline{u'v'} = \text{constant} + k\bar{v}.$$

Similarly a time-averaged form of the thermodynamic energy equation indicates the importance of the horizontal variation of zonal and meridional transient eddy heat fluxes. Observational studies of $\overline{v'T'}$ and $\overline{u'T'}$ (T being temperature) are as difficult as studies of momentum fluxes because of sampling problems in time and space, and there have been practically no estimates of the horizontal convergence of this flux. Surprisingly, too, the geographical distribution which can be obtained easily from numerical general circulation model runs has not been examined. Although, because of observing network inadequacies, the Southern Hemisphere is hardly the best region for which to compare model results with direct observations, material to do this is available in the form of the CMRC (Commonwealth Meteorology Research Centre, now ANMRC: Australian Numerical Meteorology Research Centre) 6-level/N30 general circulation model results for a normal March sea-surface temperature distribution (Simpson & Downey, 1975) and analyses of EOLE balloon trajectories in the Southern Hemisphere (Webster & Curtin, 1975). Results are shown for the 200 mb level for both the geographical distribution of $\overline{v'T'}$ (Fig. 2.2.7(a)) and the derived meridional profiles (Fig. 2.2.7(b)). These exhibit similar gross features, in particular the variation with latitude, but the longitudinal variations are not similar. Further comparative studies using more detailed numerical models for the better observed Northern Hemisphere should indicate whether the statistical features of transient eddies are satisfactorily represented in general circulation numerical models.

Although a physical basis for parameterising these fluxes in climate

---

* The vorticity equation as used in meteorology relates the rate of change of the horizontal circulation per unit area to the horizontal divergence. It is derived directly from the basic equations of motion.

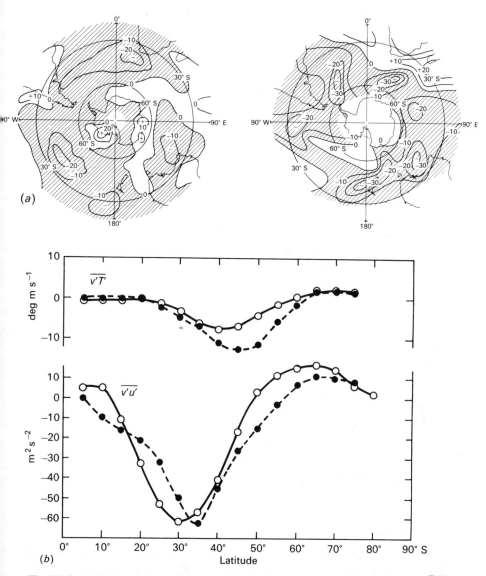

Fig. 2.2.7. (a) Patterns of meridional transient eddy heat flux (depicted in terms of $\overline{v'T'}$ (deg m s^{-1})) at about 200 mb from the ANMRC model for March (left) and from EOLE balloon data for March–May (right).

(b) Zonally averaged values of the same quantities as in (a) and of the covariance of the corresponding meridional and zonal velocity component. —○— ANMRC model (March); —●— Webster & Curtin (MAM).

models in terms of the quasi-steady state variables has been derived from considerations of baroclinic theory (Green, 1970), the fluxes are essentially due to finite-sized disturbances. Also, the lack of correspondence between, for example, regions of maximum $\overline{v'T'}$ (Newell *et al.*, 1972) in the Northern Hemisphere and regions of greatest baroclinic activity is somewhat disconcerting. For this reason empirical studies of large-scale flux:gradient relations may yet prove valuable.

One possibly useful feature has emerged from a detailed study (Tucker, 1977) of three-hourly upper wind and temperature soundings made at Laverton, Melbourne, during September–October 1967 and of standard upper air observations at Australian stations. The observed local eddy flux of sensible heat was resolved down the horizontal temperature gradient (obtained from thermal wind considerations). A distinct three-tiered atmospheric structure emerged. The results are similar to the zonally averaged scheme outlined by Newell (1964).

Because climate and particularly climatic change is concerned with longitudinal as well as latitudinal distribution of transient eddy fluxes and their divergence patterns, for the explicit general circulation models it will be important to compare model distributions with observed distributions. For statistical-dynamical models a major requirement will be a parameterisation scheme to represent transient effects in terms of steady state variables. The work referred to above has shown that vertical structure will need to be considered and that vertically integrated parameterisation schemes (e.g. Clapp, 1970) have serious limitations.

Of course other major problem areas exist, that presented by vertical velocity (both for the steady and transient states) having already been mentioned. As another example, the best parameterisations of eddy contributions are likely to contain inaccuracies which are large compared with the residual local rates of change which describe the seasonal progression of climate. Therefore it may be necessary to apply some form of computational constraint in the time integration of the equations in order to obtain stable and realistic results.

Uniqueness of the general circulation

From all that has been said so far it might seem reasonable to assume that the atmospheric general circulation represents a unique, stationary system in the sense that, given no change in external forcing, it would eventually settle down to a statistically steady state. However, following studies by Lorenz (1968), it is now believed that, even with all boundary conditions fixed, there is no certainty that the time series of atmospheric patterns would settle into a statistically steady state; and hence no assurance that the 'climate' would be unique in the sense that the statistics would be independent of the initial state. This situation, if it proves to be so, is related to the non-linearity of the dynamical and thermodynamical equations governing the behaviour of the atmosphere and is referred to as the phenomenon of 'almost-intransitivity'. If all initial states of a physical system lead eventually to the same set of statistical properties it is said to be transitive. If there are two or more sets of statistical properties and some initial states lead to one set while others lead to another, the system is intransitive. If there are different sets of statistical properties which a transitive system may assume in evolving from different initial states through a long, but finite, time span the system is called almost-intransitive. Because the atmosphere is continually subject to disturbances it might be thought that it could not be an almost-intransitive system without showing greater annual and decadal excursions than it does. But it could be that its behaviour is transitive on short time scales and almost-intransitive on long time scales. It has been suggested, for example, that the glacial–interglacial fluctuations of the last million years with their very different circulation patterns may represent the transitions of an almost-intransitive system.

Despite the spectre of almost-intransitivity and the many present gaps in our understanding of the mechanisms of the general circulation, the development of ever-more-refined numerical simulation models of the atmosphere (Chapter 5) and observational studies being undertaken within the framework of the international Global Atmospheric Research Programme promise substantial progress in the decades ahead. All the signs point to the emergence of exciting and profitable lines of research in the numerical study of climate and climatic change.

2.3 The role of the oceans

B. V. HAMON & J. S. GODFREY

The oceans influence the formation of climate through their exchange of energy, momentum and solid, liquid and gaseous matter with the atmosphere at its lower boundary. The patterns of exchange differ markedly from those over the world's land and ice surfaces, the main differences arising from the enormous capacity of the oceans to store and redistribute horizontally the incident solar energy. Almost one-half of the solar radiation absorbed by the earth–atmosphere system passes into the surface layers of the tropical oceans. Some is given up locally to the overlying atmosphere to play a part in driving the atmospheric general circulation, but a significant amount is carried off to higher latitudes and other parts of the globe by wind-driven ocean currents.

Compared with the atmosphere, the total mass of the oceans is about 280 times greater and their heat capacity is nearly 1,200 times as large

31

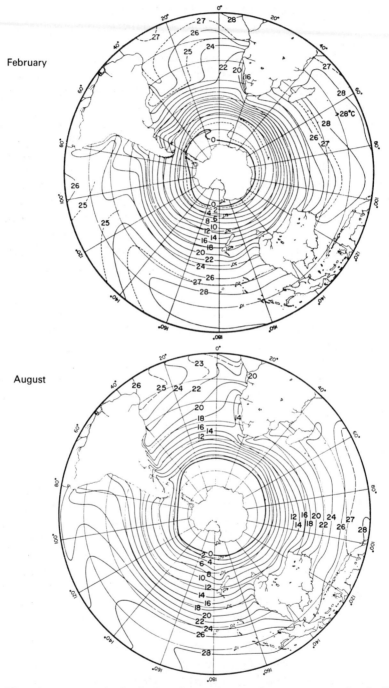

Fig. 2.3.1. Large-scale distribution of normal sea-surface temperatures for the Southern Hemisphere for February and August (after van Loon *et al.* (1972)).

(Kraus, 1972). By virtue of their very large thermal and mechanical inertia, the oceans play the role of a fly-wheel in the climatic system.

In the following pages we describe the large-scale features of the structure and circulation of the Southern Hemisphere oceans, including the major water-mass boundaries and the Antarctic pack ice. These ocean features profoundly influence the behaviour of the atmosphere and are themselves the product of interaction with the atmosphere, and accordingly some attention is given to the nature and distribution of the surface fluxes of momentum, heat and water vapour. However, it is not enough simply to describe the ocean flow patterns and temperature structure. To understand the role of the ocean in climatic change, we need to know how a change in the atmosphere will alter the ocean circulation, and how this will in turn react back on the atmosphere: and for this purpose we must know something of the internal dynamics of the ocean. Ocean dynamics is still a fairly new subject, but towards the end of the chapter we will describe some results from it that are thought likely to be important for climatology.

Structure
Temperature

Mean sea-surface temperatures for the Southern Hemisphere for February and August, roughly representative of the first half of the twentieth century, are shown in Fig. 2.3.1. Note that the western portions of the oceans are generally several degrees warmer than the eastern portions and that the summer-to-winter change is largest (around 5 °C) at latitudes 30°–40° S.

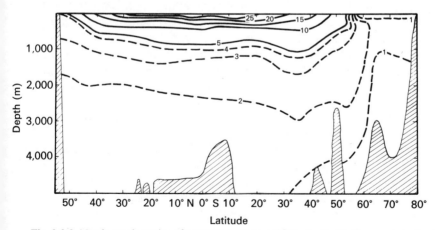

Fig. 2.3.2. North–south section of water temperature (°C) in the Pacific Ocean approximately along the 170° W meridian.

2.3   Physical basis of climate

Figure 2.3.2 shows a vertical temperature section running approximately north-south in the Pacific Ocean. In its main features it is broadly representative of any of the world's oceans. Note the thin, warm (> 20 °C) surface layer in the tropics and the almost vertical isotherms near Antarctica. This latter feature means that there is little variation in temperature as a function of depth so that, in contrast to the tropics, the stratification in high latitudes is weak.

Salinity

Salinity varies little horizontally in the open ocean and will not be discussed here except to note that its spatial distribution is related to the difference between evaporation and precipitation and that, along with temperature, it determines the density and thus the stability of stratification in the ocean.

Circulation

Figure 2.3.3 shows the main features of the surface (0–50 m approximately) circulation of the Southern Hemisphere oceans. Most of our information about surface currents has come from analysing the information obtained from the navigation of ships and from geostrophic calculations based on study of the distribution of sea-water characteristics. Maximum values for surface currents are of the order of 2–2.5 m s^{-1}. From other sources we know that, in most places, these strong surface currents decay rapidly with depth, falling to about half their surface value at 300 m below the surface. It is important to bear in mind that the surface circulation patterns shown in Fig. 2.3.3 indicate only the main long-term average features. As in the distinction between 'weather' and 'climate' in the atmosphere, the oceans show quite complex and ever-changing circulation patterns when looked at in more detail. This variability of the circulation is of particular importance to climate through the influence of the associated temperature variations on heat and moisture exchange with the atmosphere. We will return to a discussion of ocean variability later.

It may be seen from Fig. 2.3.3 that, at the surface, the currents in the tropics are mainly westward. At the western sides of each ocean basin, the currents turn poleward, becoming the narrow, strong features called 'western boundary currents' (cf. the Gulf Stream of the North Atlantic). At higher latitudes, currents are generally eastward. A broad, weak equatorward flow in the middle and east of each basin, at mid-latitudes, completes the circulation; this return flow is not conspicuous in Fig. 2.3.3, because direct wind influence apparently cancels this equatorward movement in the thin surface mixed layer. However, we can summarise this description of the mean near-surface flow by saying there is an

34

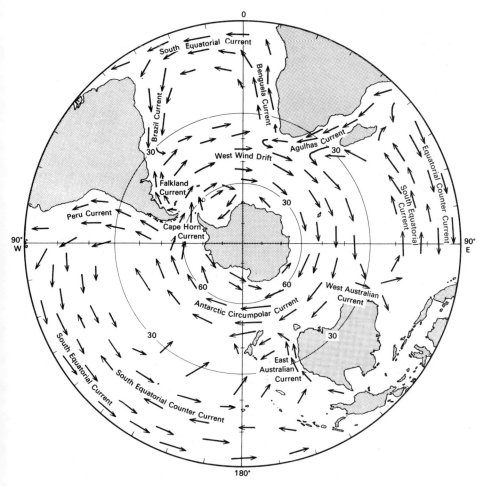

Fig. 2.3.3. The main features of the surface circulation of the Southern Hemisphere oceans.

anticyclonic (anticlockwise in the Southern Hemisphere, clockwise in the Northern Hemisphere) gyre in each main ocean basin. In the Southern Hemisphere, the Antarctic Circumpolar Current or West Wind Drift is a special feature, with no counterpart in the Northern Hemisphere. Each of these current systems is described in a little more detail below.

Equatorial currents

In both the Atlantic and Pacific Oceans, the westward South Equatorial Current extends from between 10° and 20° S to north of the equator, while in the Indian Ocean it extends only a little equatorward of 10° S with a

35

2.3 Physical basis of climate

well-developed counter current between about 2° and 8° S. The equatorial currents are relatively shallow and, in the Pacific, a strong subsurface counter current (the Cromwell Current or Equatorial Undercurrent) with its core at a depth of only 50–200 m lies directly along the equator (Tsuchiya, 1970). A broadly similar, though less well documented, feature is found also in the Atlantic (Gordon, 1975) and Indian Oceans.

Western boundary currents

Of the three western boundary currents of the Southern Hemisphere, only the Agulhas Current is comparable in strength to the Gulf Stream of the North Atlantic and the Kuro Shio of the North Pacific. The Agulhas Current flows in a general southwesterly direction along the continental shelf off the southeast African coast between 25° and 40° S; it is, in all seasons, the fastest-flowing and warmest of the Southern Hemisphere western boundary currents. Where the Agulhas water merges with the West Wind Drift, exceedingly steep temperature gradients occur and large eddies develop (Darbyshire, 1964). The East Australian Current has generally been thought of as a single, strong, narrow, southward-flowing current paralleling the continental slope between about 25° and 35° S. It is, in fact, a rather complex and variable system characterised by a series of large anticyclonic eddies (diameter 300–500 km) and strong surface currents up to 2 m s^{-1} (Hamon, 1965; Boland & Hamon, 1970). Little information is available on the Brazil Current.

Equatorward currents

Of the equatorward-flowing currents, the Peru (or Humboldt) Current is notable for its latitudinal extent, its coldness (mainly due to upwelling) and the fact that, rather than consisting of a broad equatorward flow in the central and eastern parts of the ocean, it is concentrated in a quite narrow zone near the South American coast. It is complex and characterised by subsurface and surface countercurrents. Like the Peru Current, the Benguela Current in the South Atlantic is characterised by strong upwelling with the coldest water within 150 km or so of the coast between about 15° and 35° S (Fig. 2.3.1). The West Australian Current, on the other hand, is broad and weak and characterised by the absence of significant upwelling.

Antarctic Circumpolar Current

The Antarctic Circumpolar Current has probably the largest volume transport of any current system of the globe. The entire circumpolar ring of

36

water from latitude 40° S to approximately 63° S is subject to a general eastward drift with maximum velocities near and somewhat equatorward of the Antarctic Convergence or Polar Front (Ostapoff, 1965). The Circumpolar Current is some 21,000 km long and its axis varies in latitude by nearly 20°, mainly in response to the bottom topography (Gordon, 1973). It is believed to extend with relatively little attenuation to the sea floor passing through narrow passages in places, paralleling submarine ridges for long instances in others and becoming more diffuse as it passes over broad, flat basins. In places, the current may be multi-axial and there is evidence of substantial fluctuations of its volume transport with time.

Deep circulation

The movements of water in the deep ocean (2,000–5,000 m) are not well known or understood. It seems clear, however, that these deep waters are formed by sinking of surface water near Antarctica (mainly in the Weddell Sea) in winter. Water-mass properties in the Southern Ocean suggest that the meridional circulation pattern is as indicated in Fig. 2.3.4, which also shows the probable relationship of this circulation to the major frontal zones of the region.

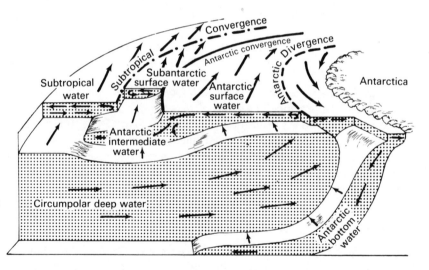

Fig. 2.3.4. Schematic depiction of the meridional distribution of water masses and meridional circulation in the Southern Ocean.

2.3 Physical basis of climate

Oceanic fronts
Antarctic Convergence

The Antarctic Convergence has been one of the most intensively studied features of the Southern Ocean. Its most outstanding characteristic, from the point of view of effects on the atmosphere, is a sharp southward drop of sea-surface temperature, often of several degrees within a few kilometres. First identified as a circumpolar feature by Meinardus (1923) and subsequently mapped by Deacon (1937) and Mackintosh (1946), it became known as the Meinardus Line, the Oceanic Polar Front and, following Deacon (1937), the Antarctic Convergence. During ten years of *Eltanin* cruises in the Southern Ocean 1962–72 (Capurro, 1973), the Antarctic Convergence was studied in considerable detail and its location mapped on the basis of both surface and subsurface data. The position given in Fig. 2.3.5 is from Gordon & Goldberg (1970).

Subtropical Convergence

The Subtropical Convergence is a second temperature discontinuity, some ten degrees of latitude north of the Antarctic Convergence. It is a much more variable feature than the Antarctic Convergence and frequently displays a streaky or ribboned structure. The mean position given in Fig. 2.3.5 is from Deacon (1966).

Other frontal zones

Two other frontal zones of the Southern Ocean are of particular interest. The Antarctic Divergence is a zone of upwelling at the boundary of the West Wind Drift and the Antarctic coastal East Wind Drift. It is a diffuse and rather variable feature. The Australasian Subantarctic Front, on the other hand, appears as a persistent sharp temperature discontinuity located a few degrees of latitude north of the Antarctic Convergence in the Australasian sector.

Pack ice

Figure 2.3.5 shows also the approximate extent of the Antarctic pack ice at the time of maximum extent (September–October) and minimum extent (February–March). Prior to the advent of meteorological satellites, knowledge of the seasonal behaviour of sea-ice was based primarily on early ship surveys (e.g. Mackintosh & Herdman, 1940) and maps depicting the mean location of the ice edge on a monthly basis are given by US Navy Hydrographic Office (1957) and Tolstikov (1966). Early satellite studies (e.g. Knapp, 1967) drew attention to the variability of the ice cover and

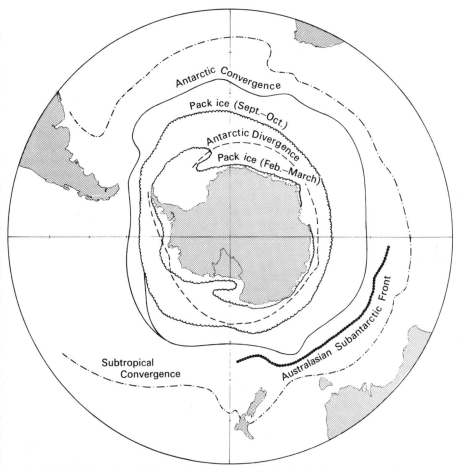

Fig. 2.3.5. The location of major surface discontinuities in the Southern Ocean. The position of the Antarctic Convergence is based on Gordon (1967) for the South Pacific and on the location of the salinity minimum at 200 metres (Gordon & Goldberg (1970)) for the Indian and Atlantic sectors. The Subtropical Convergence is as given by Deacon (1966); the Antarctic Divergence by US Navy Hydrographic Office (1957); the Australasian Subantarctic Front by Gordon (1973) and Zillman (1970); and the mean extent of the pack ice at the time of maximum coverage (September–October) and minimum coverage (February–March) by Tolstikov (1966).

the persistence of open water areas, even in winter, within the pack-ice limits. Since 1967 a variety of satellite imagery, including microwave imagery (Ackley & Keliher, 1975) has become available for observing the variation of sea-ice around Antarctica on a large scale. The recent data confirm the main features of the earlier pictures: that the ice boundary expands most rapidly from May to July; that it changes little from July to November and that melting proceeds rapidly in December and January.

2.3 Physical basis of climate

Ocean–atmosphere interaction

It is through their interaction with the atmosphere by exchange of momentum, heat, moisture and carbon dioxide that the oceans exert their dominant influence on climate. The exchanges actually take place through the agency of small-scale processes right at the sea–air interface but their effects are evidenced in the larger-scale behaviour of both media over a wide range of time periods. The atmosphere exerts windstress upon the ocean surface and drives ocean currents which transport relatively warm water slowly to higher latitudes and bring cold water to more tropical regions. The oceans give up heat and moisture where the overlying atmosphere is colder and drier and, in turn, influence the development and evolution of the winds.

There are numerous physically based methods for measuring or estimating the fluxes of momentum, heat and moisture across the sea–air interface, most of them rather difficult (Pond, 1975), often involving special equipment and thus restricted to specific sites and time periods. In the context of large-scale ocean–atmosphere interaction, the exchanges can be conveniently represented in terms of 'bulk' parameters of the sea–air interface. Thus the momentum flux into the ocean (τ), the sensible heat flux from the ocean to atmosphere (H) and the evaporation rate (E) are given by

$$\tau = \rho C_D u_a^2, \tag{2.3.1}$$

$$H = \rho c_p C_H (T_s - T_a) u_a, \tag{2.3.2}$$

$$E = \rho C_E (q_s - q_a) u_a, \tag{2.3.3}$$

where ρ is the density of air and c_p its specific heat, $(T_s - T_a)$ and $(q_s - q_a)$ are sea–air differences of temperature and specific humidity, u_a is the wind speed near the surface and C_D, C_H and C_E are non-dimensional bulk transfer coefficients for momentum, heat and water vapour. Though the values of C_D, C_H and C_E actually depend on a number of factors not always known, and certain approximations must be made in applying the formulae to climatic data, useful estimates may be obtained for the world oceans by using the empirically derived approximation $C_D = C_H = C_E = 1.5 \times 10^{-3}$.

Wind stress

The wind stress is a vector quantity and may be represented in terms of its zonal and meridional components. To obtain long-period mean values of the wind stress it is necessary to compute the eastward and northward components of the stress from individual winds and average these.

Maps of long-period mean wind stress for the oceans have been derived by various authors (e.g. Hellerman, 1967, 1968) and used as a forcing function to infer the time-averaged pattern of surface currents.

Sensible heat flux

In most parts of the world, sensible heat passes from the ocean to the atmosphere most of the time, the temperature of the sea surface being, on the global average, a little less than 1 °C warmer than the air a few metres above. It is evident from Equation (2.3.2) that the heat flow to the atmosphere is greatest where strong cold winds blow over much warmer ocean. With heat flow from ocean to atmosphere, the atmosphere becomes less stable and deep convection is favoured. The heating of the atmosphere may, either directly or indirectly, act to generate or maintain atmospheric circulation systems of larger scale. At the same time the heat loss from the sea surface (once any thermocline* is eliminated) induces an unstable stratification and deep convection in the sea.

When heat flows from the atmosphere to the ocean as when relatively warm air passes over regions of wind-induced upwelling of cold water, a stable stratification is induced in both air and sea, inhibiting convection and tending to limit the heat exchange and temperature changes to layers close to the surface. Global patterns of sea–air heat exchange are influenced by the location and strength of warm and cold currents, the disposition of land and sea, distribution and density of sea-ice cover, the location of oceanic fronts and so forth.

Evaporation

Evaporation from the oceans is the main source of moisture for the atmosphere and thus, through the subsequent release of latent heat of condensation, the primary energy source for the global atmosphere circulation. The average evaporation rate is about 100 cm/yr. The range is between about 10 and 150–200 cm/yr depending mainly on latitude with the most intense evaporation taking place in the Trade Wind belts of the tropical oceans. The actual evaporation rate at any given place and time depends very markedly on season and weather, being greatest, as evident from (2.3.3), where winds are strong and humidity of the air is low.

Heat budget of the ocean

The heat balance of the ocean at any point is (except for a few minor terms that will be neglected) given by

$$R - LE - H = S - T, \tag{2.3.4}$$

* A thermocline is a layer of water with a more intensive vertical temperature gradient than in the layers above and below it.

41

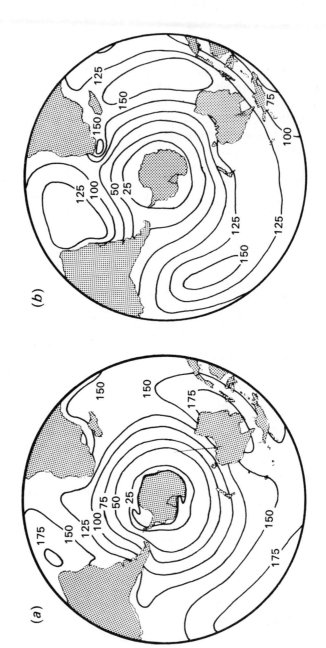

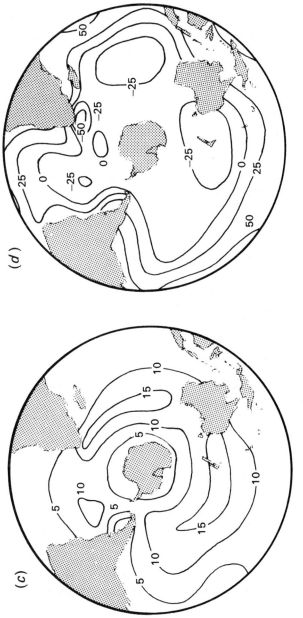

Fig. 2.3.6. Annual mean patterns of ocean-surface heat budget components (W m^{-2}) from Zillman (1972a) interpolated from the atlases of Budyko (1963) and Tolstikov (1966): (a) net radiation R, (b) latent heat flux LE, (c) sensible heat flux H, (d) heat balance $R - LE - H$. Note that the evaporation rate E in centimetres per year can be obtained by multiplying the LE values (W m^{-2}) by 1.32.

where R is the net radiation income (equal to the absorbed solar short-wave radiation minus the net upward long-wave radiation from the sea); LE is the heat expended in evaporating water from the oceans (and thus equal to the product of the evaporation rate E and the latent heat of evaporation L); and H is the sensible heat flux from the ocean to the atmosphere. On the right-hand side, S is the rate of storage of heat within the ocean column and T the rate of heat transfer away to other parts of the ocean.

In middle and high latitudes, the heat budget experiences a pronounced annual cycle dominated by the solar radiation. The oceans store heat as the surface layers warm up during the summer and the heat is given up in winter. Over a full year the net storage of heat is quite small and consideration of annual values of the surface heat balance, $R-LE-H$, indicates the regions of excess heating (mainly in the tropics) from which heat is carried by the mean ocean currents, the transient eddies and slow meridional overturning to make up the deficit in other (mainly polar) regions.

Figure 2.3.6 shows the broadscale annual patterns of the surface heat balance components for the Southern Hemisphere oceans based on the atlases of Budyko (1963) and Tolstikov (1966). Note particularly the regions of very high evaporative heat loss by the low-latitude oceans and the location of the regions of maximum heat loss to the atmosphere in the central and western parts of the major ocean basins. The net heat balance pattern as shown is very uncertain though, in its gross features, it finds some support from independent evaluations by other authors (e.g. Albrecht, 1960; Privett, 1960). Fig. 2.3.7 shows the meridional variation of the various heat-budget components and the poleward heat flow within the oceans required for balance.

The contribution of the various mechanisms to the horizontal heat transfer in the ocean and the roles of individual current systems are poorly known, particularly in the Southern Hemisphere.

Ocean variability and climate

It is evident from the discussion of ocean–atmosphere interaction and the role of the oceans in the global heat budget that an understanding of the variability of sea-surface temperature is central to questions of climatic variation. There are a number of dynamic processes which can cause sea-surface temperature to vary, on time scales from a few months to hundreds of years; here we shall discuss some of them.

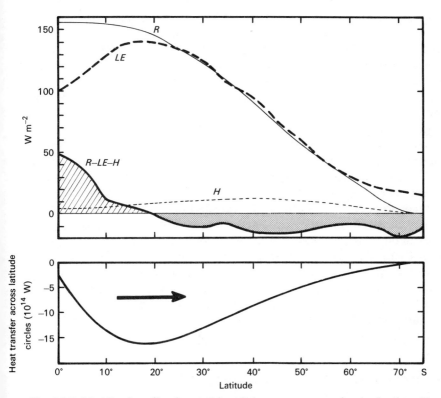

Fig. 2.3.7. Meridional profile of annual heat balance components for the Southern Hemisphere oceans and the poleward heat transfer in the oceans required for balance (based on the same data sources as Fig. 2.3.6; from Zillman (1972 *a*).

Passive response of the surface mixed layer to atmospheric exchange

In regions of the ocean with slow surface currents, a simple one-dimensional model of the upper hundred metres or so of the ocean can successfully predict sea-surface temperature variations over periods of up to one or two months. A recent example is that of Denman & Miyake (1973). A model of this kind assumes that sea-surface temperature changes passively, in response to purely local weather conditions; in each time-step of the model, allowance is made for the exchange of solar and thermal radiation between air and sea, and where necessary for the latent heat loss and sensible heat exchange. It is assumed that the only sources of mixing are convection (whenever surface water becomes denser than the rest of the mixed layer) and mixing by wind stirring.

Order of magnitude estimates suggest (Gill & Niiler, 1973), that such

45

a model would reproduce fluctuations in sea-surface temperature, with time scales of up to a few months, over the bulk of the Southern Hemisphere oceans; however, throughout the oceans, oceanic advection must become important on a long enough time scale. Suggestive evidence for this comes from Namias (1970 *a*). He prepared three-monthly average sea-surface temperature maps for the northern North Pacific (35°–55° N), and for the southern North Pacific (20°–35° N), between 1947 and 1966. Lagged correlations between the two time series showed rather little correlation beyond a few seasons, except in one case – when winter sea-surface temperatures in the southern North Pacific were correlated with winter sea-surface temperatures in the northern North Pacific, 2 years earlier. The correlation in this case was significant beyond the 1 % level. Namias proposed a very interesting explanation for this phenomenon; he suggested that in the northern North Pacific, in winter, a mixed layer of great depth is formed by vigorous wind mixing, so that the temperature anomaly for that season is imprinted on a very large water volume. In the next few seasons, this water is overlain by lighter, warmer layers, and does not show up in the sea-surface temperature; but two winters after the initial event, a large fraction of this water has drifted around to the southern North Pacific. Winter winds then stir some of the 2-year-old water back to the surface, to recreate the original anomaly.

Variability of tropical and temperate currents

Ocean currents all shift their positions, either due to their own intrinsic instabilities or to fluctuations in wind and heating. The resulting advective effects on sea-surface temperature are already known to be climatologically important in certain parts of the ocean. For example, the Kuro Shio sometimes moves sharply away from the coast of Japan and stays there for some years, leaving a wide region of cold water next to the shore; this has a marked effect on Japanese climate (Stommel, 1965). In the North Atlantic, Gulf Stream meanders appear to be very important south of Newfoundland, where unusually large sea-surface temperature anomalies (2 °C) often cover an area of about 10^6 km²; Ratcliffe & Murray (1970) studied the relation of these anomalies with European weather, and found them to have a statistically significant influence.

The East Australian Current also fluctuates greatly, and this is probably responsible for the observed large variability in sea-surface temperatures over the whole Tasman Sea (e.g. Trenberth (1975 *a*)). The connection between ocean dynamics and sea-surface temperature is suggested by Fig. 2.3.8, which shows a typical pattern of 'dynamic height' in the Tasman Sea; the figure may be read in two ways. First, wherever high dynamic heights are found (as in the centre of the eddy southeast of Sydney), the

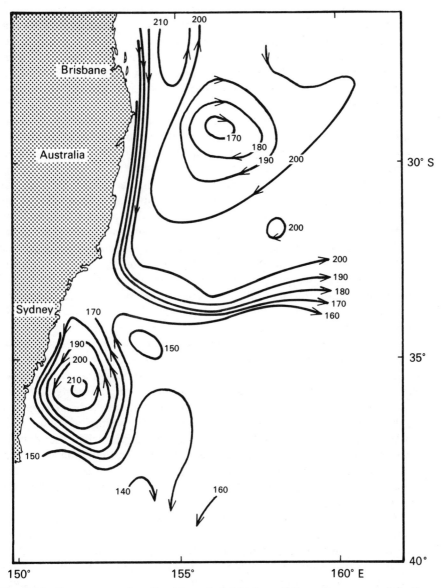

Fig. 2.3.8. Contour map of surface dynamic height (essentially surface pressure) for 8 November–4 December, 1960. The contour interval is the equivalent of 10 cm in height of the sea surface (CSIRO, 1963).

water is several degrees warmer than in surrounding low dynamic height regions, down to several hundred metres. Secondly, the figure may be regarded as an isobar map: the East Australian Current, and flow around the eddy, run parallel to the dynamic height contours. Like isobar patterns in the atmosphere, these dynamic height patterns vary drastically from one month to another. When dynamic heights are higher than usual in the Tasman, the East Australian Current flows further south, and sea-surface temperature is warmer.

It is probable that, when the data become available, variability in ocean currents will be found to have important effects on long-term sea-surface temperature changes throughout the ocean; we therefore need to understand what causes currents to change their positions.

The main driving force for the ocean currents is the wind stress; this sets up an 'Ekman transport' within the top hundred metres or so of the ocean. Due to effects of the earth's rotation, this water transport is to the left of the wind in the Southern Hemisphere, and to the right in the Northern Hemisphere. Inspection of mean wind patterns (or mean surface atmospheric pressure patterns such as that shown in Fig. 2.2.1 (*b*)) suggests that this will pile water up in the central regions of the semipermanent anticyclones and will also pull deep cold water to the surface along the equator and the eastern oceanic boundaries. The main ocean currents develop in response to the wind-induced upwelling and downwelling. A theory of the ocean's time-dependent response to this driving function was developed by a number of workers in the 1940s and 1950s; an elegant synthesis of their ideas is found in Lighthill (1969). Lighthill treated ocean currents as frictionless, linear perturbations of a uniformly stratified ocean; he was then able to predict a number of features of the response of the Somali Current to the onset of the South West Monsoon, using only simple analytic theory. While observations suggest that the current reverses in days after the wind pattern changes, rather than in several weeks as Lighthill predicted (Düing & Szekielda (1971), Leetmaa (1972)), much of the discrepancy could probably be accounted for within the context of Lighthill's linear, frictionless formulation.

If the ocean were truly linear and frictionless, any given disturbance would excite a well-defined pattern of waves. No other data would be required but the bottom topography and the average vertical distribution of density in order to determine these waves and hence to determine completely the ocean currents that would develop in response to a given wind pattern. While the linear theories have certainly had some semi-quantitative successes in explaining both steady-state flow patterns and variability in the ocean, they may unfortunately have important limitations. The main reason is that eddies with a length scale of order 100–200 km are found throughout the subtropical ocean with velocities typically ten times the mean flow. Although these eddies are small compared to ocean size, they may nevertheless influence large-scale ocean dynamics (e.g. Gill *et al.* (1974)).

Furthermore, if linear theory is somewhat inaccurate in the bulk of the ocean, it is certainly inadequate in the western boundary currents. For example, warm eddies continually form, move southwards and decay within the East Australian Current (e.g. Fig. 2.3.8). Dynamic study of them is possible, because eddies that were similar in length scale, vertical structure and other properties developed spontaneously in the western boundary current of a numerical model (Bryan & Cox, 1968 *a*, *b*; Godfrey, 1973 *a*, *b*). The outflow from the model current was subject to shifts in position of order 1,000 km. These shifts were due to the instability process, rather than to changes in wind and heating; similar internal oceanic processes may help to explain the observed changes in sea-surface temperature in the East Australian Current. Fluctuations of the Agulhas and Brazil Currents may also create climatologically important sea-surface temperature anomalies in certain areas.

Equatorial flows

The equator is a special place, in the ocean as well as the atmosphere, because the Coriolis parameter* changes sign there. Sea-surface temperature variations at the eastern end of the equatorial Pacific are particularly severe, and Bjerknes (1969) has suggested that these anomalies are connected with the equatorial atmospheric circulation in a positive feedback mechanism. Bjerknes' hypothesis is that when eastern equatorial Pacific water is colder than usual, high atmospheric pressures develop above it, driving an easterly wind down the equatorial pressure gradient. On either side of the equator, and due to the earth's rotation, this wind creates an Ekman drift away from the equator. This results in cool water from 100 to 200 m depth being brought to the surface at the equator. The upwelling tends to maintain the cold anomaly that gave rise to the easterly winds in the first place. Warm anomalies similarly feed back to accentuate weak easterly winds. Warm anomalies have the local name of 'El Niño', and their effect on Peruvian marine and bird life can be drastic.

Ekman drift, and the associated upwelling, may not be the only oceanic phenomenon contributing to development of El Niños. If the equatorial easterlies slacken, the warmer waters in the western equatorial Pacific will tend to surge eastward; the surge should propagate via equatorial 'Kelvin' waves, which move east at about 100–200 km/day (Gill & Clarke, 1974; Godfrey, 1975). Similar ideas have been used by O'Brien & Hurlburt (1974)

---

* The Coriolis parameter (proportional to the angular rotation rate of the earth and the sine of the latitude) is a measure of the effects of the earth's rotation on the motion.

to model Wyrtki's (1973 a) observations of an equatorial jet across the Indian Ocean, and Wyrtki's (1973b) observations of sea levels at near-equatorial Pacific islands suggest eastward-moving disturbances with speeds of about 100 km/day. Observations of El Niños are so far confined almost exclusively to the ocean surface; so the depth distribution of temperature during warm anomalies is not known. However, the region has been under increasing scrutiny in recent years (e.g. Wooster & Guillen, 1974; Ramage, 1975). While many discrepancies between theory and observation need to be reconciled before sea-surface temperatures can be adequately predicted, new developments are occurring rapidly.

The Southern Ocean

It is known that temperature anomalies of considerable magnitude are associated with variability in the Antarctic Circumpolar Current and the shifts in location of various oceanic fronts. There has been considerable discussion in the literature regarding the possible migration of the Antarctic Convergence itself. Deacon (1963) asserts that it is practically stationary, reported positions of the Antarctic Convergence in a given longitude generally falling within 100 km of the mean.

If so, variations in sea-surface temperature due to shifts of the Antarctic Convergence may be too small to be of climatological importance. Since there are strong indications that the path of the current is controlled by bottom topography (Gordon & Bye, 1972), this may explain the stationarity of the Antarctic Convergence.

Bryan & Gill (1971) ran an idealised numerical model of the Antarctic Circumpolar Current and found that its transport and other properties were sensitive to the geometry of the Drake Passage; wherever a circle can be drawn at a given latitude and depth that lies within the ocean all round the world – and this only occurs in the Drake Passage – there can be no net geostrophic transport across the circle, because the pressure gradient along it must integrate to zero. Bryan and Gill emphasised the importance of this constraint, and suggested that it was responsible for the existence of the Antarctic Convergence. Nevertheless, the study of the Circumpolar Current remains difficult theoretically, and is (not surprisingly) hampered by lack of data. It is hoped that the work of the International Southern Ocean Study (ISOS) over the next few years may help to correct this situation. The position of the Subtropical Convergence is much more variable than the Antarctic Convergence (Deacon, 1963), and it may be more profitable to regard it as composed of the southern end of the three Southern Hemisphere western boundary currents, rather than as a single entity.

Antarctic pack ice

A final ocean area whose fluctuations may be of climatological importance is the region next to the Antarctic Continent, where pack ice forms. Budd (1975 c) has studied the relation between maximum pack-ice cover in a given year, obtained from satellite records since 1967, and annual average air temperature at a number of observation stations around the Antarctic Continent. The two quantities appear to be closely correlated, a 1 deg C fall in air temperature at a given station corresponding to an increase in pack-ice extent of about 2.5 degrees of latitude opposite that station. Over the 18-year period 1956–74, patterns of annual mean air temperature are available around the continent; generally, there is a tendency for an increase in air temperature in one sector of the Antarctic to be compensated by a decrease elsewhere. This is also true of the variations in maximum pack-ice cover, a pattern that Budd suggests may be related to the Southern Oscillation (Chapter 4).

A strong case for the importance of the effect of variations in Antarctic pack ice on global climate has been presented by Fletcher (1969). Ice formation affects the density structure by raising the salinity of the unfrozen water and it exerts a profound influence on the surface heat budget by changing the albedo and effectively shutting off moisture and heat flow to the atmosphere. On the longer time scales it modulates the heat transport associated with bottom water formation and replacement with inflowing warm water. The cause–effect relationships between the pack-ice extent and atmosphere behaviour are extremely complex. Using satellite-borne scanning microwave radiometer data for 1973–4, Ackley & Keliher (1975) examined the time variation of the pack-ice cover in four sectors around Antarctica and its relation to the atmospheric circulation. They found that the largest part of the area within the ice edge is less than 80% ice-covered even during the coldest part of the year, probably in response to convergence and divergence in the atmospheric forcing fields and to ocean currents, waves and swell. An observed rapid decrease in pack-ice extent in the Bellingshausen Sea during the winter period of 1973 was found to correlate with a nearly real-time adjustment by the atmosphere to the change in its heat loss caused by the removal of the ice. The authors conjecture that the atmosphere responded to the relative heat gain associated with the removal of ice by an increase in storm activity which then contributed to maintaining the low ice cover. On seasonal time scales, Schwerdtfeger & Kachelhoffer (1973) find a highly significant relationship between the position of the pack-ice border and the latitudinal band of maximum frequency of cyclonic vortices.

2.3   Physical basis of climate

Importance of oceanography in climate research

It will be clear from this section that dynamic oceanography – the science of relating observed patterns in ocean circulation back to their causes in surface fluxes of heat, water and momentum – is still in its infancy. Ocean fluctuations affect the world climate, and vice versa, in a variety of different ways; the subtlety and complexity of the ocean as a component of the world's climate system is only now beginning to be appreciated.

3. The long-term climatic record

3.1 Cenozoic climates: Antarctica and the Southern Ocean

L. A. FRAKES

Data on past climates of far southern latitudes have been derived both from land evidence on the Antarctic continent and from the more accessible record preserved in sediments of the Southern Ocean. They are of highly variable nature and can be interpreted with varying degrees of confidence. The result is that there are a few quite safe and widely accepted bench-marks along the evolutionary trend of southern palaeoclimates and a large group of facts for which interpretations are conflicting or uncertain. In this paper an effort is made to differentiate these categories, in part through the methods by which data are gathered and their interpretations generated.

It is necessary to reach back 25 million years (m.y.) or more into geologic time (Fig. 3.1.1) to set the stage for the beginning of the present glacial cycles. The bulk of reliable and detailed information from geological sources covers only the interval of the last 4.5 m.y. largely because of the limited resolving power of most age-dating techniques. Happily, however, this is the most important era in considering the effects of immediate climate change on man.

A significant lesson to be learned from the history of these 4.5 m.y. is that palaeoclimatological events observed in one part of Antarctica seldom correspond precisely in time with similar events elsewhere on the continent, nor indeed with events on either hemispheric or global scales. It may be that local or short-term changes in climate obscure the geological remnants of the regional and long-term evolution of climate; or that dating is not accurate enough; or that climate change proceeds over the globe in a diachronous fashion, so that the effects of cooling, for example, are felt first in polar regions and last in the equatorial zone. The latter case will be examined in some detail while dealing with global correlation problems.

Climates in the early Cenozoic

A significant event in the climatic history of high southern latitudes took place when glaciers reached the sea for the first time. Temperatures in

53

3.1 The long-term climatic record

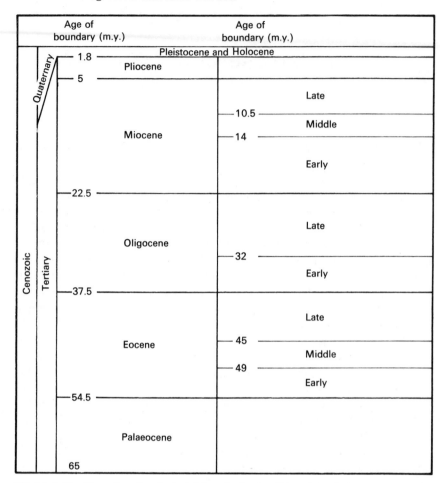

Fig. 3.1.1. The hierarchy of geological time units for the Cenozoic period, 0–65 m.y. before present (BP). Age of boundaries are based on Berggren (1972 a).

the oceans and the atmosphere apparently were low enough to permit this about 26 m.y. ago (late Oligocene) in the Ross Sea region of Antarctica (Hayes *et al.*, 1975). Possibly this was the culmination of a long build-up of ice which began with a well-established and marked drop in global temperatures in the early Oligocene ($\sim$ 37 m.y. ago). The warm oceans of preceding epochs may have provided the abundant precipitation needed to build glaciers at high latitudes (e.g. Frakes & Kemp, 1972).

   There is no evidence that sea-level glaciation occurred previously in the Ross Sea although future work in such a poorly explored region could establish a case. Glacial marine sediments on the floor of the Ross Sea (Fig. 3.1.2) contain stones dropped by floating ice, and non-glacial strata

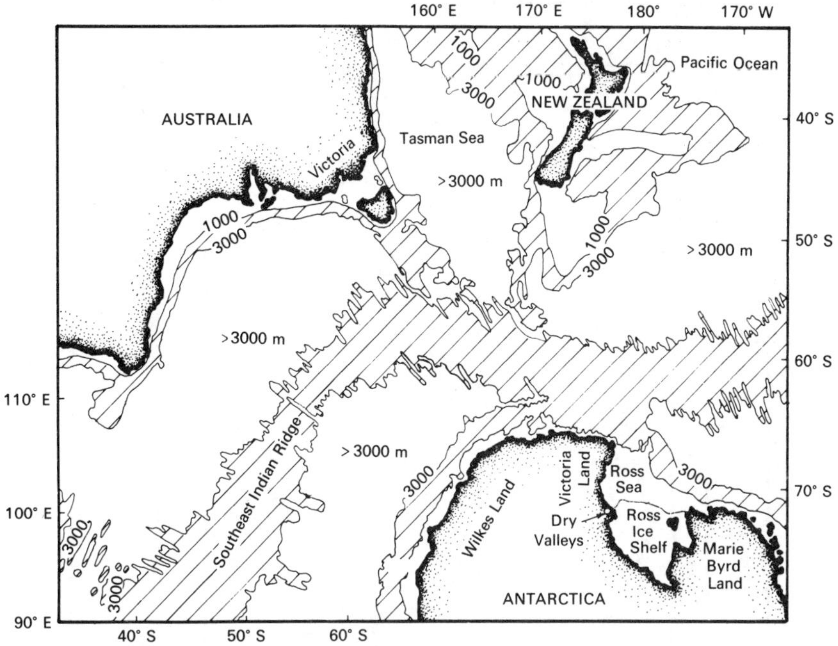

Fig. 3.1.2. The Australian–New Zealand sector of the Southern Ocean and Antarctica. Depths between 1,000 and 3,000 m are shown hatched.

beneath them have been dated by radiometric means as 26 m.y. old. Further, the glacial sediments contain fossil spores and pollen which indicate that forests grew in the region at the time, although plant remains younger than these are known only from relatively low-latitude areas of the Antarctic Peninsula. Thus, the land plants of Antarctica seem to have been almost entirely eliminated at about the same time that ice reached sea level (Kemp, 1975) and we can assume expansion of the ice to the limit of the continent at this time or soon thereafter.

Two lines of evidence raise the possibility that sea-level glaciation took place before the Oligocene; both have been subjected to criticism in the literature. First, sand-sized quartz grains occur in Eocene sediments of the southern Pacific (Geitzenauer *et al.*, 1968; Margolis & Kennett, 1971) and surface-texture features of the grains resemble those from grains ground by Pleistocene ice. They may instead represent relict structural or even depositional characteristics of entirely different environments (Setlow & Karpovitch, 1972). Secondly, the textures of Eocene volcanic rocks (hyaloclastites) in Marie Byrd Land indicate that they were erupted subaqueously, perhaps beneath melted ice (LeMasurier, 1972). If so, about 3,500 m of ice blanketed West Antarctica during part of the Eocene epoch. It has been suggested by Mercer (1972) that the radiometric dates may be

55

3.1 The long-term climatic record

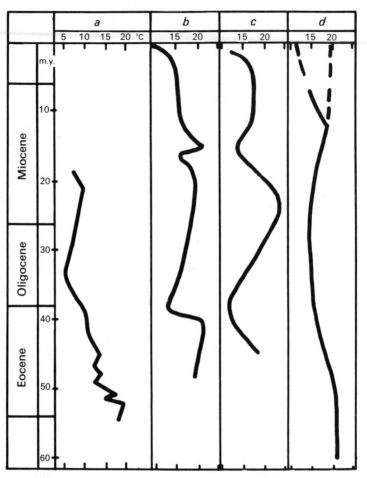

Fig. 3.1.3. Palaeotemperature curves (°C) for the Cenozoic derived from oxygen isotope ratios. Curves (a), (b), (c) and (d) are for the Tasman Sea (Shackleton & Kennett, 1975 a), New Zealand (Devereaux, 1967), Victoria (Dorman, 1966) and the northwest Pacific (Douglas & Savin, 1971) respectively.

unreliable and that the volcanic piles may have been built on the sea floor and subsequently uplifted.

An estimate of sea-water palaeotemperature can be gained through study of the isotopic composition of fossil shells, particularly those of microfossils such as foraminifera. The lighter isotope of oxygen ^{16}O is more concentrated in polar ice (relative to ^{18}O) than it is in sea water due to different rates of evaporation and hence precipitation. Variations in the $^{18}O/^{16}O$ ratio (expressed as δ values) in fossil materials therefore are taken to represent variations in the amount of ice bound up at the poles and these are frequently related to palaeotemperatures by means of volume

56

calculations for ice and sea water. The application of such studies to palaeoclimates was pioneered in the Southern Hemisphere by Dorman (1966) for Victoria, Devereaux (1967) for New Zealand and more recently for the Tasman Sea area by Shackleton & Kennett (1975 a). The most striking feature of all these isotope palaeotemperature curves (Fig. 3.1.3) is the tremendous drop in sea-surface temperatures indicated for an interval near the Eocene–Oligocene boundary (about 37.5 m.y. ago). Shackleton and Kennett's curve suggests a sharp fall of 4–5 °C for the southern Tasman Sea at this time. Frakes & Kemp (1973) have summarised the numerous geological and biogeographical data which substantiate this rapid change in climate on a global scale.

The fact that the Eocene–Oligocene temperature drop preceded the late Oligocene sea-level glaciation in the Ross Sea by about 10–12 m.y. suggests that the interval was a time of accumulation of polar ice. Alternatively, growth of the ice to sea level took place contemporaneously with the temperature drop but in another sector which remains undiscovered.

The middle Cenozoic

During much of the Oligocene and the early Miocene, climatic conditions were relatively stable and characterised by relatively small fluctuations which built slowly toward a climatic optimum. This part of the record is poorly known for high southern latitudes because of a paucity of data from Antarctica and because Southern Ocean materials of this age are known only from cores obtained during deep-sea drilling. Two types of evidence are available:

(i) drilled cores from the Ross Sea disclose apparently uninterrupted deposition of glacial marine sediments throughout the interval (Hayes & Frakes, 1975); and

(ii) the oxygen isotope curve for the southern Tasman Sea reveals little variation between the early Oligocene and the beginning of the Miocene, followed by a warming in the early Miocene and a subsequent substantial cooling continuing to the present (Shackleton & Kennett, 1975 a).

These facts lead to the conclusion that once ice reached sea level in the Ross Sea, it seems to have remained there and did not retreat sufficiently to expose much of the coast to resedimentation or to non-glacial erosion. Moreover, ice accumulated in sufficient quantities during the later half of the Miocene to affect strongly the isotopic ratio in sea water. The middle and late Miocene ($\sim$ 15–5 m.y. ago) was thus a time of substantial ice build-up on Antarctica.

Several lines of evidence indicate that climate underwent a remarkably significant change in the interval between about 5 and 3 m.y. ago. First, glacial marine sedimentation in the Ross Sea was interrupted, probably

3.1   The long-term climatic record

by an expansion of the ice sheet nearly as far as the edge of the continental shelf. This implies an advance of at least 150 km if, previous to the advance, the ice front lay at about its present position. The evidence for growth of the ice sheet includes the planing off of older sediments as seen in a regional unconformity; oxidation of the topmost sediments and a concentration of large boulders left by the ice at the surface; and trough-like depressions left on the sea floor by advancing lobes (Hayes & Frakes, 1975). Earlier, Chriss & Frakes (1972) had pointed out the probable morainal origin of the elevated rim at the edge of the continental shelf, a feature which probably belongs to this advance as well. Dating of the expansion of the ice sheet is based on microfossil determinations made during leg 28 of the Deep Sea Drilling Project and falls within the interval between the early Gilbert and the middle Gauss (5–3 m.y. ago). Glacial marine sedimentation in the Ross Sea thus had resumed by 3 m.y. ago, probably as a result of withdrawal of the ice and concomitant rise in sea level due in part to a pronounced warming during much of the Gauss palaeomagnetic epoch.

Oxygen isotope curves and related foraminifera diversity curves provide a second line of evidence. Shackleton & Kennett (1975 a, b) illustrate a lengthy cooling from isotopic data beginning within the middle Miocene, and the late Miocene portion is even more obvious in the diversity curves of Margolis & Kennett (1971) and Kennett & Vella (1975). From isotopic data on the Challenger Plateau a substantial build-up of ice on Antarctica occurred between 4.7 and 4.3 m.y. ago. This represents the most refined date available for the expansion of the ice sheet. However, palaeonto-logical correlation of the strata with New Zealand reference sections suggests this interval falls within the latest Miocene rather than the earliest Pliocene, as indicated by the time scale utilised here, that of Berggren (1972 a).

Third, shifts in sedimentation patterns in the Southern Ocean also indicate a marked cooling about 5.5–3.5 m.y. ago. At present, southeast Indian Ocean sedimentation is characterised by carbonate oozes north of the Antarctic Polar Front and siliceous ooze south of it. This seems to result in part from ecological control on distribution of micro-organisms having external frameworks (tests) of $CaCO_3$ and SiO_2 by the Polar Front, and in part from lack of carbonate materials below the carbonate compensation depth (Frakes, 1975). This critical depth, the CCD (Berger, 1972), is thought to vary in response to global temperature conditions and thus, changes in the regional distribution of ancient carbonate sediments can be used to estimate the relative magnitudes of temperature changes. Deep-sea drilling results show that carbonate sedimentation extended much farther south during the early Cenozoic, that they moved north at about 2–3 cm/yr as the southeast Indian Ocean widened in the middle

58

Cenozoic, and that the realm of carbonate sedimentation expanded greatly northward to about its present limits in the interval 5.5–3.5 m.y. ago (Kemp *et al.*, 1975). The extent of the shift is on the order of 300 km but it cannot be more precisely timed because of a lack of good palaeontologic control.

The significance of this event is enormous, considering that the cold surface waters which characterise the region south of the Polar Front zone today, must have expanded to cover nearly one-third more sea than they previously occupied.

A fourth indication of climatic change at this time comes from the apparent lack of ice-transported pebbles from the Transantarctic Mountains in Ross Sea sediments younger than the earliest late Miocene (~ 10–8 m.y. ago). Pliocene sediments in the drilled holes contain fragments seemingly all derived from West Antarctica but older Miocene units at a drill site near the Transantarctic Mountains contain an abundance of pebbles derived from diabasic igneous rocks known from the margin of the East Antarctic Plateau. Since some uncertain time in the late Miocene, most ice reaching the Ross Sea is therefore considered to have flowed out of Marie Byrd Land; earlier flow included some from the Plateau. Thus, in the late Miocene, ice was substantially thicker and conditions were probably much more extreme than during the Pliocene. It is possible, however, that flow from the East Antarctic Plateau was shut off by an event totally unrelated to climate – increased elevation of the Transantarctic Mountains barrier.

It is probable that a marked warming took place after these events. Much of the early Gilbert palaeomagnetic epoch appears to have been characteristically warm, judging from the Challenger Plateau curve and from positions of the sea-ice front determined by Weaver (1973). Variations in the abundance of warm- and cold-water silicoflagellates in the southeast Indian Ocean define a very warm phase at about 4.25–3.95 m.y. ago (Ciesielski & Weaver, 1974).

From Antarctica itself the oldest definite evidence of glaciation comes from tillites and glacial striae which underlie volcanic rocks dated by K/Ar as between 7 and 10 m.y. old. Rutford *et al.* (1972) suggest that glaciation here, in the Jones Mountains in West Antarctica near the Bellingshausen Sea, is at least as old as 7 m.y. and probably older. In the Ross Sea region, the valley cutting by Taylor and Wright upper glaciation V has been dated by K/Ar as older than 4 m.y. by Denton *et al.* (1971). Since only minimum ages are known it is uncertain whether Taylor V glaciation corresponds in age to a continent-wide cold phase pre-dating 7 m.y. ago, or to the culmination of ice accumulation within the 3–5 m.y. interval which is recorded in the Ross Sea drill sites.

The sequence in the Dry Valleys is clearer for succeeding glacial events. Spillovers of ice from East Antarctica apparently took place at 3.5–2.7

3.1 The long-term climatic record

m.y. ago (Taylor IV), 2.1–1.6 m.y. ago (Taylor III), between 1.6–1.2 m.y. and > 49,000 years ago (Taylor II) and 34,800–9,490 years ago. In addition the Ross Ice Shelf has expanded and become grounded on the floor of the Ross Sea at least four times in the interval since 1.2 m.y. ago. Denton *et al.* (1971) conclude that the last expansion was synchronous with the Wisconsin (Wurm) glaciation and that earlier expansions of the Ross Ice Shelf were thus probably contemporaneous with the major glaciations in the Northern Hemisphere. The record from the Weddell Sea, on the other hand, seems both more variable and more complex (Anderson, 1972) with at least six periods of dry base (cold) glaciation over the same interval.

It is tempting to relate the valley carving stage, Taylor V, in the Dry Valleys to the build-up of ice which led to extension of the Ross Ice Shelf 5–3 m.y. ago. Indeed, if the two events can ever be proven to be contemporaneous, this would constitute a strong case for the Ross expansion originating by means of a glacial surge (see Section 5.3). However, the counterargument is that the presence of the greatly thickened, and thus higher Ross Ice Shelf adjacent to the Dry Valleys would probably not allow the deep valley-cutting exhibited during Taylor V glaciation. Since the only constraint on dating of Taylor V is a minimum age of about 4 m.y., there is a strong likelihood that valley carving preceded the ice shelf expansion, and further that spillover ice from East Antarctica was responsible for transport of diabasic pebbles from the Transantarctic Mountains during the late Miocene, i.e. before about 5 m.y. ago. The fact that more recent Ross Sea and Taylor Valley glaciations have been out-of-phase suggests that they may have had different sources and this is indicated for the earlier ice-shelf expansion as well, by pebble provenance. In summary, the Taylor V valley carving probably represents a late Miocene expansion of ice from East Antarctica while the Ross Ice Shelf expansion corresponds to an ice build-up in Marie Byrd Land in the early Pliocene.

Further evidence bearing on continental glaciation comes from the distribution of ice-rafted debris in cores from coastal drill sites (Hayes & Frakes, 1975). Off the northern Victoria Land coast beyond Cape Adare ice-rafting began in the early part of the late Miocene, that is, approximately contemporaneously with rapid ice build-up as indicated by the isotope curves. Since this area lies offshore from the drainage basin for the Adelie Coast ice (Giovinetto, 1964), it is not to be expected that events would necessarily coincide with those taking place in the Western Ross Ice Shelf drainage basin, which contains the Dry Valleys. However, the two basins share a common drainage divide located immediately north of the Dry Valleys, thus allowing the speculation that Taylor V valley carving was synchronous with ice reaching sea level near Cape Adare. A large build-up of ice near the divide would affect both areas similarly, in this case at about 10–8 m.y. ago.

Drilling results offer evidence on the advent of ice-rafting for other sectors of the continent. For the eastern part of the Wilkes Land drainage basin (longitude 140° E) rafting is not indicated before the early Pliocene while in the western part of the basin near 105° E, ice reached sea level as early as the early Miocene (Hayes & Frakes, 1975). West of the Antarctic peninsula, rafting by icebergs apparently first took place during the early Miocene (Hollister *et al.*, 1974) and in the Scotia Sea rafting did not begin until the late Miocene (Barker *et al.*, 1974).

The Pliocene and Quaternary

Fluctuations in climate of high southern latitudes are much better known over the last 5 m.y. than in the earlier Cenozoic. Young strata are exposed near the Antarctic coast and use can be made of geomorphology in deducing past events. Also, age-dating techniques offer greater resolution on geologically recent materials, whether taken from land or sea floor. Finally, events from much of the last 100,000 years or so can be deduced from study of ice cores.

Radiometric methods have been used in dating lava flows interbedded with glacial tills in the Antarctic Dry Valleys. K/Ar determinations yield dates for four episodes of glaciation here. On the other hand, many sea-floor sediments can be dated by means of their palaeomagnetic polarity and supportive palaeontology. Workers frequently sample deep-sea cores at 10 cm intervals and calculate running averages over sets of three determinations; boundaries of polarity epochs or of short-term events are correlated with the earth's record of magnetic reversals, as known from study of lava flows which have been dated by K/Ar. The effective limit of palaeomagnetic stratigraphy is thus about 5 m.y., because for that age the precision error of K/Ar dating (usually about 10%) does not allow discrimination of epochs shorter than about 500,000 years. Polarity records back to about 15 m.y. ago are known from sediment cores but their correlates among lavas cannot be dated precisely.

The amount of carbonate in a deep-sea sediment can be used as a direct estimate of surface water temperature. It is assumed that dissolution of carbonate in the form of tests of deceased micro-organisms is more efficient with progressively colder sea water whether dissolution takes place in the water column or on the sea floor. Complicating factors to the strict control of carbonate abundance by temperature include the variable diluting effects of non-carbonate sediment and the rate of both carbonate and non-carbonate biogenic productivity in surface waters. Six palaeomagnetically and palaeontologically dated cores were analysed by EDTA titration to yield the Southern Ocean carbonate abundance curve in Fig. 3.1.4. In the cores seventeen carbonate-poor cold episodes are recognised

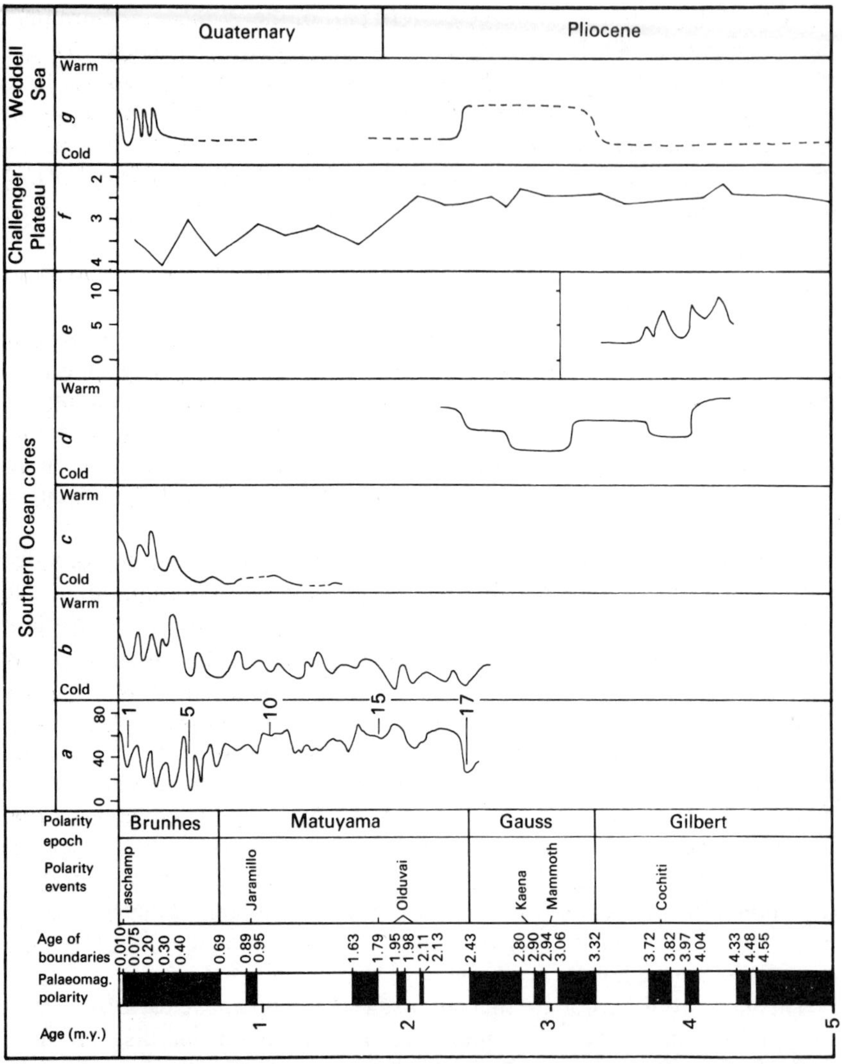

Fig. 3.1.4. Palaeoclimatic curves for the Southern Ocean over the last five million years. Curve (*a*) is based on percentage carbonate abundance data from study of Eltanin cores 34–14, 35–6, 16–15, 38–9 and 36–34. Numbers in curve (*a*) refer to the sequence of cold troughs. The other curves are based on studies of (*b*) radiolarians and foraminifera (Keany & Kennett, 1972), (*c*) nanofossils (Geitzenauer, 1972), (*d*) palaeo sea-ice front in the Southern Ocean (Weaver, 1973), (*e*) silico flagellates (Ciesielski & Weaver, 1974), scale in °C, (*f*) oxygen isotope ratios, δ values (Shackleton & Kennett, 1975 *b*), and (*g*) sedimentary data (Anderson, 1972). The time scale at left is based on the record of normal and reversed polarity of the earth's magnetic field.

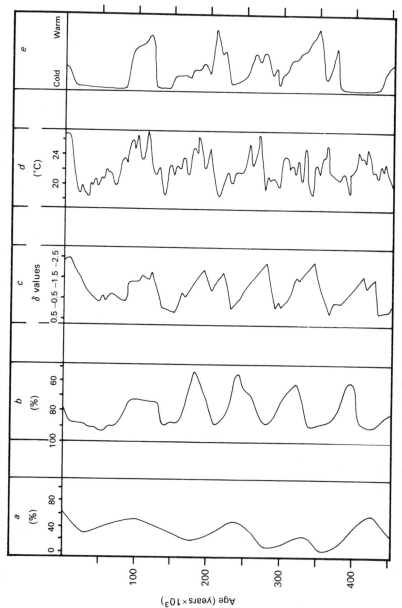

Fig. 3.1.5. Palaeoclimatic curves for the last 500,000 years. Curve (a) is the new Southern Ocean carbonate abundance curve; (b) is carbonate abundance in the eastern equatorial Pacific (Hays et al., 1969); (c) is the deviations of the oxygen isotope ratios from a standard in the Caribbean due to Emiliani (1966a); (d) is a temperature curve derived from oxygen isotope ratios in the Caribbean, by Imbrie et al. (1973); (e) is a Caribbean foraminifera curve due to Ericson & Wollin (1968).

63

3.1   The long-term climatic record

over the Brunhes–Matuyama interval (from $t = 0$ to $t = 2.43$ m.y.). Trends in the carbonate curve are as follows:

(i) in the Brunhes epoch (from $t = 0$ to $t = 0.69$ m.y.) cold intervals are more regular, more closely spaced, and more marked than in the Matuyama;

(ii) the Matuyama appears to be generally warmer than the Brunhes; and

(iii) there appears to be a general warming trend over the last 450,000 years.

The regularity of the seven Brunhes temperature cycles versus those of the Matuyama is borne out by curves depicting the variability of temperature-dependent species of micro-organisms over the same interval (radiolarians and foraminifera (Keany & Kennett, 1972); calcareous nanofossils (Geitzenauer, 1972)) and by sedimentological data from the Weddell Sea (Anderson, 1972). Further, the shape of all curves in Fig. 3.1.5 indicates rapid transition to warm periods but irregular and much more gradual cooling changes. There is a suggestion as well that, on this scale of observation, the next cooling trough will not be reached until about 70,000 years in the future, though the beginning of cooling is imminent (Fig. 3.1.5).

The small period of Brunhes climatic cycles relative to the Matuyama ones is also supported by the curves of Keany and Kennett, and Geitzenauer. From carbonate data these average at about 110,000 years for the Brunhes versus about 170,000 years for the Matuyama, while the data from radiolarian/foraminifera yield average cycles of 116,000 and 170,000 years, respectively. However, Brunhes cycles in the Southern Ocean appear to be less completely represented than those defined by Caribbean and equatorial Atlantic oxygen isotopes in Fig. 3.1.5 (average period in both is about 85,000 years), perhaps because of variable sedimentation rates or eroded sediments in the Antarctic cores.

A third characteristic of all southern curves is the large amplitude of climatic fluctuations in the Brunhes relative to the Matuyama. This suggests that although later oscillations of temperature may have been regular they also have been profound. A corollary of this conclusion is the notion that the next cooling may be expected to be of substantial magnitude.

Whether the Brunhes was in general cooler than the Matuyama is problematic. The carbonate curve presented here and radiolarian variations observed by Bandy *et al.* (1971) suggest a cooler Brunhes, but the work of Keany and Kennett, and of Geitzenauer point to the opposite conclusion. On a global basis, there is good evidence that the north and equatorial Atlantic were colder during the Brunhes than the Matuyama

(Ruddiman, 1971) although conditions in the Arctic during the Matuyama are interpreted with difficulty (Hunkins *et al.*, 1971).

A warming trend over the past 450,000 years is suggested in the carbonate curve (Fig. 3.1.5) and in the Keany and Kennett curve (Fig. 3.1.4). This is not supported by other curves in Fig. 3.1.5. The lack of any discernible trend in the oxygen isotope curves particularly casts doubt on the reality of a general warming. It is possible, however, that trends in carbonate dissolution and microfossil variations result from water column changes which significantly affect high-latitude sites without observable concurrent effects elsewhere. For example, if both trends are caused by dissolution deep in the water column, the causative factor could be a general oceanic warming which did not involve substantial melting of polar ice.

In the foregoing discussions, reference to palaeotemperatures or palaeotemperature trends from microfossil materials has been in terms of surface water conditions, as analyses are generally performed on planktonic organisms which lived in the near-surface regime. The remains of these organisms as they accumulated on the sea floor and were eventually covered, necessarily represent a somewhat different assemblage from the living one because of differential dissolution of robust versus fragile calcium carbonate tests. Accordingly, many palaeontological studies on such specimens actually are concerned with total dissolution effects of the water column, including exposure time on the sea floor, rather than strictly surface temperatures. Conversely, carbonate abundance in oceanic sediments is a function not only of vertical temperature range in the water mass but also of the rate of supply from surface waters (i.e. productivity). Careful selection of sites with reference to water-mass boundaries can resolve some of these problems.

Global correlations

Here we examine the correspondence in time of climatic events in Antarctica and the Southern Ocean with those observed elsewhere. The generalised scheme of events for high southern latitudes is as follows:

~ 37 m.y. ago – A sharp drop in surface ocean temperatures (4–5 °C).

~ 26 m.y. ago – Glaciers reach sea level in the Ross Sea.

~ 15–5 m.y. ago – A pronounced drop in surface ocean temperatures. Rapid build-up of ice in East Antarctica. Valley-carving stage in Transantarctic Mountains (Taylor V).

5.5–3 m.y. ago – Antarctic Polar Front shifts 300 km northward. Ross Ice Shelf expands to edge of continental shelf.

4.7–4.3 m.y. ago – Substantial build-up of ice in polar regions.

3.1 The long-term climatic record

4.25–3.95 m.y. ago – Marked warming of surface ocean temperatures (8–10 °C). Possible melting of the ice sheet in West Antarctica.

3.95–3.35 m.y. ago – Lengthy and marked cooling.

3.35–0.7 m.y. ago – Fluctuations in ocean surface temperatures (period = $\sim$ 170,000 years).

0.7 m.y. ago–present – Wider fluctuations in ocean surface temperatures (period = $\sim$ 110,000 years).

There is abundant evidence on a global basis that much of the early Cenozoic was unusually warm and humid. The early Oligocene temperature drop seen in the southern isotope curves is recorded as well in a curve for the northwest Pacific (Douglas & Savin, 1971). What is notable, however, is that for these considerably lower latitudes, the change is not only less severe but it is spread over a greater time span encompassing much of the Oligocene epoch. Similarly, while palaeontological data in California and Germany indicate a significant though gentle climate change at about this time (Durham, 1950; Schwarzbach, 1968) floral groups in high-latitude Alaska show a striking and abrupt shift (Wolfe & Hopkins, 1967). From the isotope curves one can surmise that isotopic changes do not reflect instantaneously the full impact of ice build-up at the poles on a global basis, though the initial effects may be rapid enough. In order for changes in isotopic ratios to be stabilised over the earth, a lengthy period of mixing is implied.

It is suggested here that cooling of high-latitude surface waters may proceed over a substantial time before effects are discernible in the equatorial zone. This results both from reduced effects of high-latitude cooling through low-latitude insolation – we would expect temperature effects to be less in low latitudes – and from the greater difficulty in observing small changes, in assemblages of fossils, for example. While these arguments do not apply to isotopic changes they perhaps explain the general lack of evidence for early Oligocene climate change in low latitudes.

The growth of glaciers to sea level 26 m.y. ago, on the other hand, is not paralleled by late Oligocene climatic cooling elsewhere. Most isotope curves are featureless over this interval, for example. One can conclude from this that sea-level ice was not unique to this time and must have existed previously, perhaps since the early Oligocene climatic deterioration. Alternatively, the Ross Sea may have provided special conditions suitable for sea-level glaciation on a relatively local scale which was not sufficiently large to affect isotope ratios in the sea. In most other sectors of Antarctica ice did not reach the sea until well into the Miocene.

Geological evidence from Antarctica indicates a rapid build-up of ice in the middle and late Miocene and is supported by trends of the isotope curves in both hemispheres. Ice apparently reached sea level in the later

66

part of the middle Miocene in the Gulf of Alaska and this is substantiated by changes in fossil floras indicating a 7 °C drop in July temperatures (Bandy *et al.*, 1969; Wolfe & Hopkins, 1967), and by the earliest known mountain glaciation on North America, in the Wrangell Mountains of Alaska (Denton & Armstrong, 1969). At somewhat lower latitudes of northwestern Europe (Netherlands), floras typical of subtropical climates in the late Miocene were followed by types more characteristic of present higher latitudes in the Pliocene (van der Hammen *et al.*, 1971). The marine climate of mid-latitude California, as reflected in the distribution of mollusks, began a progressive deterioration during the middle Miocene (Addicott, 1969). Uncertainties in age-dating are such that the relative timing of these events is not certain, but the Californian cooling appears slightly younger than the Alaskan one, as would be expected if cooling were progressive over latitude.

The expansion of the Ross Ice Shelf is assumed to have taken place somewhat later and in the interval 5–3 m.y. ago, perhaps concurrently with the marked shift in position of the Polar Front. Since cold fluctuations after the warm interval at 4.25–3.95 m.y. ago are relatively minor it is likely that these events took place immediately earlier, and in the 4.7–4.3 m.y. interval of ice build-up recorded in the isotope curves. It is possible though not yet proven that the mechanism of surface cooling in the Southern Ocean was the sudden large increase in the abundance of sea-ice and bergs associated with expansion of the ice sheet. A glacial surge (Wilson, 1964; Hollin, 1965) is thus not unlikely. The very warm phase which followed at 4.25–3.95 m.y. ago possibly resulted from melting of this ice, and possibly caused a marine invasion of Wright Valley (Webb, 1972; Ciesielski & Weaver, 1974).

Elsewhere evidence to support these trends is scant. This sharp cold spell followed by a pronounced warming has been suggested by Ciesielski & Weaver (1974) to be sufficient to melt much of the West Antarctic ice sheet. Sea-level fluctuations should therefore record a marked and sustained rise. As with other lines of evidence for this interval of time, age-dating of terrace levels is not generally accurate. Further, few areas have been so intensively studied from a palaeoclimatic viewpoint for this interval of time as has the Southern Ocean. Verification of the coincidence of events assumed above therefore is needed on a global scale.

The lengthy cooling phase in the late Gilbert (3.95–3.35 m.y. ago) similarly is not presently correlated with any specific event elsewhere, though it may correspond to an early glacial deposit of Argentine Patagonia (Mercer *et al.*, 1975; Ciesielski & Weaver, 1974). There are a few studies of sufficiently detailed dating from elsewhere which support the two cool phases in the Gauss palaeomagnetic epoch (4.0–3.6 and 3.2–2.4 m.y. ago) noted by Weaver (1973). In the Labrador Sea minor ice-rafting apparently

3.1 The long-term climatic record

took place at about 3.0 m.y. ago (Berggren, 1972b) and glacial marine sedimentation began about 3 m.y. ago in the Arctic (Voring Plateau (Talwani et al., 1975)). The time around 2.6 m.y. ago was suggested for the initiation of temperate latitude glaciation by Devereaux et al. (1970) on the basis of oxygen isotope variation in New Zealand and this is supported by the isotope curve of Shackleton & Kennett (1975b) and by Berggren (1972b) on heavy ice-rafting near Labrador. In this case there is good correspondence of Northern and Southern Hemisphere data for a major event.

Climatic fluctuations for the Brunhes and Matuyama palaeomagnetic epochs in the Southern Ocean suffer from not being as precisely dated as those from the eastern equatorial Pacific and the Caribbean. Workers on low-latitude sites have applied radiocarbon, radiometric methods based on thorium, palaeomagnetism and, more recently, oxygen isotope stratigraphy to yield a detailed account of changes through the Brunhes and part of the Matuyama epoch (Shackleton & Opdyke, 1973; Emiliani & Shackleton, 1974). However, in the absence of more elegant techniques, the biostratigraphic techniques have been perfected to a high degree of resolution and have been particularly useful in conjunction with palaeomagnetic stratigraphy. Moreover, carbonate abundances are shown here to correspond well with the earlier palaeontologic diversity studies to give a coherent picture of late Cenozoic climates.

Conditions during the Matuyama as illustrated in Fig. 3.1.5, can be compared with palaeontological diversity curves for the equatorial zones of the Atlantic (Ruddiman, 1971) and the Pacific (Hays et al., 1969). Possibly because the curves actually represent an unknown parameter (or perhaps several) rather than strictly surface temperature, there is little correspondence among them. The most striking divergence is in major-peak periodicity – about 170,000 years for the Southern Ocean versus 120,000–130,000 for equatorial materials. Additionally, there is only poor agreement between the Atlantic and Pacific trends, though periodicities are similar. Some of these discrepancies may be real, reflecting basic differences in climatic history for the regions concerned but alternatively, some may derive from poor dating, variable sedimentation rates in the cores, or possibly incomplete sequences.

For the Brunhes epoch, the situation is somewhat happier: the Southern Ocean isotope curve for the last 300,000 years (Shackleton, this volume, Fig. 3.2.3) agrees well with the Southern Ocean carbonate curve. However, divergences appear elsewhere – for example, the two lengthy cold intervals centred on about 50,000 and 420,000 years ago in the Caribbean foraminifera curve coincide with warm intervals in the Southern Ocean carbonate curve on Fig. 3.1.4. This would tend to support the conclusion reached by Ruddiman (1971) and Keany & Kennett (1972) that climatic cycles operate out of phase between low and high latitudes, except for the fact that the cold interval at 420,000 years ago from Caribbean foraminifera is also out of phase with the Caribbean isotope curve. This is true as well for the new isotopic time scale illustrated by Emiliani & Shackleton (1974) for Caribbean cores.

The most obvious disagreement in these curves is the occurrence of only five warming peaks in the Southern Ocean carbonate curve, as compared with six in all other curves in Fig. 3.1.4. Yet, the Southern Ocean carbonate and microfossil diversity curves correspond almost exactly in their spacings as well as their relative amplitudes (Fig. 3.1.5). More precisely, the agreement among all curves is limited to the last cycle in the series, i.e. the last 110,000 years or so. Does this represent a shift from an older out-of-phase relationship to an in-phase one? If the answer is yes, a recent change in the mechanism of climatic cycles is indicated.

In summary, a few climatic events of considerable magnitude seem to be represented in both hemispheres:

(i) the dramatic cooling of surface waters in the early Oligocene;

(ii) extensive ice build-up at high latitudes in the middle and late Miocene;

(iii) intensive ice accumulation during the middle Gauss epoch; and

(iv) the last 110,000-year cycle.

Significantly, the analogues of some major events from high southern latitudes remain unrecorded in the Northern Hemisphere:

(i) the earliest known sea-level glaciation in the late Oligocene;

(ii) shifting of the Polar Front and expansion of the Ross Ice Shelf in the late Miocene; and

(iii) extended cold and warm phases during the Gilbert and Gauss epochs.

Many of these 'problems' may cease to exist once the new and sophisticated stratigraphic techniques are applied to Southern Ocean core material of early Brunhes and Matuyama age. But there is an equal need for detailed studies on high northern latitude sediments of Gauss and older age.

3.2 Some results of the CLIMAP project

N. J. SHACKLETON

CLIMAP (Climate: Longrange Interpretation, Mapping and Prediction) is a multi-institutional project for the study of climatic change. It is funded in the United States under the IDOE programme (International Decade of Ocean Exploration), and the member institutions include the Lamont–Doherty Geological Observatory of Columbia University, Brown University, Oregon State University, the University of Maine and the University of Rhode Island. In addition there are several national and international corresponding members, of whom the author is one. The work I shall comment on here is therefore the result of co-operation among a large number of people.

Climates of 18,000 years ago

A dramatic CLIMAP contribution is the construction of a surface temperature map of the oceans at the height of the last glacial, about 18,000 years ago (CLIMAP Project Members, 1976) (Figs. 3.2.1 and 3.2.2). This aspect may be of particular interest because, although it was not originally designed with this purpose in mind, the reconstruction provides a primary set of boundary conditions with which to test both the robustness and the veracity of numerical atmospheric models. In order to be of use for this purpose, information additional to the sea-surface temperatures have been added. For continental areas estimates were made of altitude above glacial level (taking into account ice sheets up to 3 km in altitude) and of continental albedo on a 4° latitude by 5° longitude grid over the entire surface of the earth. The new boundary conditions have been used in a Minz–Arakawa two-level atmospheric circulation model and the results of that application have also been published (Gates, 1976). In addition, a compilation has been made, mainly from the published literature, of a set of test data consisting of estimates of climate within the continents, by which the results of model predictions may be assessed.

Sea-surface temperatures on the CLIMAP reconstruction were estimated using the methods of statistical treatment of microplankton assemblages from the sea bed first described by Imbrie & Kipp (1971). Foraminifera, radiolaria and coccoliths have been used, as no one group is abundant in all areas. Analytical error for each estimate is about ±1.6 °C and varies slightly from one region to another. Faunal variations down a core at a single site reflect the magnitude of local environmental changes, including climatic ones; the beauty of the CLIMAP approach, and the reward for the great investment of money and manpower that it has involved, is that the distribution of faunas, and hence environments, can now be mapped in space at selected times in the past. About a decade ago McIntyre (1967) first pointed out that the distribution of coccoliths in glacial sediments of the Atlantic delineated the former position of the Gulf Stream; mapping the distribution of foraminiferal assemblages has indicated exactly the same feature. The mappable former distribution of coccolith-free sediment (deposited beneath polar water) of the polar foraminiferal assemblage, and of ice-rafted debris all portray the same disposition of surface isotherms. Thus confidence in the map stems not

Fig. 3.2.1. Sea-surface temperatures, ice extent, ice elevation (m) and continental albedo, α (%), for southern winter 18,000 years ago with sea level at -85 m. From CLIMAP Project Members (1976). A, $\alpha > 40$; B, $30 \leqslant \alpha \leqslant 39$; C, $25 \leqslant \alpha \leqslant 29$; D, $20 \leqslant \alpha \leqslant 24$; E, $11 \leqslant \alpha \leqslant 19$; F, $\alpha \leqslant 10$.

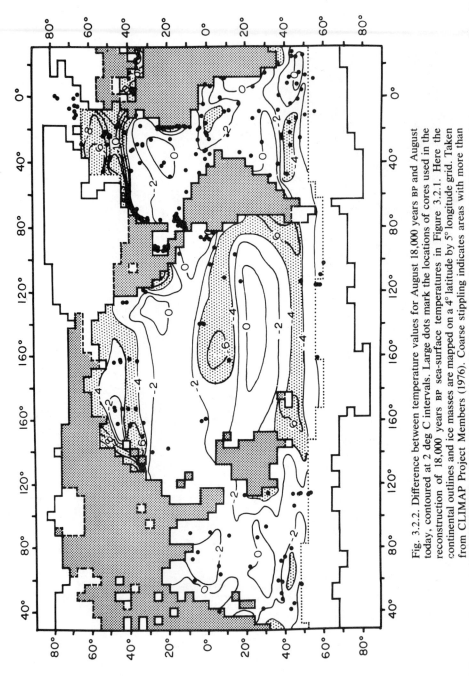

Fig. 3.2.2. Difference between temperature values for August 18,000 years BP and August today, contoured at 2 deg C intervals. Large dots mark the locations of cores used in the reconstruction of 18,000 years BP sea-surface temperatures in Figure 3.2.1. Here the continental outlines and ice masses are mapped on a 4° latitude by 5° longitude grid. Taken from CLIMAP Project Members (1976). Coarse stippling indicates areas with more than 4°C decrease from present.

only from the statistical confidence limits attached to the temperature estimates: much more, it arises from the coherent map that results.

Initially, it was necessary to perform down-core analysis in every area in order to obtain a stratigraphic framework of world-wide validity into which the temperature estimates could be placed. It emerged that the most widely applicable stratigraphic scheme was provided by oxygen isotope records. As ice sheets accumulated on the continents of the Northern Hemisphere, they removed from the oceans a certain amount of isotopically light water, i.e. water depleted in oxygen-18. The result was a slight enrichment in oxygen-18 in the remaining ocean water, an enrichment that is recorded in the oxygen isotopic composition of calcareous foraminifera preserved in deep-sea sediments. In any sediment core this record of isotopic change may be perceived by making oxygen isotopic analyses of the foraminifera. Roughly speaking, as one proceeds to sample a core from top to bottom, the last glacial maximum will be represented by the youngest deposits which give an isotopically positive peak. Thus the primary stratigraphic sequence was obtained from the oxygen isotope record (Shackleton & Opdyke, 1973). A number of other stratigraphic techniques were used as well, and quite apart from the results of the palaeoclimatic reconstructions, CLIMAP has made a significant contribution to the stratigraphy of Quaternary marine deposits.

Southern Hemisphere climate changes

From the distribution of samples shown in Fig. 3.2.2 it can be seen that coverage of the Southern Hemisphere for the reconstruction is somewhat patchy. Certainly the isotherms are very poorly controlled in some areas. But the fact is that the map facilitates understanding of climatic change in the Southern Hemisphere, as illustrated by examining a down-core record.

Figure 3.2.3 shows measurements of three climatically sensitive parameters in a sequence from the subantarctic Indian Ocean (43° 31′ S, 79° 52′ E). The record is a result of measurements from core RC11-120, which contains sediments extending back to about 300,000 years ago. We statistically estimated the summer sea-surface temperature (T_s), determined percentages of the radiolarian *Cycladophora davisiana*, and measured oxygen isotopic composition in *Globigerina bulloides* (from Hays *et al.*, 1976, and Hays *et al.*, in press). The important and immediate conclusion that one can draw from Fig. 3.2.3 is that temperature changes in the Southern Hemisphere (T_s) occurred essentially synchronously with events in the Northern Hemisphere (δ ^{18}O, measuring Northern Hemisphere ice). The abundance of *C. davisiana* is thought to have been controlled by changes in the salinity structure in Antarctic waters to the

73

3.2 The long-term climatic record

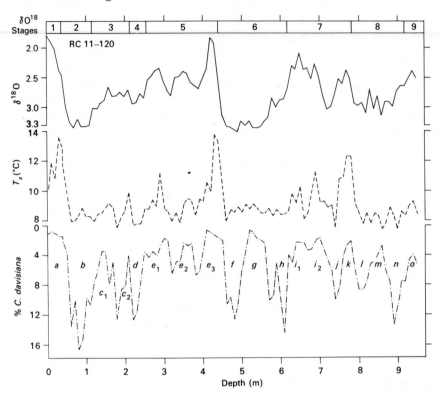

Fig. 3.2.3. Oxygen isotope, estimated summer sea-surface temperature and *C. davisiana* records in subantarctic core RC11-120, which spans about 300,000 years and accumulated at 3.2 cm per thousand years. Notice that Stage 5 contains three distinct peaks about 20,000 years apart (denoted e_1, e_2 and e_3 in the *C. davisiana* record). *C. davisiana* abundance is expressed as a percentage of the total counted radiolarian fauna. Taken from Hays *et al.* (in press).

south of the coring site (this species is not used in the temperature estimating equations). Interpretation of the *C. davisiana* data is thus more complicated, but whatever the precise variable estimated, it is clear that changes in the mass of Antarctic waters were also essentially synchronous with those in the north.

Originally the data from Southern Hemisphere cores were collected simply to provide the stratigraphic base for Fig. 3.2.1, so that encapsulated in Fig. 3.2.1 is our discovery that the 18,000 year point in the glacial records of the Northern Hemisphere coincides approximately with maximum cooling in the Southern Hemisphere. However, the records proved to be very valuable for another reason, which accounts for the fact that we have carried them further back than the 18,000 year point.

The record shown in Fig. 3.2.4 comes from an equatorial Pacific core

74

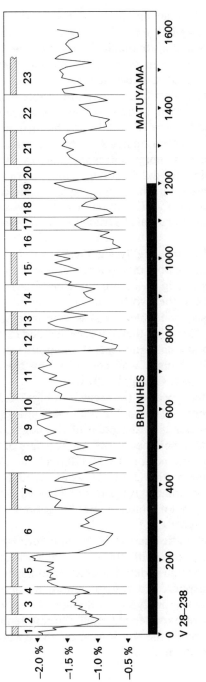

Fig. 3.2.4. Oxygen isotope record in equatorial Pacific core V28–238, which spans about 850,000 years and accumulated at 1.7 cm per thousand years. Notice that within Stage 5 there is little variation. Taken from Shackleton & Opdyke (1973).

3.2 The long-term climatic record

(1° 01′ N, 160° 29′ E) that accumulated at a rate of about 1.7 cm per thousand years. This core has proved a very useful standard sequence because it can be calibrated chronologically through the presence of the Brunhes–Matuyama palaeomagnetic reversal at about 700,000 years ago (Shackleton & Opdyke, 1973). Comparing Figs. 3.2.3 and 3.2.4, significant differences can be seen. In particular, there are visible fluctuations about 20,000 years apart in the subantarctic which we know are real, but which are obscured in the Pacific core (Fig. 3.2.4) by bioturbation (disturbance by marine organisms). Spectral analysis of the numerical record in Fig. 3.2.4 reveals the presence of energy corresponding to a 40,000-year cycle. Work on the records in Fig. 3.2.3 (Hays *et al.*, in press) confirms the presence of spectral energy in the same frequency band in the oxygen isotope record, and also in the T_s and percent *C. davisiana* records. In addition, spectral energy is present at a frequency corresponding to a cycle of about 22,000 years.

These cycles are of course highly significant for they correspond to periodicities in the changing obliquity and precession in the earth's orbital geometry. The records in Fig. 3.2.3 are sufficiently good to confirm the fact that the ratio between the two frequencies is about 1.8 as expected, a test that is particularly valuable because it is independent of small uncertainties in dating. This and other tests are detailed in our paper (Hays *et al.*, in press), which strongly supports the hypothesis that the timing of Pleistocene climatic changes was controlled by the earth's orbital changes (Milankovich, 1938; Chappell, 1973).

There is one additional aspect of the records in Fig. 3.2.3 that is worthy of attention. Both at about 40 cm and at about 440 cm, the peak in T_s is reached significantly earlier than the point of full deglaciation indicated in the ^{18}O record. We have obtained a radiocarbon date of 9,400 years for the most recent temperature peak in RC11-120, and the same feature has been observed and dated in other cores (Hays *et al.*, 1976). This is rather earlier than the usual estimates for the so-called hypsithermal in the Northern Hemisphere, a point which cannot be ignored when examining Southern Hemisphere records.

In summary, recent climatic changes in the Southern Hemisphere have been essentially synchronous with those in the Northern Hemisphere during the Pleistocene, though in detail Southern Hemisphere temperature led the Northern Hemisphere ice volume record by about 3,000 years (estimated by cross-spectral analysis). Although no part of the ocean in the Southern Hemisphere experienced such extreme temperature variations as those that occurred in the North Atlantic, there were areas of quite substantial temperature anomalies.

3.3 Climatic and topographic changes from glaciological data
D. JENSSEN

Oxygen and hydrogen isotope ratios such as $^{18}O/^{16}O$ or D/H in deposited snow and ice are controlled by many factors, including the ocean temperatures where the water was initially evaporated, its experiences before its final sublimation, and the manner of its final deposition. However, as has been shown (e.g. by Dansgaard *et al.*, 1969), the isotope ratio reflects closely the climatic temperature at the time and place of deposition.

On the polar ice sheets the surface temperature decreases not only with latitude but also with surface altitude (see e.g. Budd & Morgan, 1973). Thus the isotope ratios of ice flowing outward from the interior of an ice sheet will correspond to lower temperatures simply due to the higher surface elevation, even with steady climatic conditions. Further, any changes in ice elevation through time may cause isotope ratio variations similar to climatic change effects.

Two questions now arise: how important relatively are elevation change effects, and how can they be separated from true climatic fluctuations in the isotope ratios?

Elevation change effect

To estimate this effect it is necessary to find the height of the ice surface, at the time of deposition, for each parcel of ice making up the core. For this purpose, use will be made of the glacier flow model of Budd & McInnes (1974) as applied to a flow line in East Antarctica. In calculations covering many hundreds of thousands of years this model has produced a record of calculated surface elevations and average ice velocities with numerous rapid advances of the ice coupled with surface lowering, followed by gradual restoring of a shorter and thicker ice-sheet profile (see Section 5.3). During the latter periods the ice particles tended to move mainly along the vertical, whereas during the advances or 'surges' the ice for a short time moved quasi-horizontally over considerable distances. Typical trajectories are shown in Fig. 3.3.1. Of twenty-five data points in the simulated core (vertical line) twenty originated within 300 km of the core, so that 60–70% of its ice is of relatively recent origin.

Now let the relation between the surface elevation change in time and the temperature at the surface be linear:

$$T' = T^0 + \gamma(E' - E^0), \qquad (3.3.1)$$

where T is temperature, E is elevation, γ measures the temperature change produced by a unit change in elevation and the superscripts refer to any

77

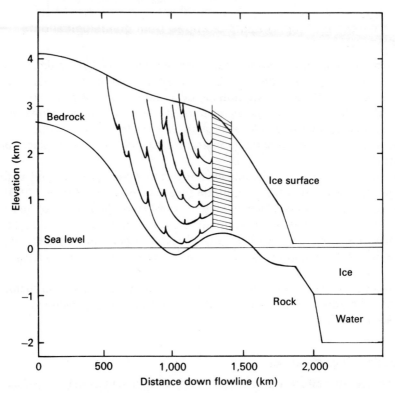

Fig. 3.3.1. Computed trajectories in a cross-section of the Antarctic ice sheet. The cross-section is along a flowline from about 1,850 km inland to the coast, and simulates an observed flowline roughly from Vostok station (81° S, 78° E) to Mirny (67° S, 93° E). The trajectories are for 25 parcels now in a vertical line through the ice about 500 km from the coast (or 1,350 km from the central ice divide at which the flowline begins). The lines show the history of these 25 parcels and their initial elevations (which will correspond to the height of the ice surface in the past). Thus, for the second lowest parcel in the core, the ice surface when it was deposited was much lower than the present surface, whereas for the sixth lowest parcel, the ice surface was much higher at deposition. The ladder-like structure at the end of the particle paths indicates rapid flow of the ice in recent times.

time (t) and a reference time (0). Since the oxygen isotope ratio δ can be assumed to be another linear function of temperature

$$T = a + b\delta,$$

where a and b are constants, δ and E are linearly related.* Hence the elevations can be used directly to portray the changes in δ produced by time variation in the ice-sheet profile.

Typical results are shown in Fig. 3.3.2(a) for three simulated cores,

---

* The δ value is defined by

$$\delta = \{[(^{18}O/^{16}O)_{sample}/(^{18}O/^{16}O)_{standard}] - 1\} \times 10^3.$$

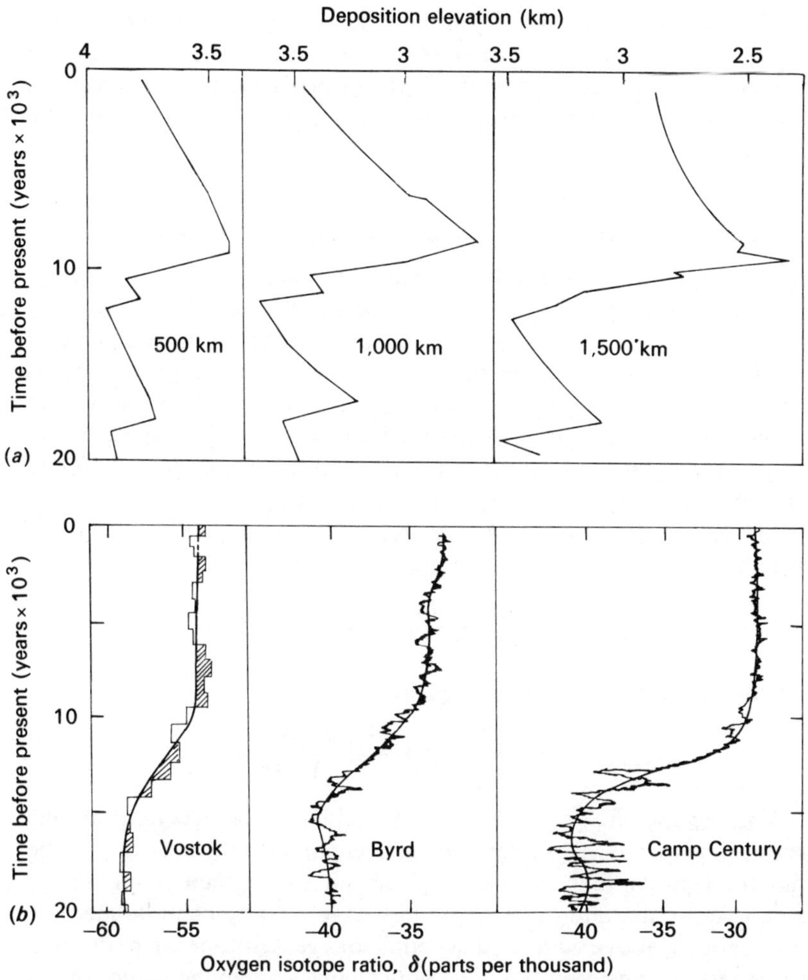

Fig. 3.3.2. Simulated oxygen isotope ratio curves for constant climate (*a*), and observed ratios in three different cores (*b*). For the curves in (*a*) the text shows there is a linear relation between the height above sea level of the ice when it was deposited and oxygen isotope ratio. Because of the uncertainties in the constants of that relationship, the elevations have been used. The three simulated curves are for ice in vertical columns 500, 1,000 and 1,500 km, respectively, from a central ice divide in Antarctica (see Fig. 3.3.1 also). In (*b*) the three curves are for, respectively, Vostok and Byrd stations in Antarctica, and Camp Century in Greenland; from Barkov *et al.* (1975).

located respectively 500, 1,000 and 1,500 km from the highest point of the ice sheet. Using a plausible value of $\gamma = 0.006$ °C/m and the usual value of $b = 0.8$ °C/(1‰ change in δ),† then the computed δ variation in the core nearest the coast is almost 5‰. This should be compared with the

† ‰ signifies parts per thousand.

79

3.3   The long-term climatic record

observed variation of about 10‰ in the Camp Century core (Dansgaard *et al.*, 1971) in Fig. 3.3.2(*b*). The observed isotope profiles for the Byrd and Vostok cores are also shown there and clearly have many similarities with the profiles computed from elevation changes in steady climatic conditions.

Separation of elevation from climatic changes in the δ-record

Additional information is needed to separate these two effects in the measured isotopes: this is provided by measurements of the total gas content within any core section. It has been pointed out by Raynaud & Lorius (1973) that the volume of the gas entrapped within the ice is dependent on the temperature and pressure at 'close-off' (transformation of firn into ice). But pressure is related to both elevation and temperature; hence, given gas contents and isotope ratios in the same core, it should be possible to separate elevation and climatic changes.

Jenssen (1977*a*) developed a theory and model in which this was done, showing that the elevation change (ΔE) is given by

$$\Delta E = \frac{R}{g}\left[\frac{(a\bar{\delta}+b)}{\bar{V}}\Delta V + a\Delta\delta\right]$$

and climatic temperature change is given by

$$\Delta T_{\text{climate}} = \left(1+\frac{\gamma R}{g}\right)a\Delta\delta + \frac{\gamma R}{g}\left(\frac{a\bar{\delta}+b}{\bar{V}}\right)\Delta V,$$

where R is the specific gas content for dry air; g is the acceleration due to gravity; $\bar{\delta}$, $\bar{V}$ are averages over two vertical points in the core of isotope ratio and total gas content, respectively; $\Delta\delta$ and ΔV are their changes over the same two points; and other quantities have already been defined.

The equations above will hold whether oxygen isotope or hydrogen/deuterium ratio values are used. The only changes are in the values of the constants a and b.

Figure 3.3.3 left shows the measured hydrogen isotope ratios and total gas content at Camp Century, both plotted for selected points in the Camp Century core. The isotope record is more detailed than represented here, but the gas content is not. For reasons of consistency, therefore, the data as presented to the model are as shown in the figure.

Figure 3.3.3 (right) gives (*a*) the deduced climatic temperatures – with elevation effects eliminated – for three values of the temperature/elevation constant γ (see equation (3.3.1)), and (*b*) the deduced elevation changes, all as functions of depth. No attempt has been made to fit a time scale to the depths but, on the scaling of Dansgaard & Johnsen (1969), 1,100

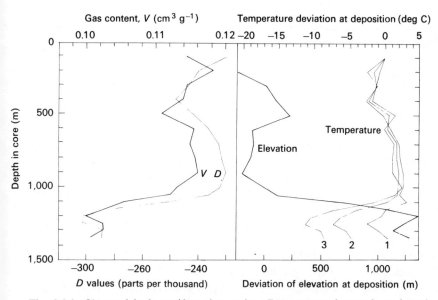

Fig. 3.3.3. Observed hydrogen/deuterium ratios, D (parts per thousand), and total gas content, V (cm^3g^{-1}), for Camp Century are shown on the left as functions of depth in the core. Data from Raynaud & Delmas (1977). On the right are deviations from the present of the temperature and height at deposition of the ice in the Camp Century core as deduced by the model of this paper. The three temperature curves correspond to three different values for the ratio between temperature change and elevation change at the surface of the ice with time. This ratio is, for the three curves respectively, 0.01, 0.006 and 0.003 °C/m. Thus for curve 2, the ice now at a depth of 1,200 m was deposited 1.3 km higher, with a temperature about 6.5 °C colder, than the present surface of the ice.

m correspond to roughly 10,000 years BP and 1,250 m to very roughly 20,000 years BP.

The three temperature curves agree reasonably closely to a depth of about 1,000 m and all show cooler temperatures before that time; but at earlier dates the three curves exhibit significant differences. In the absence of data or firm theory which would yield an unambiguous value for γ, it is impossible to say just what the climatic change has been.

The dangers of interpreting oxygen isotope ratio fluctuations simply as climatic temperature changes are thus obvious. To allow in addition for the effect of downstream ice transport on both temperature and surface elevation requires a model of the ice flow. Given such a model a better estimate of the climatic temperatures can be made, and the changes in elevation of the ice surface (as distinct from that of ice in the core) can be determined. These corrections depend essentially on the past dynamics of the ice, and further computer models, which already exist, must be used (see Section 5.3).

3.4   Quaternary climates of the Australian region
D. WALKER

The region today

Australia and the island of New Guinea (Fig. 3.4.1) have been united into a single land mass during those periods of the Quaternary, that is the greater part of it, when the sea stood more than 20 m below its present level. It is therefore admissible to treat the whole, extending from the equator to 44° S, as a single unit for present purposes. Nevertheless, the variety of latitudes, topographies and geological histories presented a wide range of contrasting environments throughout the Quaternary as at the present day.

The greater part of the uplift which formed the mountains of New Guinea took place late in the Tertiary although some vertical movement still continues. The highest peaks (e.g. Mt Carstensz, 4,883 m) bear small snow fields and glaciers and the altitudinal forest limit lies at about 3,800 m a.s.l. The climatic equator sweeps south and north over these mountains annually bringing rain from all directions; only rarely do evaporation and runoff exceed precipitation long enough to cause drought stress on the vegetation. The mean annual sea-surface temperature around New Guinea is about 28 °C and the mean annual air temperature falls about 0.5 °C with each 100 m altitude. The forests of these mountains contain an inheritance from Gondwanaland as well as major Malesian and Pan-tropical components and some more specifically Australian plants.

The lowland bulge of southern New Guinea is built of low plateaus and the alluvium of the island's short but powerful rivers. Although its climate is now modified by the shallowly flooded Arafura Sea, which restricts the extent of seasonally (May–November) dry lowlands to about 60,000 km², during large parts of the Quaternary it must have shared broadly the same climate as northern Australia.

South to about the Tropic of Capricorn, the continent of Australia receives practically all its rain between December and March and from monsoon winds. Except along the coastal fringe the total quantity is small, its incidence is erratic and evaporation under the prevailing high temperatures is relatively great so that most of the region is effectively arid, covered with desert, salt-bush, grassland and shrub communities. Along the western and northern coasts on-shore winds create a discontinuous fringe of consistently wetter conditions which maintain patchy eucalypt forests, some of them deciduous. The main modification, however, is in the east where the wet Trades, impinging on the coastal highlands, provide the highest annual rainfalls of Australia (e.g. 3,600 mm at Innisfail). The wettest enclaves support rain forests with counterparts

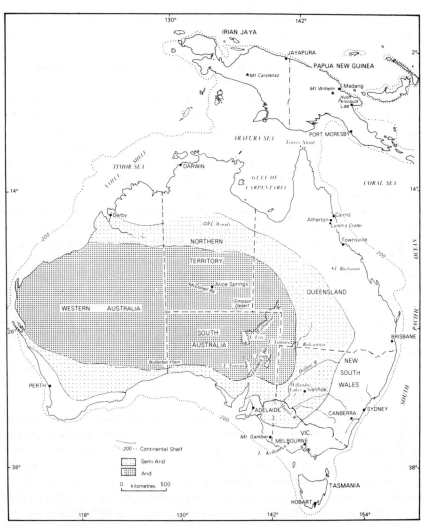

Fig. 3.4.1. Australia and New Guinea showing some of the places mentioned in the text (modified from Bowler *et al.*, 1976).

in New Guinea and more characteristically Australian wet sclerophyll (eucalypt) forests but effective precipitation falls off so rapidly westward that savanna woodland and shrubland are generally only 100 km inland and themselves give way to grassland less than 600 km from the coast.

The zonal high-pressure cells which migrate eastward across the continent south of the Tropic allow the occasional penetration of unstable tropical air bringing occasional rains to those latitudes. During winter the same area may receive rain from occasional incursions of southern

83

maritime air. There is therefore a broad belt from the west coast to about 140° E in which the small amount of rain that falls may do so at virtually any time of the year.

Southwestern and southeastern Australia have dependable winter rains, and the coast of the Great Australian Bight between these areas receives such rain as it gets during the same season. In the east of the continent, winter rain penetrates north as far as 28° S, which, together with the southern excursion of the Trades in summer, ensures a more-or-less even distribution of rainfall throughout the year in the eastern half of the state of New South Wales. Nevertheless, the total average annual rainfall falls to about 500 mm and shrublands and salt-bush steppe replace forests and savanna only 400 km from the east coast.

The latitudinal variation of temperature ranges between annual means of 15 °C at Melbourne and 28 °C at Darwin. The hottest month in Darwin (November) experiences an average temperature of 30 °C and the coldest month in Melbourne (July) one of 10 °C. The temperature regime is, of course, significantly modified by humidity and by topography. In the arid inland, a summer day reaching 40 °C may be followed by an almost freezing night, whereas as far south as 37° S on the east coast, even winter frosts are rare at sea level. On the Snowy Mountains of New South Wales and in Tasmania, snow lies for several months each winter down to 1,700 m and 900 m a.s.l. respectively.

Advantages and limitations

Australia and New Guinea present a number of strikingly unusual natural features stimulating to climatological study: a maritime equatorial mountain chain high enough to harbour glaciers today, a large desert in a dominantly oceanic hemisphere, winter, summer, and year-round rainfall regimes and a more-or-less continuous, largely montane, eastern coastline running north and south through about 40 degrees of latitude. On the other hand, it lies in latitudes too low to have been directly affected by the Quaternary expansions of polar ice or for its own mountains to have served as nucleii for extensive continental glaciation. But this has left the continent free from glacio-isostatic movements rendering the interpretation of sea-level changes about its margin more straightforward.

Other disciplines have now begun to establish facts which demand palaeoclimatological contexts for their best understanding. Amongst these are the unexpected antiquity of man in Australia ($>$ 40,000 years), the great age of agriculture in the New Guinea highlands ($>$ 6,000 years), the patterns and discontinuities in plant and animal distributions, major hydrological changes in large parts of the continent and the increasingly evident importance of arid centres in affecting world climates during colder periods.

Quaternary research is young in Australia; little had been done before the last decade. Even now, the number of workers is distressingly small. Without extensive sequences of boulder clays and associated interglacial materials, Australia lacks the most convenient basis for the stratigraphic, climatic and broadly chronologic division of the Period. And even in the intensely, and probably repeatedly, glaciated parts such as Tasmania and the Irian Jaya mountains most of the evidence is erosional rather than depositional and therefore difficult to interpret in detail. As today, so also during much of the Quaternary, most of Australia must have carried a relatively small biomass, decomposition of which was rapid after death. These factors, combined with the paucity of perennially flooded basins of accumulation, have rendered organic deposits the exception rather than the rule, limiting the orthodox applications of some techniques (e.g. pollen analysis) familiar in the Northern Hemisphere. On the other hand, the areal extent of some Quaternary features sometimes implies a uniformity of physical and biological conditions over areas the size of England or bigger and lends a certain confidence to their interpretation and the assessment of their regional importance. This is particularly the case in the study of Quaternary soils developed in Australia's vast areas of slow inorganic accumulation and of the stratigraphy of lake fills and associated deposits in geographically widespread, now semi-arid, places. The relative stability of much of Australia's coastline and the rather steady tectonic uplift of the Huon coast of Papua New Guinea offer unusually attractive circumstances for studying problems of world-wide, and therefore climatic, significance which have only recently begun to be exploited.

A number of general difficulties in interpreting stratigraphic, bio-stratigraphic and geomorphic data in palaeoclimatic terms apply with particular force in Australia and New Guinea. For example, in a continent where the occasional but intense event, such as a 'flash flood', today may have more profound geomorphic effects and hydrological repercussions than slower but long-maintained processes, the importance of separating intensity and frequency in the factors contributing to a Quaternary deposit is very apparent. Indeed, since certain combinations of these, particularly in regard to rainfall, typify some Australian climates today, it is important not only to recognise their different products but also to try to quantify them in the past. This can be particularly difficult in the absence of close chronological control. Most Australian Quaternary deposits do not contain large quantities of material suited to orthodox radiocarbon dating and others, such as buried soils, are unlikely to yield consistent palaeomagnetically-based chronologies. In some instances (e.g.

the use of soil carbonates for radiocarbon dating) special technical developments have been justified but a sufficient chronological framework will for long limit the climatic reconstruction of Quaternary Australia.

Other difficulties arise in the interpretation of biological, particularly palynological, data. The pollen morphological survey of Australian and New Guinean plants is still far from complete, even at the generic level. Nevertheless, enough is now known to confirm that, even with recourse to electron microscopy, many species and even genera are not separately identifiable. Coupled with a lack of ecological information on individual species (autecological information) hardly credible to Northern Hemisphere scientists and the reasonable suspicion that some of the most important taxa (e.g. *Eucalyptus*) have been evolving actively during the Quaternary, this imposes severe difficulties in finding living analogues for fossil pollen assemblages and attributing climatic implications to them. Even when coincidence in the geographic limits of pollen-analytically identifiable taxa at the present day allow the more-or-less confident comparison of a temporal boundary in the past with a spatial one in the present, the separation of temperature and rainfall effects (let alone those associated with ecological adaptation to changes in soil characteristics and fire frequency) can be daunting. In some cases, such as the Papua New Guinea mountains, it is reasonable to suppose that water supply has never limited plant growth and that climatically induced vegetation changes have been largely temperature dependent. But elsewhere, such as at the rainforest–savanna boundary in north Queensland, the plants react directly to temperature but more particularly to the availability and effectiveness of soil moisture to which both precipitation and temperature (through evapotranspiration) contribute. Under these circumstances reconstruction of the courses of the separate climatic elements is hazardous and can only be successful where there is convincing evidence from other sources for the stability or independent direction of change in one of them. Even so, the complexity of the taxa identified (families or genera) and the poverty of autecological information call for restraint on the part of the investigator, even when the dominant climatic factor can be identified. This is because the larger the taxonomic unit the wider its climatic tolerance and the more varied its migration rates and its competitive ability in different circumstances. Under all but exceptional circumstances it is only possible to attribute a range to the climatic factor selected to account for a fossil pollen assemblage; 'single line' renderings of climatic change from palynological sources must always smack of dubiously based confidence.

Climate affects other natural phenomena indirectly as well as directly. In Australia the indirect effects are well exemplified by incidence of fire. There is plenty of evidence that a section of the Australian flora is adapted not only to periodic drought, but to fire, and that its past dis-

86

tribution will therefore have been strongly affected by fire frequency and intensity. Natural fires are, to a degree, climate-dependent which has both advantages and disadvantages in climatic reconstructions from pollen-analytical studies coupled with comparable analysis of the occurrence of fine charcoal in stratified deposits. But man-made fires have a greater range and higher frequency than natural ones and have rather different ecological effects. Moreover, man himself practising non-agricultural or subsistence agricultural economies (such as Aboriginal Australians and New Guineans respectively) moves and burns partly at the dictate of climatic change. Evidence of fire in the past, although undoubtedly having some climatic significance, calls for careful interpretation.

Palaeoclimatology was retarded for decades in Australia not only through a lack of personnel but by a reluctance to adapt techniques to the materials available and to ask of these materials questions which they were suited to answer. On a continent where travel was arduous and maps sketchy, there was an understandable bias amongst colonial scientists favouring the application of techniques and ways of thought developed for different situations to whatever material presented itself on their new doorsteps. This led to a number of fragmentary investigations, usually unrelated to any general problem, and inconclusive results. Although such miscellaneous observations still accumulate, the general improvement of our knowledge of the natural history of Australia and New Guinea, easier travel and good aerial photographic coverage have resulted in a remarkable improvement in the formulation of problems and the development of appropriate techniques during the past twenty years. It is from these relatively few but penetrating investigations that the outline of Australian Quaternary palaeoclimatology must be drawn. Because of this and the size of the region it is still too early to generalise about its past climates. The best that can be done is to address a few simple problems and derive what answers we can from investigations in specific areas.

The biggest climatic fluctuations

Whereas the major shifts in Quaternary climates of North America and Europe were first defined from the sequence of glacial and interglacial deposits, such deposits in Australia are still being investigated, are confined to southeastern Australia, and relate mainly to the last glaciation. On the Huon Peninsula of Papua New Guinea, however, there is a 'staircase' of coral reef terraces raised by tectonic movements from which sea-level changes throughout almost the last half million years have been derived (Chappell, 1973, 1974a; Bloom et al., 1974).* It is a reasonable assumption that sea levels were highest when ice volumes were smallest,

---

* The specific references in this section have been selected and inserted by the editors. More extensive references will be found in Bowler et al. (1976).

as a result of the warmest world climates. Also the times of high sea levels during the last 250,000 years, established by these data, correspond well with those of highest ocean surface temperatures derived from oxygen isotope analysis of microfossils (foraminiferan remains) in deep-sea cores of the Caribbean and the equatorial Pacific (Chappell, 1974 b).

The pollen-analytical record from crater-lake deposits on the Atherton Tableland in North Queensland, probably spanning the last 70,000 years, is divisible into at least four sections, of contrasting climatic implications (Kershaw, 1974 and 1976). From Lake George, near Canberra, sedi-mentological and pollen-analytical investigations suggest periods of low water level when the lake was surrounded by herbaceous vegetation interspersed with others during which cool temperate forest bordered a full lake (see Section 3.5). The exact implications in terms of temperature and precipitation might be open to several interpretations but there is no doubt that major climatic fluctuations occurred during the last 100,000 years.

Fragmentary though they are, these data confirm that Australia did not escape the major climatic fluctuations of the Quaternary Period although their expression is often obscure.

The climate of the full glacial

Several recent investigations have begun to throw light on the climate of the later part of Wisconsin–Weichsel time.

In the mountains of New Guinea about 2,000 km^2 were glaciated and the snowline appears to have lain between 3,400 m and 3,650 m a.s.l. Glacial retreat began 15,000 to 14,000 years ago. Pollen-analytical data from outside the glaciated area suggest that mean annual temperatures were at their lowest, between 7 and 11 °C below present, between 18,000 and 16,000 years ago but that before then the cold need not have been so intense (Bowler *et al.*, 1976). Indeed, for about 1,500 years after 27,000 years ago temperatures need only have been 1.5 °C below present (Fig. 3.4.2).

Around Lynch's Crater in North Queensland, currently suffering a rainfall of about 2,500 mm annually, open sclerophyll woodland grew between 38,000 and about 9,000 years ago where rainforest now flourishes. It is difficult to separate the effects of temperature changes from those of rainfall changes in this instance but there can be no doubt that, during this long sclerophyll phase, effective precipitation was probably less than half its present level.

In Lake George, Fig. 3.4.3, the water stood high and fresh from sometime beyond the limit of radiocarbon dating until about 22,000 years ago and cool temperate forest grew around its shores. This has been

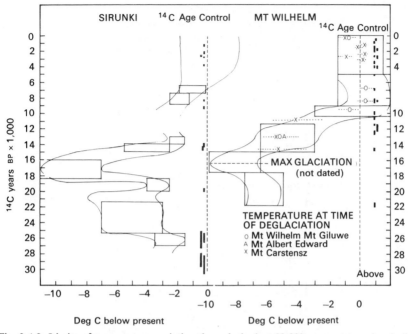

Fig. 3.4.2. Limits of temperature variation through the last 28,000 years at two sites in the highlands of New Guinea derived from pollen analysis, with estimates of temperature indicated by glacial stages on four mountains (from Bowler *et al.*, 1976).

interpreted to imply a mean annual rainfall perhaps lower than today's, coincident with summer temperatures about 3 °C below present. From about 22,000 to about 14,000 years ago the lake level was lower. Evidence on the nature of the surrounding vegetation at the time is exiguous but it certainly was not forest. It must therefore be tentatively concluded that the climate was drier but substantially colder than before or since. In the nearby mountains of New South Wales, periglacial landforms (including solifluction mantles and blockstreams) suggest that frost action in the soil and weathering mantle was particularly active for a period after about 35,000 years ago (Caine & Jennings, 1968; Costin, 1972) and are variously taken to indicate mean annual temperatures between 6 and 10 °C lower than the present (e.g. Galloway, 1965; Costin, 1971). These events seem to have heralded the one glaciation for which there is clear evidence in the Snowy Mountains (Galloway, 1963), the maximum for which is placed somewhat before 20,000 years ago. The data from Lake George (Fig. 3.4.3) and the Snowy Mountains clearly indicate that by 35,000 years ago temperatures were lower than at present and that this refrigeration had intensified by about 20,000 years ago. There is conflict, however, about the length of this last cold period which only future work, aimed parti-

89

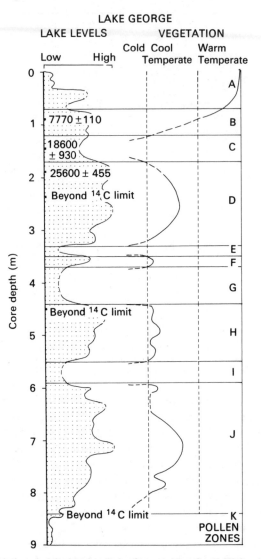

Fig. 3.4.3. Variations in lake level at Lake George, New South Wales, based on the analysis of pollen and algae, together with the indications of temperature derived from the pollen of terrestrial plants (from Bowler *et al.*, 1976).

cularly at isolating the effects of lowered temperature from those of lowered rainfall, can resolve.

In Tasmania, the last major glaciation covered about 5,000 km² (Derbyshire, 1972) and periglacial conditions were widespread, their products being recorded down to 600 m a.s.l. or even lower. This implies mean annual temperatures as much as 5 °C below present whilst other evi-

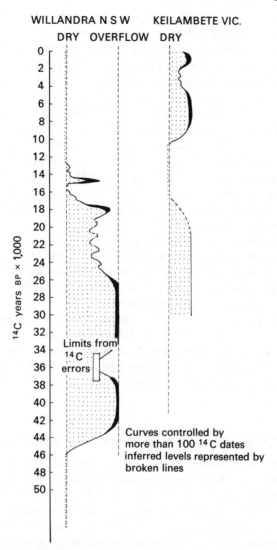

WILLANDRA N S W   KEILAMBETE VIC.

DRY   OVERFLOW   DRY

Limits from ^{14}C errors

Curves controlled by more than 100 ^{14}C dates inferred levels represented by broken lines

Fig. 3.4.4. Movements of lake levels in the Willandra Lakes of New South Wales and Lake Keilambete in western Victoria derived from sedimentological analyses of lake deposits and from related features (from Bowler *et al.*, 1976).

dence suggests no great change in the amount of precipitation although its seasonal incidence and the geographical pattern of its distribution may have been somewhat different from those of today.

Lake stratigraphy and pollen analysis from sites near Mt Gambier (Dodson, 1975) in southeastern South Australia suggest that from a drier time before 40,000 years ago conditions as wet as the present were

91

3.4   The long-term climatic record

experienced for about 10,000 years, after which the effective precipitation oscillated until about 26,000 years ago when somewhat drier conditions set in for about 15,000 years. At Lake Keilambete (Bowler, 1970; Dodson, 1974 b), across the state border in Victoria, 10,000 years of oscillating water levels were followed between about 18,000 and 10,000 years ago by a period so dry as to permit soil formation over the lake floor deposits (Fig. 3.4.4). In spite of differences in detail, therefore, this group of sites evidences a long period of oscillating effective precipitation culminating in a period substantially drier than the present before 10,000 years ago.

The Willandra Lakes of western New South Wales seem to have been empty for many millennia before about 45,000 years ago. From then until 26,000 years ago the lakes were flooded but thereafter their levels and salinities oscillated until, after a brief flooding about 18,000 years ago, they began to dry out more consistently, periodic wind erosion under a possibly more seasonal climate leading to marginal dunes of characteristic structure and form during the early stage (Bowler, 1971, and in press).

It is evident that, at its most extreme, the Full Glacial was colder or drier than the present, or both. Where the data allow more precise temperature estimates, they indicate mean annual values between 5 and 11 °C below present. The shift towards extreme conditions became expressed in the sediments between 38,000 and 26,000 years ago depending on the site and culminated between 26,000 and 11,000 years ago. There is some suggestion that the preferred period of most extreme cold lay between 22,000 and 14,000 years ago but that comparative dryness often persisted longer.

Climatic change around the Pleistocene–Holocene transition

In many parts of the world there is strong evidence of comparatively rapid climatic change, particularly of rising temperatures, between about 12,000 and 8,000 years ago or in many cases during a shorter period within that span. Partly due to natural circumstances, but also because of the limited resolution of some techniques in the Australian context, it is unlikely that close correlation of events of this period will be achieved in Australia and New Guinea. The following account is therefore concerned with the degree and rate of shifts from the most extreme conditions of the last Full Glacial to what might be described as 'mean Holocene' conditions roughly comparable to those of the present day.

From New Guinea, evidence of glacier retreat and of vegetation changes establishes incontrovertibly that temperatures had begun to rise by about 15,000 years ago (Bowler et al., 1976; Hope et al., 1975 and 1976). Outside the glaciated area, vegetation response was rapid, mean annual temperature rises of between 4 and 9 °C being indicated by 13,000 years ago. At greater altitude on Mt Wilhelm the ice lingered in the higher valleys until about 9,000 years ago and vegetation change was correspondingly somewhat slower. Throughout the New Guinea mountains, indeed, temperatures indistinguishable from those of the present were not achieved until between 9,000 and 7,000 years ago.

On the Atherton Tableland of North Queensland stratigraphic evidence argues for an increase in effective precipitation beginning about 11,000 years ago and finding expression in the westward migration of rain-forest until about 6,000 years ago. It has been argued that the latter part of this transition period also experienced a mean annual temperature rise of about 3 °C (Kershaw, 1975 *a*, *b*).

In the Snowy Mountains of New South Wales it seems that glaciers had begun to retreat about 20,000 years ago (Costin, 1972) and had virtually disappeared by 14,500 years ago. By contrast, a positive response of the vegetation to temperature rise seems not to have occurred until about 16,500 years ago (Raine, 1974). This may indicate that starvation of precipitation was the main factor responsible for the early decay of the glaciers.

At nearby Lake George, consistent overflow levels were abandoned by 19,000 years ago at the very latest since when the water, though oscillating, has reflected drier climate than before. The climatic implications of the vegetation changes at this time are complicated by the possibility that a coincidence of temperatures high by previous standards with only modest rainfall called for a flora which was only distantly available and consequently took a long time to arrive. But there is nothing at Lake George to conflict with the notion that the persistence of Full Glacial dryness accelerated the deglaciation of the Snowy Mountains beyond the rate determined by slowly rising temperatures.

In Tasmania, ice retreat had clearly begun before 11,000 years ago and was probably complete 3,000 years later (Macphail & Peterson, 1975). During this period the eastern part of the island at least was drier than today.

In the Mt Gambier area, lake levels, low between 15,000 and 11,000 years ago, were rising during the following 2,000 years (Dodson, 1974 *a*, 1975). At Lake Keilambete, the long consistently dry period broke about 10,000 years ago giving way to 3,000 years of saline oscillatory conditions leading to a freshwater stage (Dodson, 1974*b*). If, from general evidence elsewhere, the mean annual temperature in the 15,000–10,000 years ago dry phase can be assumed to have been 3–4 °C lower than today, precipitation cannot have been more than 60% of present since evaporation kept the lake dry. From that time until the maintained flooding of the lake

3.4   The long-term climatic record

7,000 years ago, temperatures certainly rose and evaporation must therefore have increased. In order to flood the lake under these conditions, precipitation must also have risen very considerably, if erratically.

For temperate Australia, therefore, the evidence commonly indicates that the dryness of the earlier Full Glacial period persisted until 12,000±2,000 years ago. Temperatures certainly rose through several degrees during the transitional time which, to an extent, would have offset any increase in precipitation. Rainfall evidently was increasing, however, to the point where lake waters in currently subhumid areas became fresh. Further inland, as at the Willandra Lakes and comparable sites in Western Australia, the precipitation:temperature balance did not shift sufficiently to flood the lake basins. In the tropics it seems that temperatures probably began to rise about 15,000 years ago and that, in spite of this, rainfall increases were sufficient to move the rainforest border westward until about 6,000 years ago. Throughout the Australia–New Guinea region the transition from Full Glacial to 'mean Holocene' climates seems to have taken place between 15,000 and 7,000 years ago.

Climatic variation in the Holocene

Climatic events formally within the Holocene but before about 7,000 years ago can best be considered as the final stages of the transition from Full Glacial conditions. The period since then is relatively short in relation to the resolution power of methods currently applied in Australia and New Guinea. Nevertheless, there is some evidence that the climate has not remained stable during the past 7,000 years.

From the mountains of New Guinea it appears that the natural altitudinal forest limit stood higher than its present level between 8,500 and 5,000 years ago (Hope & Peterson, 1975) to the extent of implying mean annual temperatures 0.5–2.0 °C above present during that time. There is also evidence for very limited glaciation down to about 4,400 m a.s.l. on Mt Wilhelm and for greater length of the glaciers on Mt Carstensz since 5,000 years ago (Hope et al., 1976). Together these data suggest that mean annual temperature might have fluctuated as much as 2 °C above and below its present level during the Holocene and that half this amplitude was certainly attained.

The vegetation history of the Atherton Tableland suggests the possibility that highest Holocene temperatures and rainfall were reached between 6,000 and 3,000 years ago and have since fallen by about 1 °C and 50% respectively (Kershaw, 1975 a).

In the Snowy Mountains, the natural vegetation pattern has approximated to that of the present for the last 6,500 years except for a period from 3,800 to 1,700 years ago when the hygrophilous element was peculiarly ill-represented suggesting conditions drier than today. At the same time, non-sorted soil steps of periglacial origin, now inactive, developed, indicating a mean annual temperature 3 °C lower than now or stronger winds preventing the accumulation of insulating snow (Costin *et al.*, 1967).

Lake George enjoyed brief spells of high water level at about 8,000, 5,000 and 3,000 years ago. The climatic implications of the development of its surrounding vegetation are somewhat enigmatic, for the interaction of climate and fire led to the establishment of the modern warm-temperate dry sclerophyll woodland during the last 6,000 years for the first time in the hundred millennia so far investigated there.

In western Victoria, southeastern South Australia and southwestern Western Australia, rising lake levels and the pollen-analytical record attest that the earlier aridity was substantially overcome by 5,000 years ago (Dodson, 1974 a). Since temperatures are unlikely to have fallen during that period, the rising lake levels were presumably due to a real increase in rainfall. Thereafter the precipitation: evaporation regime has remained unstable until the present day, lake levels continually rising and falling. The rate at which such fluctuations may have occurred is exemplified by a 16 m fall in water level at Lake Keilambete. This took place between AD 1850 and 1969 and does not appear to have been due to European settlement.

So far as the evidence goes it seems that a climatic turning point might have been passed throughout the Australian region about 6,000 to 4,000 years ago. Immediately before then many places were effectively wetter more consistently than today. Since then, where the techniques have been capable of depicting detail, rather wide fluctuations seem to have been imposed on a marginally cooling and drying trend. However, the limitations of the techniques used are substantial. It will be interesting to see, for instance, how the greater resolution power of tree-ring studies, only just beginning in the region, increases our understanding of the real significance of the sedimentological and palaeontological evidence which has so far dominated the subject.

Conclusions

The data so far available do little more than provide pointers towards a synthesis of Quaternary climatic change in Australia and New Guinea. Not surprisingly the region partook of the refrigeration of Wisconsin–Weichsel time to about the same degree as other parts of the world, perhaps with cells of extreme cold here and there. Where temperature and precipitation effects can be separated it seems that the warming had probably begun about 16,000±2,000 years ago. Except on the far and mountainous northern fringe of the region, where only an extraordinary reduction in rainfall would significantly affect any of the main indicators, low Full Glacial temperatures seem usually to have been accompanied by less water for vegetation and lakes than in the early part of the Holocene. This must mean that, in general, low rainfall, almost certainly lower than today's, accompanied the cold. Subsequently, any tendency for further desiccation imposed by rising temperatures was erratically overhauled by increasing rainfall until the mid-Holocene, far from being a Great Arid period, proved effectively wetter than the present.

Nobody familiar with the variety of modern Australian and New Guinean climates could write or read generalisations of this order with very much confidence. In order to attain them it has been necessary to set aside much of the evidence for variety which emerges from comparisons between investigated sites and to make somewhat arbitrary decisions about the relative importance of reconstructed events. Some statements, however generally authenticated from the semi-arid and humid fringes of the continent, must be meaningless for its arid core. Yet conditions there, partially locally generated, must have been critical in determining the depth of penetration of rainbearing winds, the position of the boundary between winter and summer rain and a host of other variables which combined to produce the climatic patterns of the Quaternary. There is little hope for productive investigations in the interior itself where data are limited to such tantalising but frustrating features as the patterns of orientation of dateless dunes. Similarly the narrow and discontinuous fringes with climates suitable for rain forest, though containing more than their share of deposits, are least likely to have registered anything but the most extreme climatic changes. Nor can there be very much more to be learned from the small glaciated regions.

The most obvious and potentially rewarding development would be the extension of sedimentological and pollen-analytical investigations to lake deposits in currently semi-arid areas in eastern and northern Australia, particularly near the winter rain–summer rain boundary. It is essential, too, to adopt technologies which will provide detailed chronologies from a wider variety of sedimentary materials and which will allow the separation of temperature and precipitation effects. Against a background of data assembled from such sources, the processes implied by soil types developed during particular periods in the past, the distributions of mammalian fossils and evidence with similarly partial climatic content, could be used to extend conclusions geographically and enhance them with detail.

Such a network of data, which with the current level of effort could take two decades to collect, would allow the reconstruction of the coarse pattern of temperatures over the region and of the modifications of today's wind systems from time to time in the Quaternary past. From that point it should not be too hazardous to extrapolate to conditions which probably applied in areas lacking direct evidence of past climates.

The work of the past ten years has demonstrated that the Australian region has the materials necessary for the investigation of its Quaternary climatic history. Partly because of the nature of these materials and partly because of the geographical position of Australia and New Guinea, the elaboration of such history has a special contribution to make to world climatology. Information currently available whets the appetite and allows the more critical statement of hypotheses. Unfettered fancy is curbed and the basis for controlled investigation is laid.

3.5 Closed lakes and the Palaeoclimatic record
D. M. CHURCHILL, R. W. GALLOWAY & G. SINGH

Closed lakes exist today under a wide variety of climates ranging from hot (Tchad) through seasonally warm (western United States) and cool (Patagonia, Tibet) to cold (Antarctica). They occur in suitable closed topographic depressions where evaporation balances precipitation over the lake plus runoff from the tributary area. In Australia they occupy areas with present-day mean annual precipitation from 125 to 800 mm and evaporation from 1,000 to 2,500 mm.

Because precipitation, evaporation and discharge vary, the area of the lake fluctuates over time scales ranging from diurnal through seasonal, annual and decadal, to millennial. Lake area is related to the cross-sectional geometry of the basin, to weather and seasonal variations in the short term, and to climatic changes in the long term, with a lag which can vary from hours to many years according to the hydrologic conditions. Lake area responds to changes in the ratio between supply (rainfall and discharge) and loss (evaporation) rather than to their absolute amounts. The area of water in a closed lake can expand with a decrease in precipitation if there is a concomitant greater decrease in evaporation.

Seepage may be safely neglected for many lakes and the basic hydrologic budget can then be expressed in the simple equation

$$E \times A_l = P \times A_l + Q \times A_t, \qquad (3.5.1)$$

where E is mean annual lake evaporation, A_l is mean lake area, A_t is tributary area, Q is mean runoff per unit area of catchment, P is mean annual precipitation.

Obviously Equation (3.5.1) can be used to determine one item of a closed lake budget if the others are known. Numerous attempts have been made to deduce the precipitation during the last glacial maximum by this method, taking fossil strandlines as a guide to former lake areas (e.g. Bobek, 1937; Snyder & Langbein, 1962; Haude, 1969; Galloway, 1970). The many

3.5   The long-term climatic record

difficulties with the method, not always recognised in the past, invite augmentation and testing of geomorphological interpretations from other sources. In this section some closed lakes in southeastern Australia are used as a guide to past climates, and palaeoecological indications of the palaeohydrology of Lakes George, Gnotuk and Bullenmerri provide a multidisciplinary approach to study of climate changes over long intervals of time.

Some properties of closed lakes

Towards the humid limit of their distribution, closed lakes are generally permanent bodies of water, have low catchment:lake area ratios, receive a high proportion of their supply by direct precipitation on the surface and may overflow seasonally or in particularly wet years. In the driest parts of their range, closed lakes contain water only in exceptionally wet years (e.g. Lake Eyre) and have large catchments which contribute most of the water they receive.

Closed lakes require closed depressions which originate by tectonic, volcanic, erosional and depositional processes. Some Australian examples are related to faulting (Lake George, New South Wales) or warping (the Lake Eyre depression as a whole), some may be related to wind erosion (the Monaro lakes of the Southern Tablelands of New South Wales, Lake Eyre *sensu stricto*) others are in volcanic craters (the maar lakes of western Victoria) but the great majority in Australia are formed by depositional processes such as disruption of an early drainage system by wind-blown sand or blocking of low-gradient tributary valleys by levees on a major river.

As a consequence of their predominantly depositional mode of formation, plus the low relief of the continent, most Australian closed lakes are comparatively shallow with gently sloping shores. In such lakes a change in the water balance rapidly produces major fluctuations in lake area which reduces the opportunity for waves to build impressive shore features. Closed lakes in the dry interior of Australia rarely have prominent shorelines (Krinsley et al., 1968) unless dunes on their margins have provided locally steeper slopes.

Waters of closed lakes range from slightly brackish to supersaturated. The salinity of a lake naturally varies with fluctuations in its volume. Closed lakes are generally more saline in arid than in humid areas but there are many exceptions. While it is to be expected that salinity increases with time it is a poor guide to the age of a lake (Langbein, 1961). Salinity changes deduced from lake sediments and palaeoecology (see below) can be a guide to changes in hydrology (Flint & Gale, 1958).

All closed lakes fluctuate in level and some long records of lake fluctu-

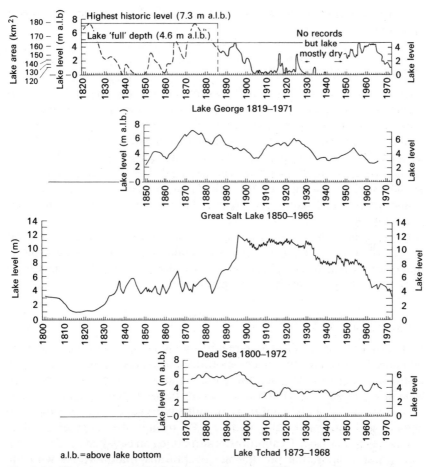

Fig. 3.5.1. Historic records of lake levels. Lake George (from Coventry, 1973), 35° 05 S, 149° 25 E. Great Salt Lake (from Whitaker, 1971), 41° 00 N, 112° 30 W. Dead Sea (mainly from Klein, 1961), 31° 30 N, 35° 25 E. Lake Tchad (from Touchebeuf de Lussigny, 1969), 14° 00 N, 14° 28 E. Lake levels reconstructed by using Nile discharge records at Aswan as a guide to supply to Lake Tchad; discontinuity at 1907 reflects discontinuity in Nile record.

ations in historic time are illustrated in Fig. 3.5.1. Lake George near Canberra has one of the best records and there is also a considerable body of climatic and hydrologic data from the district which assists in understanding the behaviour of the lake.

Records of water levels can be extended into the geologic past on the basis of such evidence as former shore lines and changes in sediment (e.g. Bowler & Hamada, 1971). Adjacent lakes fluctuate synchronously and there is increasing evidence for parallel long-term changes extending over wide regions (e.g. Butzer *et al.*, 1972). The outlines of a continent-wide

99

3.5   The long-term climatic record

or even world-wide stratigraphy of closed lakes are now becoming apparent.

Former evaporation rates

Past evaporation rate is usually estimated by applying temperatures deduced from geologic evidence to modern evaporation/temperature relationships from appropriate geographic environments. More precise estimates could be made by incorporating such data as humidity and wind speed but this in turn requires further unsupported estimates of the former climate that are rarely justified. The deduced rate of evaporation should be adjusted in the light of the calculated climates. In a dry climate, evaporation would have been greater than in a wet climate at the same temperature.

Despite the difficulties, evaporation is one of the better-defined terms in palaeohydrologic budgets from closed lakes. Current thinking suggests that mean annual evaporation rates in Australia at the last glacial maximum were about half to two-thirds the present values.

Former discharge

Estimation of former discharge requires a model incorporating precipitation amount and intensity, net evapotranspiration losses, storage capacity of the soil and surface conditions. Precipitation amount is the unknown term in Equation (3.5.1) but trial values can be used. Precipitation intensity as measured by daily rainfall is broadly correlated with precipitation amount and this relationship can be used to limit the range of possible intensities in the past. Evapotranspiration rate can be taken as 75% of the free water evaporation but the net evapotranspiration varies with the length of the period when water is available, and can thus be much lower in arid than in humid climates, and is not easily estimated for past conditions. Storage capacity of the soil and surface conditions are usually assumed to be stable through late Quaternary time although this is by no means certain, especially in areas of intermittent ground-water discharge.

In some regions but not others discharge during summers of colder phases of the Quaternary may have been at least equal to present-day winter discharge for the same precipitation. In the Yass valley near Canberra, where precipitation is about 55 mm in every month, discharge is four times greater in winter than in summer (Basinski, 1960) even though rainfall is rather more intense in the latter season. On the other hand, at Stephens Creek in semi-arid northwestern New South Wales the mean January discharge depth (1.5 mm) is twice the mean June discharge even though both months have identical mean precipitation (18–19 mm). Here, high summer precipitation intensities (rather than amount) mask any control of runoff by the lower evapotranspiration in winter and therefore modern winter discharge cannot be regarded as approximating summer runoff in glacial times.

Schumm's (1965) curves relating rainfall, discharge and temperature in the United States grossly underestimate modern discharge in semi-arid Australia and consequently their applicability to past conditions is suspect. It must be concluded that estimating former runoff is hazardous. Uncertainties in estimating past runoff can be mitigated by considering closed lakes from relatively humid areas where the runoff contribution to the budget is relatively less than in arid areas.

Former lake area

Modern and fossil shorelines indicate the extent of lakes in closed basins and the fossil shorelines can often be dated. However, using the fossil landforms to estimate palaeohydrologic budgets presupposes that a particular dated raised or drowned strandline represents a former mean lake area and is an indication of long-term hydrological (and climatic) conditions rather than a short-term stand under variable conditions. Historical records of the fluctuations in levels of Lake George and Lakes Gnotuk and Bullenmerri illustrate that such data may vary in palaeoclimatic significance with geomorphic and climatic considerations.

Lake George lies in a closed basin (932 km²), and shows no signs of subterranean discharge (Jennings *et al.*, 1964). The depth of the lake has ranged from 0 to 7.3 m over the 155 years of historic records (Fig. 3.5.1).

From Fig. 3.5.1 the maximum historic extent of Lake George is 180 km², the normally full extent is 165 km² and the mean extent is *c*. 80 km². Which area should be used in a palaeohydrologic budget? The mean area is close to the value required to balance the hydrologic budget over the years 1886–1974 if runoff ratios from the adjacent Yass catchment apply and if allowances are made for zero evaporation when the lake is dry and for a high evaporation rate when the water is shallow. In fact the lake has rarely been that mean size over the entire historic record. Lake George is 'full' not under mean precipitation and evaporation but when these factors are at about the eighth and second deciles of their respective historic ranges. Similarly the Dead Sea record (Fig. 3.5.1) indicates three separate modal levels since the early nineteenth century, each lasting a few decades and none corresponding to the mean level for the entire period.

In contrast to the record from Lake George is the consistent fall in water level since 1840 that has occurred in the twin crater-lakes Gnotuk and Bullenmerri in the Western District of Victoria (Currey, 1970). Similar

3.5   The long-term climatic record

patterns have been recorded for similar crater-lakes over a wide area of the western volcanic plains of Victoria and it is argued that the tendency is regional and with the water levels of Gnotuk and Bullenmerri differing by 42 m, they are not much influenced by ground-water movements or leakage. Today the craters thus form small internal drainage basins (Bullenmerri 873 ha, Gnotuk 702 ha) that act as natural evaporimeters at times when the lakes stand below overflow level.

From 1881 to 1974 Lake Bullenmerri has fallen 17.37 m from 162.76 to 145.39 m, an average rate of 18.68 cm/yr and Lake Gnotuk has fallen at a comparable rate of 17.78 cm/yr. The rainfall record from nearby Camperdown, 4 km away, shows no overall trend to decreased precipitation over the period of record (Fig. 3.5.2) and the loss is attributed to excessive evaporation over rainfall. Of what use are these closed basins in the task of extending our often meagre meteorological records?

The mean annual rainfall for Camperdown (Fig. 3.5.2) for the period

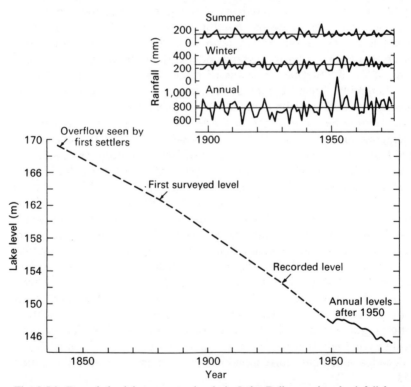

Fig. 3.5.2. Recorded minimum water levels in Lake Bullenmerri and rainfall for nearby Camperdown. (Data courtesy of State Rivers and Water Supply Commission of Victoria, Bureau of Meteorology, and Currey, 1970).
Note: Annual variation in water level is less than 1 m.

1898–1974 is 777.5 mm. The rainfall is normally distributed about the mean with a standard deviation of 135.7 mm. Recorded extremes are 511.4 and 1,253.4 mm. Since the tributary catchments are very small, discharge is a minor part of the hydrologic budget. Hence mean annual evaporation approximates the mean annual rainfall minus the mean annual change in level (−186.8 mm in Lake Bullenmerri and −177.8 mm in Lake Gnotuk). Therefore the mean annual evaporation rate for the last 75 years is 955–964 mm/yr for the area around Camperdown. Accepting evaporation in excess of precipitation as responsible for falling lake levels the explanation must be in a change among the factors that determine evaporation. Work done by the State Rivers and Water Supply Commission (J. Sutcliffe, unpublished) suggests that a rise in annual temperature since 1840 in the Camperdown area would account for the falling lake levels.

Neither the Lake George nor the Lake Bullenmerri water-level record discussed above could be deduced from geomorphic or palaeoecological evidence. Conversely not all water balance conditions over periods as short as 100 years or so will leave an environmental record, despite the fact that in some cases shorelines can form rapidly (Klein, 1961; Krinsley, 1970).

In areas where detailed lake-level records are not available, old strand-lines have to be used as evidence for hydrological conditions despite the fact that proper definition of lake area in terms of the frequency of occurrence is one of the most difficult and least studied aspects of lacustrine palaeohydrology. The difficulty is most acute in arid areas of Australia where rainfall and runoff are more variable than in humid areas. Obviously, wherever possible, palaeoclimatic interpretations of fossil landforms associated with closed basins should be tested against palaeo-ecological evidence from stratigraphically intact deposits from the lake-bottom.

Biologically meaningful periods may be defined relative to the threshold characteristics of the component species found fossilised in lake bottom deposits and hopefully some quantitative indications of the difference between climatic periods thus represented in the fossil biota can be derived. Unfortunately threshold characteristics vary between species so that different associations of species represented in the record mean that bioclimatic interpretation must be based on the ecology of different sets of organic remains present in cores of lacustrine sediments.

Cores from bottom sediments that contain fossil pollen grains of the aquatic and terrestrial vegetation provide a record of vegetation changes and, with it, an indication of environmental change of which climate is a major component. The internal drainage basins of Western Victorian crater-lakes and Lake George have long sedimentary records of such changes and palynological study of the core samples reveals the vegetation

3.5   The long-term climatic record

history of surrounding regions for tens of millennia. A hydrological test of the palynological method is offered by analysis of modern aquatic diatom communities in which the relations between species composition and water salinity are known (Tudor, 1973). The fossil diatom record of changes in water salinity is of palaeohydrological significance in a basin of internal drainage, because prehistoric water levels can be derived from the salinity of the water once the geometry of the lakes and the ratio of cyclic salt accumulation to salt weathered from the drainage basin are known. In what follows the scope of these palaeoecological techniques for analysing Quaternary climatic change are illustrated from palynological studies of core samples from Lake George and fossil diatoms from Lakes Gnotuk and Bullenmerri.

The prehistoric record from Lake George

That water levels in Lake George have varied in depth and extent, was first suggested by C. S. Wilkinson in 1886, who noted landforms indicating overflow on the western side, and recent geomorphological studies by Coventry (1976) have supported this view. His work on the abandoned shorelines suggests that lake water was 36 m deep and overflowed sometime between 26,000 and 19,000 years BP. Since then higher lake levels than present have occurred at 15,000, 5,000, 3,000 and possibly 8,000 years BP and these are closely paralleled by the pollen evidence (Fig. 3.4.3).

The mean annual rainfall of the Lake George area is 650 mm, uniformly distributed throughout the year. However, periods of erratic rainfall are not uncommon so that floods and droughts cause the lake level to rise and fall without significant long-term geomorphological or ecological legacy. Because the lake and its surroundings are drier and experience higher summer temperatures than the adjoining mountain ranges, and there are frost-hollow effects in winter, the vegetation in the catchment differs markedly from that on the mountain ranges. The vegetation around Lake George ranges from warm temperate dry sclerophyll eucalypt forests to savanna grasslands. The heavy frost in winter restricts the main periods of plant growth to autumn and spring, or to exceptional periods of good rainfall during the summer months (Slatyer, 1960). The mountain ranges, on the other hand, harbour a wet sclerophyll vegetation which demands both greater moisture and lower summer mean temperatures than those prevailing at Lake George. In the past, any change in the temperature and moisture regime at Lake George, favouring wet sclerophyll vegetation, would have allowed a quick migration of this flora from the mountain ranges. Lake George is therefore critically placed to record past changes in temperature, evaporation and precipitation. The vegetation record is

104

preserved in the bottom sediments from which continuous core samples were recovered (Singh, 1975).

Pollen from the top 9 m of the core has been analysed using absolute counts of pollen, spores and algal remains estimated per 100 cm²/yr subject to the constraints of ¹⁴C dating. Organic carbon content yielded erratic radiocarbon dates, and the three carbonate dates given in Fig. 3.4.3 require further supplementation. For a short summary of the pollen work at Lake George see Bowler *et al.* (1976).

In the 9 m section, which consists mainly of silty clay, the pollen sequence is divisible from below upwards into five wooded periods, zones J, H, F, D and A, alternating with five non-wooded phases, zones K, I, G, E and C; zone B is transitional between zones A and C (Fig. 3.4.3). Except for zone A, representing the most recent wooded interval starting about 6,000 years ago, all other wooded periods represent cool temperate forests. Zone A marks the first development of warm temperate dry sclerophyll woodlands and runs contemporaneously with a brackish water to low oscillatory phase. Supplementary evidence gained from micro-fragments of charcoal in the sediments suggests that fire activity was prevalent during all the wooded periods whereas charcoal is generally lacking for the treeless intervals (Singh, 1975).

The present-day evidence suggests that periods with brackish or shallow lakes when trees had disappeared from the Lake George area, were times of low precipitation and low temperature (approximately 10 °C lower than today's January mean). The intervening periods with cool temperate forests indicate summer temperatures about 3 °C lower than today's, but absolute precipitation was greater than before, so that tree growth continued and greater evaporation from the lake may have occurred. Runoff into the lake must also have increased to make it deeper and fresher. Nevertheless, the mean annual precipitation may still have been less than at present (cf. Galloway, 1965).

The final wooded period, of dry sclerophyll vegetation, occurs with saline and low lake water, and is interpreted as a response to higher temperatures and higher evaporation than before. It is suggested that an absolute increase in rainfall to today's values has not offset the increased evaporation and so the modern lake levels are erratic (Fig. 3.5.1).

Cool temperate zone D may have lasted from about 50,000 years BP to about 22,000 years BP. Zone C followed till about 14,000 years BP. The transitional zone B perhaps lasted till 6,000 years BP when zone A began and lasted till now (Fig. 3.4.3).

Eucalypts and myrtaceous shrubs were rare in the two lower wooded periods, zones J and H (Fig. 3.4.3). The herbaceous component of the vegetation was also scarce. The forests were probably closed, and light demanders, such as eucalypts and herbs, did not flourish during these

3.5 The long-term climatic record

periods. Unless proved otherwise through future work, the occurrence of open vegetation in zones F and D could not be regarded as features peculiar to the late Quaternary Period but are possibly associated with human disturbance.

The present warm temperate, dry sclerophyll forest represents the latest wooded period and is a clear evidence that warming took place following the last glacial episode.

The twin crater-lakes, Gnotuk and Bullenmerri

Landforms, sediments and associated features of the Western District crater-lakes indicate a history of fluctuating water levels during the late Quaternary (Bowler & Hamada, 1971; Dodson, 1974 b). These include emerged and submerged tree stumps *in situ*, emerged algal deposits (Dodson, 1974 c) and sequences of coarse and fine sediments (Bowler, 1970) all suggesting a regional (climatic) cause for the synchronous lake-level changes. In lakes Gnotuk, Bullenmerri and Keilambete these changes in lake level produced changes in water salinity that affected the composition of the diatom floras living in these water bodies.

The total amount of dissolved salts in Lake Gnotuk is not very different from that in Lake Bullenmerri, but there is contrasting salinity (averaging 59.6 g l^{-1} and 8.8 g l^{-1} respectively) because there is over six times as much water in the latter lake (Currey, 1970). There is also a striking contrast in the species composition of the diatom communities. The precise relation between diatom associations and water salinity can be defined so closely that some of the fossil diatoms in the lake sediments furnish quantitative estimates of water volume and lake level, palaeoclimatic parameters.

Tudor (1973) used linear regression of principal axis ordination (X-axis) of the diatom communities with total dissolved salts to obtain the contemporary salinity of the water at each 10 cm level in the cores from these two crater-lakes to devise a history of salinity. Radiocarbon dating of selected samples (twelve from Lake Gnotuk and four from Lake Bullenmerri) allows this information to be presented as a plot of surface water levels. The curves (Fig. 3.5.3) are corrected for salt accumulation from weathering and cyclic salt (Currey, 1970). The results agree with those obtained by Bowler (1970) for Lake Keilambete using sedimentological analysis. Lakes Gnotuk and Keilambete during some months of the year now have water salinities twice that of sea water but from 7,000 to 4,000 years ago were quite fresh as indicated by the presence of only a wide range of algal (including *Pediastrum* and *Scenedesmus*) and diatoms species with salinity tolerances of less than 1.7 g l^{-1}.

This wetter period followed one much drier and extending back to 10,000 years BP and beyond, but the detail of the earlier part of this record awaits further work.

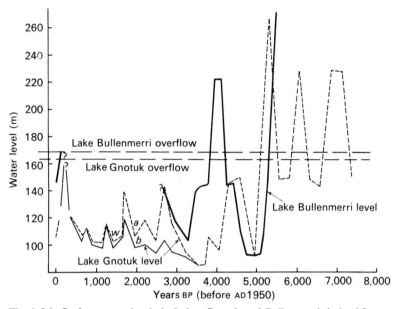

Fig. 3.5.3. Surface water levels in Lakes Gnotuk and Bullenmerri derived from salinities indicated by fossil diatom communities, supplemented by spot levels indicated from marl (calcareous clay) and submerged trees (W). '*a*' indicates the curve uncorrected for salt accumulation and '*b*' indicates the curve corrected for annual salt accumulation through cyclic salt and weathering.

Note: Water levels shown above the overflow heights are calculated as though the crater continued above its present level, in an ever-expanding cone.

Summary

Where long-term meteorological records are absent the continuous evidence of water-level fluctuations for environments characterised by closed lakes may provide the best available indication of climatic history. Historic records demonstrate that isolated prehistoric evidence of discontinuous events that are preserved as wave-built or wave-cut landforms do not necessarily coincide with significant ecological or climatic phases of long duration. Despite the difficulties associated with developing complete models of past runoff using the elevation of old shorelines as a basis for calculating palaeohydrologic budgets the method is worth pursuing. It is likely to be more reliable for closed lakes from relatively humid areas, than for open basins or those from drier regions.

Use of biostratigraphical, sedimentological and geomorphic evidence allows interrelated interpretations to be tested one against the other in the same basin. Thus the sequence of low and high lake levels might be derived from indications of changes in water salinity, bottom sediments, emerged and submerged shorelines. The palaeoclimatic implications of this evidence can be tested against palynological studies which usually

107

elucidate both regional terrestrial and local aquatic vegetation histories, that have responded to past changes of climate.

Although all the methods discussed above have not yet been used for one single closed lake basin, a multidisciplinary and interdisciplinary approach will be amongst those yielding the most significant hydrological results in areas beyond the Quaternary ice bodies of the southeastern uplands of Australia. The epoch over which such studies are relevant, however, must always be defined.

3.6   Eustatic sea-level changes and environmental gradients
H. A. MARTIN & J. A. PETERSON

The boundary conditions regarded by CLIMAP Project Members (1976) as necessary to attempt a climatic reconstruction of the 'ice age earth' (e.g. at 18,000 years BP when the main continental glaciers were at their maximum extent) were:

(i)   the sea-surface albedo and temperature pattern of the world ocean;

(ii) the extent and elevation of permanent ice;

(iii) the albedo of land and ice surfaces;

(iv) the geography of the continents (especially the shape of the coastlines during glacio-eustatic low sea level).

The reconstruction has been used in an empirical attempt to explain the cause of the abrupt and frequent climatic changes that characterised Quaternary times. This involves using glacial stage (often loosely referred to as 'ice age', cf. Section 1.1) boundary conditions in physical/mathematical models of the general circulation of the atmosphere (Gates, 1976). For land surfaces the results of the modelling can be verified if reliable estimates of past temperature and precipitation are available from the palaeoenvironmental record.

The derivation of sea-surface temperatures from dateable microplankton fossil assemblages has been described by Shackleton (Section 3.2). The temperature distribution in the surface layer of the oceans is a major climatic influence, especially in the Southern 'oceanic' (Maury, 1855) Hemisphere. The fact that Pleistocene Southern Hemisphere sea-surface temperature trends seem to have led the Northern Hemisphere ice volume record by some 3,000 years (Shackleton, Section 3.2, also cf. Young & Schofield, 1973) suggests that in the Northern Hemisphere, climatic changes were complicated by feedback processes associated with the growth and decay of huge land-based ice sheets, and that climatic histories from mid-latitude maritime Southern Hemisphere areas may be amongst those more closely related to the fundamental causes of or triggers of the ice-age climatic changes. Areas close to the Southern Ocean and without the complications arising from active fold mountain building

108

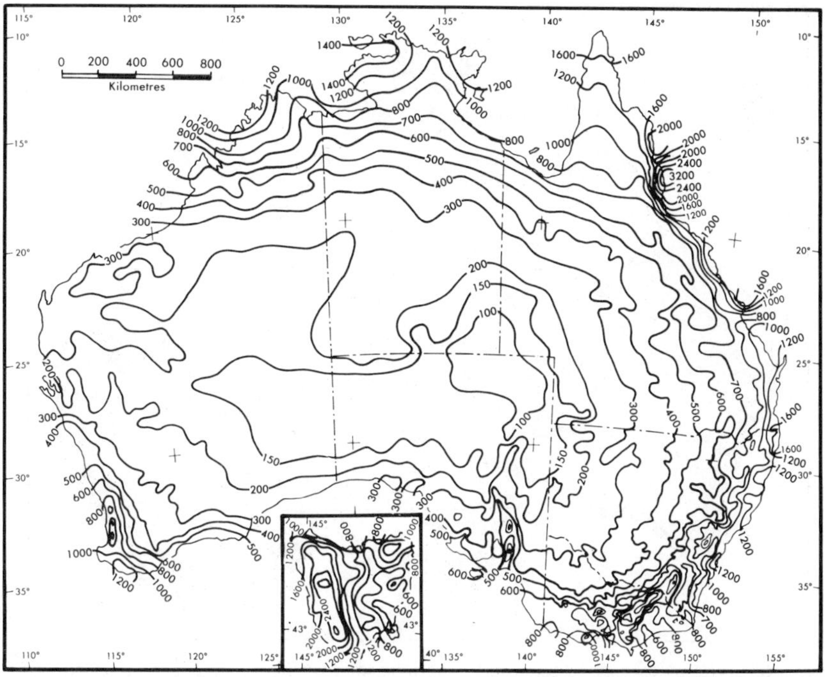

Fig. 3.6.1. The modern precipitation distribution in Australia is one of the most fundamental determinants of the slope and direction of environmental gradients. This map, from one supplied by the Australian Bureau of Meteorology, shows annual rainfall (median) in millimetres.

are probably best placed in this regard (e.g. Tasmania, coastal southern Australia and southern Africa).

The last three boundary conditions mentioned above involve palaeo-environmental interpretations from terrestrial evidence and it is this as well as new ocean core data that will be used to test and refine hypotheses embodied in the CLIMAP modelling of 'ice age earth' (Gates, 1976). Of most value will be sedimentary sequences providing long continuous records of regional rather than local significance and providing not only a temperature record to test alongside the ocean core results, but also something that cannot be derived from the deep ocean cores: a record of precipitation changes. However, deriving a history of any one component of the water balance is fraught with difficulty as illustrated by the problems associated with interpreting the record from closed lakes (Section 3.5). Moreover the great variability in precipitation in arid and semi-arid areas (most of Australia: see Figs. 3.4.1 and 4.1.3) and associated environmental instability leads to variation in types and rates of sedimentation, while erosion causes discontinuities in the depositional record which, as pointed

109

3.6 The long-term climatic record

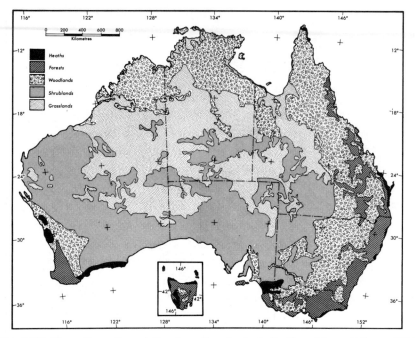

Fig. 3.6.2. The pattern of vegetation formations in Australia (from Moore & Perry, 1970) reflects environmental gradients.

out in Section 3.4 might record catastrophic events as often, or instead of, more representative ones. Hence for large areas of Australia there is a shortage of reliable land-based data to test the results of modelling. Where a palaeo-record of precipitation only exists, interpretation presents the greatest difficulties.

Climatic gradients

Where estimates of ice-age temperature and precipitation are thinly scattered and of varying reliability the use of the general circulation model might be tested by comparing the simulated surface air flow patterns with actual patterns deduced from evidence of the orientation of former precipitation gradients between the coasts and 'rain shadow' areas inland. Over much of Australia there are well-marked environmental gradients 'normal' to the coastline. These are summarised in the rainfall, vegetation, and runoff patterns (Figs. 3.6.1, 3.6.2 and 3.6.3). These climatic gradients embody the overall causes of ecological and geomorphic distribution patterns and are an expression of the atmospheric circulation pattern as reflected at the earth's surface. While it is possible to assign values for temperature and precipitation along modern environmental gradients it is

110

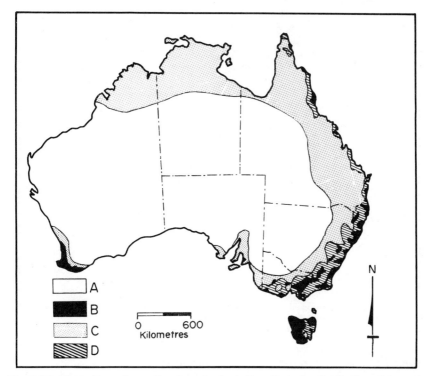

Fig. 3.6.3. The pattern of runoff regimes in Australia reflects the distribution of the main components of the water balance and echoes the general vegetation and rainfall patterns (after Ceplecha, 1971). Zone A – area where precipitation is entirely accounted for by evaporation. Zone B – area with mainly perennial runoff. Zone C – area with seasonal precipitation, seasonal moisture deficit and non-perennial runoff. Zone D – area with seasonal precipitation excess and seasonal moisture deficit and mainly perennial runoff.

more difficult to do so for glacial stages. This is not only because of interdependence of temperature and precipitation in determining the water balance, but also because glacial stage changes in these elements were unevenly distributed and no matter how firm an estimate derived from evidence at particular locations, extrapolation of values into uninvestigated regions must be very tentative.

An alternative method is to search for evidence of the direction and slope of glacial stage environmental gradients in the hope that an indication of surface circulation patterns might be obtained without the necessity to assign values to temperature and precipitation. Indeed palaeoenvironmental data that might never yield unequivocal estimates for these fundamental climatic elements can still be used to indicate the changing nature of the earth's surface during glacial stages or to test models of associated atmospheric circulations.

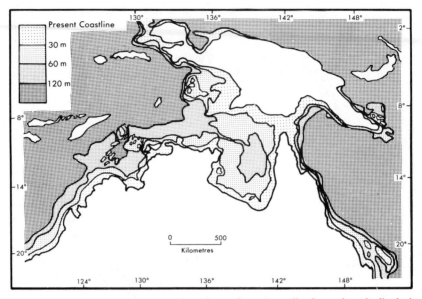

Fig. 3.6.4. The distribution of land and sea in northern Australia changed markedly during the last major eustatic changes in sea level (after Nix & Kalma, 1972). Contours show coastline corresponding to indicated drops in sea level.

Eustatic sea-level change: a complicating factor

During the glacial stages the growth of ice caps, mainly in the Northern Hemisphere, and the extension of permafrost areas delayed the runoff component of the hydrologic cycle over large areas, and sea level fell. The distance of present terrestrial locations from the sea did not increase everywhere by the same amount, however, because of irregularities in the shape of the continental shelf at −130 m (compare Figs. 3.6.4 and 3.6.5). Assuming the general configuration outlined in Figs. 3.6.4 and 3.6.5 is correct, one can propose as a working hypothesis that glacial-stage cirulation patterns and climatic gradients were parallel or sub-parallel to the present ones in Tasmania, may have been so in the region of the Great Australian Bight, and are unlikely to have displayed such simple patterns in the region of Bass Strait (Bassiana) or on either side of the glacial-age Arafura Plains or the plains of Carpentaria. These latter especially represent large areas, where terrestrial evidence of environmental history is now drowned by warm shallow seas and bottom cores will contain only discontinuous stratigraphies (including old soils) rather than the continuous microfossil accumulations of deep ocean cores.

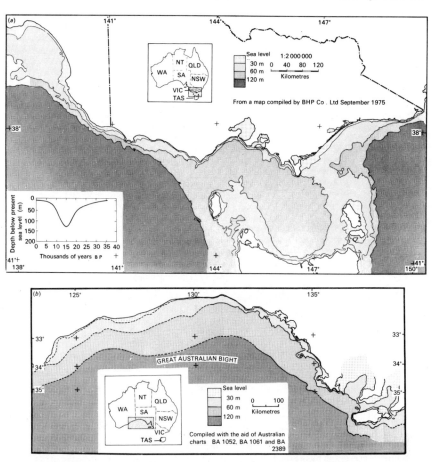

Fig. 3.6.5. Changes in land–sea distribution in southern Australia during the last major eustatic changes of sea level. Eustatic sea-level curve after Emery (1969). Contours show coastline corresponding to indicated drops in sea level.

Late-Pleistocene climatic gradient in Tasmania

Although the coarse grid of the ice-age earth model of the last glacial maximum (CLIMAP Project Members, 1976, Fig. 1) designates a large area of southeastern Australia from 28° S to the southern tip of Tasmania as having the albedo of over 40% (snow and ice), the total glaciated area was probably no more than 5,000 km². This is not much different from, and possibly less than, the sum of the total glacierised areas of New Guinea, although there are indicators of earlier and more extensive glaciations in both areas (Galloway *et al.*, 1973; Colhoun, 1975; Hope *et al.*, 1976). The unequivocal evidence for glaciation on the Australian mainland is on the highest and presently wettest mountains, and is restricted to thirteen

113

3.6 The long-term climatic record

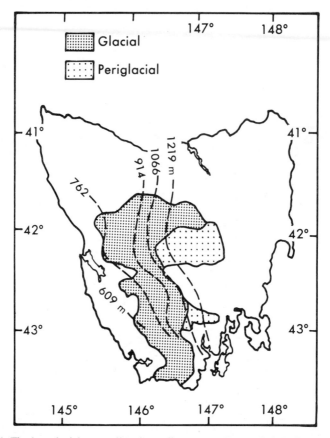

Fig. 3.6.6. The last glacial stage climatic gradient across Tasmania is indicated by a rising snow line (represented by the distribution of the lowest cirque floor levels) and the distribution of glacial stage glacial and periglacial processes (after Davies, 1965, 1967) across the Central Plateau and south west.

Fig. 3.6.7. (a) Ahlmann's (1948) curve derived from detailed glacio-climatic research, quantified the relationship between snow amount and summer temperature at the point where ice accumulation and melting are balanced on the maritime mountain glaciers of Norway. This work illustrates that, even within the maritime mountain zone, (i) positive mass balance of glaciers can be supported by a range of climates, and (ii) along the transition between the moister and drier extremes of this range, the snowline will rise according to the temperature fall necessary to offset the diminished snowfall into cirques.

Fig. 3.6.7. (b) The lowest elevation at which ice appears to have formed during the last glacial stage marks the lowest local level of the snowline. At the lower boundary of the warm infiltration ice formation zone (Shumskii, 1964) that characterises densification of maritime mountain snowfields, ice accumulation occurs in topographically favourable sites for snow accumulation (cirques) and the lower boundary of net annual accumulation marks the local elevation of the orographic snowline. Above these sites, or in areas where local topographic factors preclude cirque formation, ice abraded surfaces on plateau and summit areas

114

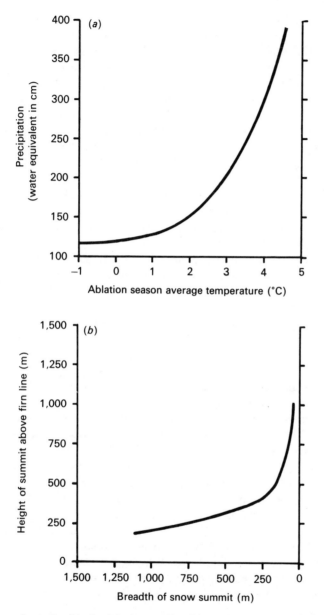

approximate the altitude of the former climatic snowline. However, the formation of summit ice caps is more likely, the broader the summit, as illustrated by Manley's curve (Manley, 1955). Although a range of climates may be found along the lower boundary of the lower maritime cryosphere, the systematic regional component of altitude variation in geomorphological indications of former snowline levels indicates the nature of the contemporary glacio-climatic gradient.

115

cirques (glacially enlarged valley heads immediately in the lee of the crestal ridges). In marginally glacial climates cirques are topographically favourable sites for snow and ice accumulation. In less marginal and more maritime glacial climates with heavy snowfall, glaciers may extend well beyond the sheltered ice accumulation area of the cirque. In Tasmania the heaviest glaciation was in the west, and lower limits of glaciation generally rise towards the east where morphological evidence suggests ice was thinner. The same applies to the cirque morphology and elevation. In the southwest they are lower, or deeper, with bedrock thresholds and overlooking end moraines down-slope well beyond the cirque. Towards the northeast the cirques are smaller and the thresholds buried by moraines (Peterson, 1968). Moreover, in the many areas where glaciation was marginal to the extent that ice only accumulated in cirques, all these features were in the lee of natural snow fences that were effective only for snow-bearing winds from westerly quarters.

A late Pleistocene glacio-climatic gradient across Tasmania was postulated by Davies (1967) (Fig. 3.6.6) who argued that the lowest altitudes of ice accumulation (as indicated by the altitudes of cirque floors) indicated that the altitude of the glacial-age snowline rose from southwest to northeast roughly parallel to the modern precipitation gradient. The idea can be tested not only with the morphological and altitudinal distribution data but also timewise because a snowline with such a gradient would cause glaciation at any given altitude to start earlier and finish later in the west. The highest cirque glaciers in the highest mountains further west and south would be the last to disappear. Deglaciation dates for some of the highest cirques were obtained (Macphail & Peterson, 1975) and the extent to which the deposits from which they came represented the period immediately before ice disappearance was tested by palynological analysis. The fossil pollen study illustrated replacement of a predominantly herbaceous late glacial flora by a predominantly arboreal post-glacier flora, suggesting that the dates were from deposits accumulated soon after deglaciation.

Beyond the limits of glaciation on the Central Plateau of Tasmania and on higher eastern mountains beyond the present 1,500 mm isohyet the climate was cold enough to cause ice to form in the soil but not snowy enough to form glaciers, even in sheltered locations. Thus as a general rule in upland Tasmania glacial landforms give way to periglacial features from west to east lending further support for the proposed glacial-stage climatic gradient.

The evidence for this glacial-age climatic gradient suggests that surface air flow patterns were not radically different from those prevailing at present. This palaeoenvironmental evidence offers a test of the Tasmanian part of the ice-age climate models, that does not require a particular temperature or precipitation value to be assigned to palaeoclimatic evidence. Glaciation can be supported by a range of climates even within the

116

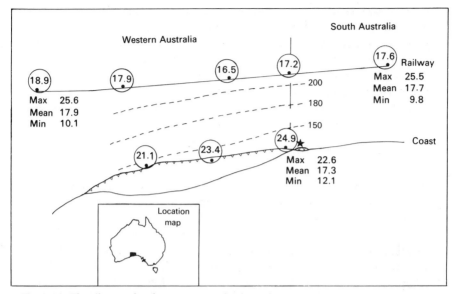

Fig. 3.6.8. The climate, showing mean annual values of rainfall (cm) and temperature (from standard period average, Australian Bureau of Meteorology) and evaporation (Martin, 1973) in the Nullarbor region (see inset map). ★ , excavation; (16.5) rainfall; --- 180, evaporation; max. 25.5, mean 17.7, min. 9.8, temperatures; ▼▼▼ Hampton Range.

maritime mountain zone. Moreover, in areas of cirque glaciation any precipitation value derived from applying the relationship summarised by Ahlmann's (1948) curve (Fig. 3.6.7(*a*)) is for topographically favourable sites for snow accumulation, and is not necessarily typical of the surrounding area (Fig. 3.6.7(*b*)).

Evidence from the Nullarbor

The Nullarbor Plain on the shores of the Great Australian Bight is an arid to semi-arid area (Fig. 3.6.8). Tertiary limestones contain caverns that accumulate and preserve deposits lacking from most areas as dry as this. The country is flat; the only topographic feature is the Hampton range, an emerged sea cliff line rising some 60 m. Otherwise, minor relief varies from 2–3 m in some areas and 5–10 m in others. Near the coast there are small patches of siliceous sands blown inland from the coastal dunes. Topography and soils are very uniform and do not have any major influence on plant distribution. Pollen has been studied from deposits in archaeological excavations. These excavations are situated under rock shelters associated with the edges of sinkholes or the escarpment of the Hampton Range. The deposits consist mostly of wind-blown dust and contain mostly wind-blown pollen, artifacts, animal bones and charcoal. Three sites were examined, and the deepest deposit, at Eucla, 540 cm, is

3.6 The long-term climatic record

over 20,000 years old by radiocarbon dating. Each deposit contains abundant pollen which could be identified with plant species growing in the region today, but the quantitative relationships of the different pollen types varied through the depth of the profile. Interpretation of pollen counts depends upon an understanding of the present-day vegetation and its environmental controls. The pollen in surface deposits beneath known vegetation types provides the means of interpreting the pollen counts in terms of the vegetation.

The vegetation

The vegetation of the Nullarbor region can be divided into three zones, viz. the coastal eucalypt scrub, the arid scrub and the inland shrub steppe. These zones are shown on Fig. 3.6.9.

The eucalypt scrub zone is found nearest the coast and the dominants are about ten species of eucalypts and one species of tea-tree. In the most favourable habitats, they grow to small tree size, but further inland their height decreases and the space between individual plants increases. Eucalypts and tea-tree belong to the family Myrtaceae and their pollen is very similar (see Fig. 3.6.10). The shrub layer is discontinuous and contains many species, including saltbush and bluebush. These last two belong to the family Chenopodiaceae and the pollen is very uniform in this family (see Fig. 3.6.10).

In the arid scrub zone, the eucalypts and tea-tree are less common and the shrub layer is much denser. Saltbush and bluebush (chenopods) are

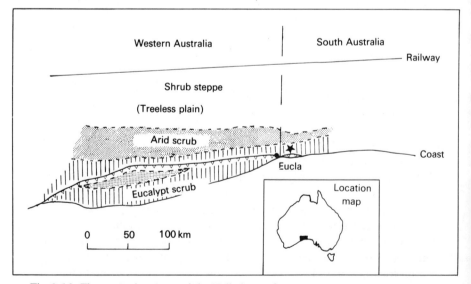

Fig. 3.6.9. The vegetation zones of the Nullarbor region.

118

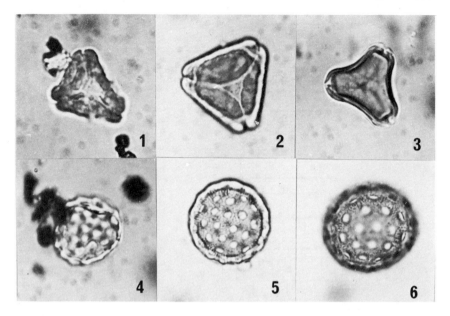

Fig. 3.6.10. The pollen types. 1, fossil Myrtaceae-type, from the bottom of Eucla excavation. 2, eucalypt; 3, tea-tree; 4, fossil chenopod-type, from the bottom of the Eucla excavation; 5, saltbush: 6, bluebush. Magnified 1,000 times.

much more abundant here. There is a wealth of other species but most of them are low pollen producers and very little pollen from them is recovered. This applies especially to *Acacia* species which are a conspicuous part of the vegetation but rarely constitute more than 1–2% of the pollen count. In many respects, the arid scrub zone is intermediate between the eucalypt scrub and shrub steppe zones.

The shrub steppe or treeless plain is the furthest inland. Trees and large shrubs are extremely rare here, and the shrub layer is mostly saltbush and bluebush.

Most of the plant species are found in the three zones, but their abundance varies considerably. Thus eucalypt species and tea-tree are common near the coast, becoming more infrequent with distance from the coast. The chenopods (saltbush and bluebush) are sparse near the coast becoming more common further inland and they are the major component of the vegetation on the treeless plain. These two important plant groups show opposite trends in their distribution patterns.

It is clear that the rainfall gradient is the major factor controlling plant distributions. As discussed previously, the other environmental factors are relatively uniform and there is no evidence that they produce the zonation seen in the vegetation. The salt content of the soils in the region is fairly high, judging from the halophytes (salt-tolerant vegetation) present, but it only determines local patterns and not the major vegetation zones.

3.6 The long-term climatic record

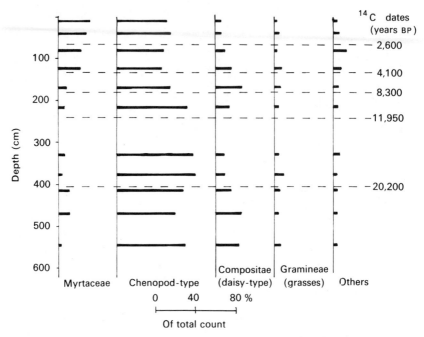

Fig. 3.6.11. The pollen counts of the Eucla excavation with radiocarbon dates.

The pollen counts

Figure 3.6.11 presents the pollen counts for the Eucla excavation and the radiocarbon dates through the profile. Clearly, the chenopod pollen type is the major group, with a maximum for the period 20,000 to about 11,000–10,000 years BP. Then follows a decline to about 4,000 years BP, after which the values are somewhat erratic. The Myrtaceae pollen type is a smaller group, and older than 20,000 years BP, there is a minor peak. The period 20,000 to 10,000–9,000 years BP registers consistently low values, and there is a continual increase from about 10,000–9,000 years BP to the present day. The other groups do not show any consistent trends and are relatively minor. Clearly, the two major plant groups show the same pattern of variation in the pollen counts as that seen in the vegetation, i.e. where the Myrtaceae group is larger, the chenopod group is smaller and vice versa. Consequently, the ratio of Myrtaceae to chenopod pollen has been plotted and the curve is shown in Fig. 3.6.12. The values of surface sample ratios from the three vegetation zones allow interpretation of the profile in terms of the vegetation.

Figure 3.6.12 is interpreted in the following way. Older than 20,000 years BP the pollen ratios are widely scattered and the reason for this is not known. From about 20,000 to 10,000–8,000 years BP, treeless plain occu-

120

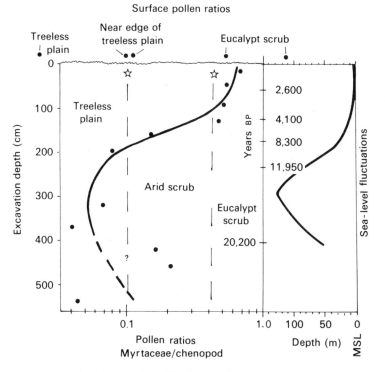

☆ Approximate limits of vegetation zones

Fig. 3.6.12. The pollen ratios of the Eucla excavation and the generalised curve of world-wide sea-level fluctuations (from Emery, 1969). The surface pollen ratios have been used to determine the approximate limits of the ratios from the different vegetation zones.

pied the Eucla area. In the period 10,000–8,000 to 5,000–4,000 years BP, there is a marked change to the eucalypt zone and from 5,000–4,000 years BP to the present, there has been relatively little change, with perhaps only a small increase in the density of the eucalypt cover.

This interpretation of the pollen counts is supported by a variety of evidence, including the records of animal bones and of artifacts, all of which are in substantial agreement (Martin, 1973).

Climatic interpretations

These postulated changes in the vegetation could not have occurred unless the climate was suitable, hence they imply a climatic change. Thus the period 20,000 to about 10,000–8,000 years BP was more arid with perhaps a rainfall of about 16–18 cm/yr, the values for the treeless plain today. From 10,000–8,000 to 5,000–4,000 years BP was a time of increasing rainfall, and after 5,000–4,000 years BP, the rainfall has been at much the same level as today, viz. about 25 cm/yr. But on this flat featureless terrain, distance from the coast determines the rainfall, and the coastline has changed its position in the last 20,000 years, in concordance with world-wide eustatic adjustments. The generalised changes in sea level (from Emery, 1969) are shown in Fig. 3.6.12 also, plotted to the same time scale. Clearly, the sea-level and pollen-ratio curves are parallel. Thus 18,000–15,000 years ago, when sea level was at its lowest, the coastline was some 160 km south of its present position, near the edge of the continental shelf. At this time, the treeless plain extended south to the Eucla district. In the subsequent period, as the rise in sea level brought the coastline closer to the Eucla site, the vegetation zones must have shifted, no doubt keeping pace with the changes in the coastline and in rainfall. Today, the Eucla excavation is about 8 km from the coast, and well within the coastal eucalypt zone.

This work illustrates the effect of changing sea level on gradients of climate and vegetation. Stations on the coast today would have had a more continental climate at times of low sea level. This effect would have been most marked where topography (of land and continental shelf) is flat, as it is in the Nullarbor region, where a slight change in sea level would have caused a relatively large movement of the coastline.

Difficulties arise from attempts at estimating glacial-stage temperature and/or precipitation through a comparison with modern plant distributions.

Unless scope can be found to use geophysical or geochemical methods to obtain a firm estimate of palaeotemperature (as is obtained for ocean core material) or precipitation regimes, interpretation of the evidence from any one stratigraphic level requires the whole range of probable combinations of the elements to be assigned. The ecologically more important variable in the Nullarbor region is precipitation and the palaeo-environmental record of changes in this element is more a direct function of changing distance from the sea because of eustatic sea-level change, than it is an indication of the overall difference between glacial stage and modern climates.

Despite these complications a glacial-stage climatic gradient similar to that prevailing at present is indicated as a useful working hypothesis for the Nullarbor region.

Ice-age environmental gradients: northern Australia and Bassiana

The distance from the sea of various locations at and above present sea level in western Tasmania would have changed by approximately the same amount for any given eustatic rise or fall of sea level during the late Pleistocene, assuming no major tectonic deformations during those times.

The same probably applies within the Nullarbor region. The comparatively short and uniform distance retreated by the Bismarck and Coral Seas during the glacial stages would not have of itself affected climatic gradients very much, but the exposure of the Arafura Shelf, Torres Strait and Carpentaria would have quite changed the moisture content of the southeast Trades passing across these areas towards New Guinea (Fig. 3.6.4).

At present there is a seasonal contrast in circulation and associated precipitation patterns between the south- and north-facing 'halves' of New Guinea. The southeast Trades blow across the warm shallow Arafura Sea during the middle months of the year and bring a wet season to coastal and mountain areas facing the south coast. At the same time the north coasts and hinterlands experience their drier season. The reverse occurs during the 'northwest' season around the summer solstice. The south-facing mountains of western New Guinea would in that case have been drier during glacial times than today and certainly drier than the contemporaneously glacierised mountains in eastern New Guinea. Field evidence supports this hypothesis.

At Mt Jaya (4,883 m) in Irian Jaya, the outermost moraines of the last glaciation are higher than at Mt Wilhelm (4,509 m) in Papua New Guinea. Deglaciation seems to have begun later at Mt Wilhelm and so comparatively drier conditions on Mt Jaya are indicated. Nix & Kalma (1972) suggested that progressive eustatic flooding of the Arafura Shelf after 15,000 years BP would have caused a marked increase in precipitation on the southern coasts in the west of New Guinea. Initial support for this is found in the advance of Mt Jaya ice 12,500 years BP and 11,000 years BP (Hope & Peterson, 1975). No such glacier fluctuations are recorded from eastern New Guinea (Galloway *et al.*, 1973) nor any other tropical mountains (Hope & Peterson, 1976).

It would appear that in New Guinea the present climatic gradients are not everywhere a guide to those of the glacial stages. Obvious complications to the use of modern patterns as a guide to the reconstruction of glacial-stage circulation patterns in the New Guinea region include the fact that decreasing sea-surface temperatures and sea levels would have diminished or precluded the local formation of tropical cyclones. Moreover, there is the unknown status of any local circulations, such as occur today (Brookfield & Hart, 1966), and the possibility of unknown variation in the terrain-level atmospheric circulation at altitudes well above sea level, to add to the uncertainty. However, the early retreat and subsequent anomalous glacial advances in the Jaya Mountains suggest that the southeast Trades were at least as persistent as they are today.

The flooding of Bassiana during the Flandrian transgression between 14,000 years BP and 10,000 years BP would have quite changed the pattern of any climatic gradients that were governed by distance from the sea (Fig.

3.6.5). However, the late Pleistocene history of Victoria is not well known and very little evidence for either glacial-age climatic elements or gradients is available. Increased continentality of parts of the uplands would mean that periglacial limits might be lower than would be expected in comparison with maritime western Tasmania. Galloway (1965) has compared evidence from the Snowy Mountains in nearby New South Wales with data from Tasmania and concluded that the glacial age mean temperature drop was the greater in the Snowy Mountains. Unfortunately, enclosed depressions that yield vegetation histories are less common in periglacial than in formerly glaciated areas, and most bogs that are found in the uplands will probably be raised bogs perhaps dating back to the onset of warmer and wetter climates probably no more than 10,000 years ago. It is worth noting that if the climatic amelioration that melted the glacial-stage Northern Hemisphere ice caps was a global warming, there may have been a millennium during which the climate of the southern mid-latitudes became warmer while glacial sea level (and precipitation?) remained near its lowest.

Summary

The verification of simulated glacial-stage air flow patterns by models of the general circulation is offered by certain palaeoenvironmental data that may not yield or do not require interpretation for glacial-stage temperature or precipitation estimates. Palaeoenvironmental records that reflect temperature and precipitation changes with distance from the sea, rather than the overall difference between glacial stage and modern climates, can still be used to aid verification of glacial-stage general circulation models.

3.7   Abrupt events in climatic history
H. FLOHN

Gradually, the fascinating climatic history of the last few hundred thousand years is emerging from ignorance and becoming quantitative knowledge (Joint Organizing Committee for GARP, 1975; US Committee for Woldstedt, 1969; Lamb, 1972a; GARP, 1975). Obviously the global climate is subject to vacillations between two near-equilibrium states – Lorenz (1968, 1975) has coined the term 'almost-intransitive' for this behaviour. Increasing evidence has shown that such fluctuations may take place quite abruptly, on a time scale comparable to human life: the consequences of such an event for a world with 4×10^9 human inhabitants (soon increasing to 8×10^9 or 15×10^9) must be catastrophic.

The Alleröd–Younger Dryas fluctuation

The peak of the last ice age (Würm–Wisconsin) occurred around 18,000 years BP. The subsequent melting of continental ice-domes of the Northern Hemisphere was completed by about 8,000 years BP (Fennoscandia) and finally 6,500 years BP (Laurentide ice). Disappearance of the Fenno-scandian ice was nearly contemporaneous with the disintegration of the Hudson Bay ice sheet into two or three separate residuals, and a catastrophic incursion of the sea into Hudson Bay (Ives *et al.*, 1975). After 6,500 years BP the climatic optimum (Atlanticum, Hypsithermal) developed and sea level rose eustatically about 2–3 m higher than today. During this period, sea-ice in the Canadian archipelago disappeared, including the northern coasts of Greenland and Ellesmere Island, with similar effects in the European and Siberian Arctic (Vasari *et al.*, 1972).

The retreat of the continental ice was by no means uniform. Several glacial readvances happened apparently simultaneously (Mörner, 1973) and were accompanied by climatic variations all over the globe. Three readvances occurred after about 12,500 years BP, about 11,900 years BP, and the last (Younger Dryas) after about 10,900 years BP. These were separated by two warm interstadials: Bölling (around 12,200 years BP) and Alleröd (around 11,400 years BP). The two full oscillations of the 'Alleröd Sequence' together covered less than 2,000 years, and the variation in mean annual temperature reached 5–6 °C. Both warm interstadials have been accompanied, in the Colombian Cordillera (5° N, 2,850 m), by humid phases (van Geel & van der Hammen, 1973) in contrast to the Würm III peak and the Younger Dryas, which were both arid. Of special interest are the high (400 m!) level fluctuations in Lake Kivu (Degens & Hecky, 1974), which are shown in Fig. 3.7.1.

The catastrophic readvance of the ice after the Alleröd destroyed 'wholesale' full-grown forests. This, together with marked cooling at Camp Century, Greenland (76° N) and Byrd Station (West Antarctica, 80° S) (Johnsen *et al.*, 1972) and desiccation in Colombia seems to have occurred within a time span of 200–350 years. Similar time spans are reported for warming at the beginning of the Alleröd and after the Younger Dryas. It should be realised, however, that vegetation changes need more time than true climatic variation. Similarly, a thermophilous insect fauna from near Birmingham was affected rapidly by climatic change. In one deposit a typical northern fauna was found at a layer dated 10,025±100 years BP, while 10 cm higher no Arctic element survived, at an age of 9,970±110 years (Osborne, 1974). Taking these data literally, this 'very rapid' warming occurred within a time span of 50 (±100) years.

Within the complex Alleröd sequence the warm phases coincide with high humidity in the African Tropics (Butzer *et al.*, 1972). This is also true

125

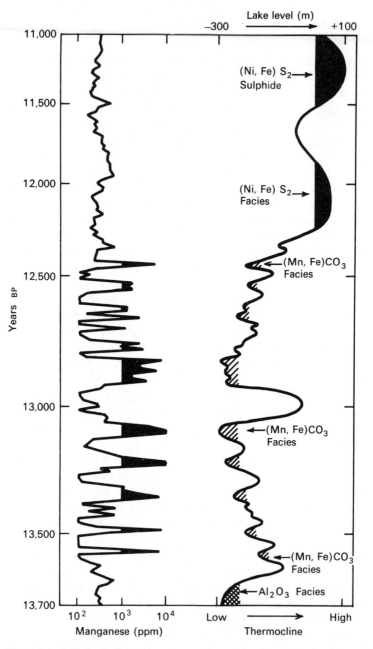

Fig. 3.7.1. Lake-level fluctuations (m) and associated bottom chemistry of Lake Kivu (2° S, 29° E), 13,700–11,000 years BP. After Degens & Hecky (1974).

for the Rajasthan desert (Singh *et al.*, 1974); here a multiregression technique leads to an estimate of the annual rainfall amounting to 300–400% of the present rainfall (Bryson, Section 7.3).

In Australia there is also evidence for a humid phase together with the final warming (Walker, Section 3.4) but little evidence of a complex Alleröd fluctuation.

The Alleröd complex was accompanied by a rise of sea level, from about −50 to about −35 m (Mörner, 1973). At −38 m the Bering Strait reopened and Arctic waters again entered the Pacific. The latter had not been as cold as the North Atlantic, where 90% of the meltwater from the continental ice-domes formed a cool, low-saline, surface layer. During and after the peak of the glacial phases, Atlantic surface water was less saline than the Pacific, not as a cause (Weyl, 1968), but as a consequence of glaciation of the northern continents. This cool layer may have contributed to final undernourishment of ice-domes and to the widespread aridity (Adam, 1973, 1975).

Further examples during the Pleistocene

Four other examples of abrupt cooling at several distant places during the last 100,000 years have been quoted (Flohn, 1974*a*); only one case (near 55,000 years BP) coincides with the beginning of a glaciation (Würm I). The beginning of the last (Würm III) glaciation (about 25,000 years BP) must also have been nearly instantaneous (Ives, 1956, 1957; Lamb & Woodroffe, 1970). The famous 'Blitz' (Flohn, 1974*a*, Fig. 1) of Camp Century (Dansgaard *et al.*, 1972) is now subject to intense study; the estimated age of 90,000 years BP is still uncertain. But from other evidence (Barbados, Matthews 1973; New Guinea, Bloom *et al.* 1974) three rather rapid fluctuations are known about 130,000–80,000 years BP; in one of these cases the development of full glaciation, with eustatic lowering of sea level of about 80 m took place within 5,000 years. This is roughly equivalent to an average annual storage of 20 cm water equivalent over a glaciated area of about 30×10^6 km². This is not unrealistic, if one assumes that the annual storage at the centre may be as low as 5 cm/yr, while it may rise at the outer margins to 50–60 cm/yr, similar to present accumulation in Antarctica. Two of these coolings may also coincide with a double rapid transgression of polar water in the Atlantic from north of latitude 62° N down to latitude 44° N, dated (with some uncertainty) at between 145,000 and 125,000 years BP (McIntyre *et al.*, 1972). Several abrupt transgressions over a distance of 2,000 km are indicated. Also, detailed palaeobotanical records from Macedonia (Wijmstra, 1975) show at least 16 coolings and warmings, some of them very rapid, during the last 400,000 years.

3.7   The long-term climatic record

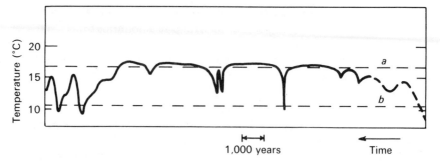

Fig. 3.7.2. Temperature trend for July during the Holstein Interglacial in northern Germany. The absolute age is around 210,000 years BP., and a relative time scale is indicated. '*a*' denotes recent prevailing July temperatures, and '*b*' that at the northern forest boundary. After Müller (1974).

In the Holstein Interglacial of northern Germany – probably equivalent to Point 7 in Emiliani's revised palaeotemperature curve (Emiliani & Shackleton, 1974), with an absolute age around 210,000 years BP – H. Müller (1974) investigated an extended series of thin annual layers – verified by the seasonal pollen sequence – revealing several sharp coolings. In one case the catastrophic destruction of a thermophilous *Corylus–Taxus–Alnus* forest occupied less than 100 years, after which a subarctic *Betula–Pinus* forest recovered. Figure 3.7.2 (simplified from Müller's Fig. 6) gives a schematic interpretation of the (summer) temperature variations together with the relative time scale. The two subarctic phases lasted only 300–400 years; the total duration of the Holsteinian has been estimated at 15,000–16,000 years.

Judged from these events, the frequency of such abrupt (apparently aperiodic) climate processes, with a time scale near 100 years, appears to be near one per 10–20 millennia.

Many Pleistocene fluctuations resemble a flip-flop mechanism, with abrupt changes between two opposite climatic states. A full continental glaciation needs, for building-up as well as for decay, not more than 5,000–8,000 years: this is definitely less than any component cycles of the Milankovitch mechanism. However, in an arid phase, a dusty atmosphere absorbing radiation and emitting heat can initiate and accelerate the decay processes (Flohn, 1974*a*). Existence of arid conditions during the last glaciation is now beyond any doubt. This aridity coincided with advective cooling of the ocean surface and suppressed evaporation, while the tropical continents, albeit also cooling, remained warm with a nearly constant surface net radiation.

The possibility of 'abortive' glaciations (Flohn, 1974*a*) – i.e. rapid widespread coolings of several degrees centigrade lasting some centuries, but not long enough to initiate a full glaciation – merits further discussion.

128

Another example is the 'Little Ice Age' with its peak around AD 1600–1700, characterised by thin ice-sheets and perennial snowbanks at the plateaus of Baffin Island (Andrews *et al.*, 1975). Several knowledgeable scientists have suggested (but not yet published) a marked spectral peak near 100,000 years BP (Kutzbach & Bryson, 1974), coinciding with the Milankovitch time scale, but incoherent with the frequency of these short-living processes. This enigma may perhaps be solved in assuming a selective role for the earth's orbital elements, producing a positive or negative feedback (Kellogg, 1973) towards a full-grown glaciation.

Possible causes of abrupt climatic events

The climatic system consists of several subsystems – atmosphere, ocean, snow, ice and soil – with different response times and many degrees of freedom. Even with a constant energy source, such a system must operate with much interannual variability, like any other self-regulatory system.

Internal feedback processes with a storage time of weeks or months (e.g. an early snow cover or upwelling of surface waters) and covering areas in the order of 10^5–10^6 km^2, can alter significantly the regional heat budget. Processes of this kind – together with positive (or negative) feedback – control, for periods of several months, the tropospheric circulation and the tracks of travelling eddies. Varying frequency of such anomalies characterises climatic fluctuations, but only for intervals of the order of 10–100 years; longer time series usually produce a spectrum with indefinite and inconstant peaks (Kutzbach & Bryson, 1974). Is climate completely random? Or what external sources may add or subtract energy to or from the climatic system, thus causing abrupt climatic 'changes' with a lifetime of 10^2–10^4 years? Three groups of external causes will be considered:

(i) solar events;
(ii) frequency of explosive volcanic eruptions;
(iii) glacial surges of very large ice-sheets (Antarctica).

Variations of the solar 'constant' with a time scale of 10^2 years and above are unknown. Direct (Kondratyev & Nikolsky, 1970) and indirect (Lockwood, 1975) measurements in the visible part of the spectrum seem to indicate some 11-year variability, probably below 1%. Remarkable evidence for strong variations of the solar activity during the 'Little Ice Age' has recently been given (Schneider & Mass, 1975). Computations on comprehensive climate models (Wetherald & Manabe, 1975) demonstrate strong effects and should provide a stimulus towards regular surveillance of solar radiation by satellites.

The possible role of volcanic activity has been discussed especially since Lamb's catalogue was published. Volcanic particles (mostly 0.1–1 μm) in the stratosphere absorb and backscatter part of the sun's radiation (Lamb,

129

3.7 The long-term climatic record

1970). Absorbed radiation heats stratospheric layers with but little effect on the ground. The loss of back-scattered radiation thus produces surface cooling (McCracken & Potter, 1975; Peyinghaus, 1974) even if the signal may not be distinguishable from noise in many cases (Landsberg & Albert, 1974). Convincing evidence for a simultaneous occurrence of volcanic activity and glacial events has been recently presented (Bray, 1974 a, b, and Section 5.5).

From Lamb's catalogue (period AD 1500–1968) the frequency of volcanic eruptions exceeding a given 'dust veil index' (DVI) is derived; to this the eruption of Fuego (November 1974) has been added. For eruptions with DVI = 1,000 the average recurrence time is 21.3 years (Table 3.7.1). Assuming a Poisson distribution of independent rare events, one can estimate the frequency, with which eruptions cluster in a period of 10 or 25 years. Table 3.7.1 shows the recurrence time of clusters of n volcanic eruptions for these periods and for different DVI values.

Table 3.7.1. *Recurrence time in years of clusters of n volcanic events (Poisson distribution)*

	10-year period			25-year period	
	DVI ≥ 1,000 $m = 21.3$	500 9.01	300 5.46 years ($n = 1$)	≥ 1,000	≥ 300
$n = 3$	925	133	61	300	150
4	7,870	480	133	1,025	130
5	83,700	2,200	365	4,360	145
6	1.07×10^6	11,680	1,195	22,300	190
7	15.9×10^6	73,500	4,570	0.13×10^6	290
8	—	0.53×10^6	20,000	0.89×10^6	505
9	—	4.35×10^6	98,000	6.2×10^6	1,000
10	—	—	0.54×10^6	—	2,170
11	—	—	—	—	5,210
12	—	—	—	—	13,660
13	—	—	—	—	38,640
14	—	—	—	—	0.12×10^6
15	—	—	—	—	0.38×10^6

m = average recurrence time, data from Lamb (1970)

Checking the assumption of a Poisson distribution against observations (Fig. 3.7.3), it is obvious that the real frequency distribution is not random (Lamb, 1970); small and large frequencies occur more frequently than in a Poisson distribution. Apparently a kind of 'infectious' process exists; nearly simultaneous eruptions alternate with periods of weak volcanic activity. The Poisson-predicted frequencies (Table 3.7.1) of clustering volcanic activity are definitely too low, and the recurrence time of simultaneous eruptions is shorter (about one-half) than predicted.

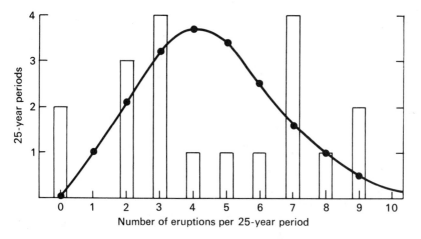

Fig. 3.7.3. Frequency of observed volcanic eruptions per 25-year period, based on Lamb's (1970) catalogue for 1500–1974 (dust veil index ⩾ 300), compared with a Poisson distribution (smooth curve).

The residence time of volcanic particles in the stratosphere is not constant. Apart from the tropical Hadley cell (Reiter, 1975), only isentropic exchanges within the tropopause gaps (jet-streams) are instrumental in removing particles from the stratosphere. These processes are rare and inefficient in polar regions. In most seasons a stratospheric circulation is developed above the poles (Hesstvedt, 1964) (Fig. 3.7.4). Impressed by repeated visual observations of strong dust well above the tropopause in the interior Arctic (notably after eruptions of Agung and Fuego), Flohn (1974 *b*) has proposed a longer residence time in the polar stratosphere (perhaps 3–5 years) compared to the global average of 1–1.5 years. A near coincidence of major eruptions – such as in the period 1750–70 or 1810–35 – can thus produce a prolonged and intensified stratospheric dust layer above the polar cap, with significant consequences for the thickness of the Arctic ice and for the air above during summer.

The idea of instability of the Antarctic ice-dome (Wilson, 1964) arrived as a real surprise at the scientific scene. It may also have been to the author's surprise, that four years later one of his basic assumptions – melting processes at the bottom of the ice – had been confirmed by the first deep Antarctic ice core (Gow *et al.*, 1969). While in East Antarctica only small, widely-dispersed meltwater lakes have been observed (Oswald & Robin, 1973) the ice cover of West Antarctica (based below sea level) could be in an unstable state (Hughes, 1973, 1975). Evidence exists of a collapse of the West Antarctic ice during the last 12,000 years which may continue (Denton, 1974). This process is related to a retreat of the grounding-line of the Ross Ice Shelf (Hughes, 1975). It may be suggested

131

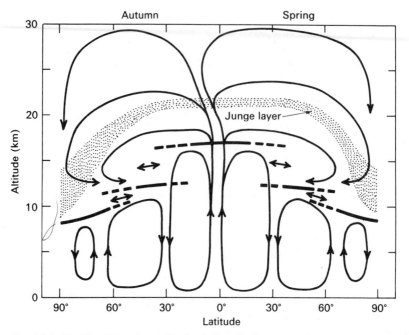

Fig. 3.7.4. Meridional (north–south) circulations and transports in autumn and spring, with tropopauses and the Junge layer of stratospheric dust indicated.

that the actual world-wide sea-level rise of 1.2 mm/yr – inconsistent with a positive mass-budget of Antarctica – can be partly attributed to this retreat. Assuming a retreat of 400 m/yr, an average thickness and width of shelf-ice at this line of 800 m and 800 km respectively, yields an annual increase of floating shelf-ice of $0.32 \times 800 \simeq 250$ km³/yr, equivalent to a global sea-level rise of 0.7 mm/yr.

The author has speculated (Flohn, 1974 *a*, Fig. 4) how an Antarctic surge in the Weddell Sea–Atlantic section could eventually produce, by ice–ocean–atmosphere interaction, a quasi-permanent cold low over the Gulf of Mexico. This could act as an anchor to a tropospheric summer trough along the east coast of North America, as a necessary prerequisite for a permanent snow cover at Labrador–Ungava, which could serve as a nucleus for a North American ice-dome. However, the estimated magnitude of an Antarctic surge used by Flohn (1974 *a*), is apparently too high. If we assume instead a surge of the order of 1×10^6 km³, 80% of which could break up into tabular icebergs covering 4×10^6 km² with a thickness of 200 m, and 20% into small ice floes, covering more than 20×10^6 km², the climatic effect should be not too much different. Such a surge would produce a sea-level rise of 2.5 m. Hughes (1975) has collected evidence for such an event happening about 110,000 years ago, closing the warm

132

Eem–Sangamon interglacial period. The Antarctic ice sheet and the subantarctic drift-ice – 18×10^6 km² of which is freezing and melting seasonally – should be monitored by satellites as a warning system against initiation of an abrupt (and possibly disastrous) climatic event. Antarctic ice surges are discussed further in Section 5.3.

Inquiry into the initiation of a glaciation

It seems self-evident that the nucleus of a new glaciation of the northern continents should develop at the same location where the decaying remnants of the last glaciation have been observed. However, this assumption may be challenged with the indication that during its development the ice centre may have migrated. Without further arguments (Ives *et al.*, 1975), we may discuss a model with the first assumption.

The initiation of a new glaciation must start with a sufficiently large (synoptic scale) snow cover surviving one summer (Flohn, 1974 *a*; Adam, 1973, 1975). In this case, the next winter adds to its thickness, and the probability of persisting rises from year to year. In a boreal forest, an effective permanent snow cover could be built up in a few decades (Flohn, 1974 *a*). Then a hardly destructible feedback process (Kellogg, 1973) begins to produce, within some millennia, a gigantic ice-dome until an equilibrium state is reached. The evolution of such a nucleus on the vast tundra plateaus of Baffin Island would probably need the same time. Here a cooling of 2 °C would be sufficient without increase of precipitation, since the plateaus of north central Baffin Island are only 200 m below the present snowline. Such a permanent snowbed would act as another anchor for a quasi-stationary trough along the eastern coast of North America during summer, i.e. during that season when it actually disappears.

This would lead immediately to cooling and increasing snowfall over the Ungava plateau (latitude 53–60° N), where a similar snow cover would form. For this climatic evolution an initial cooling of about 2 °C is needed, only lasting during, say, 50 years.

The same arguments could be given for Fennoscandia (Ives *et al.*, 1975) with a rather similar orography. But from a meteorological viewpoint this is unnecessary: the European ice-dome was much smaller and disappeared earlier. Most probably its formation had started slightly later, following the formation of an upper trough west of the Norwegian coast, actually a frequent cold-season feature interpreted as a quasi-stationary Rossby wave downwind of the East American trough.

Some of the earlier conclusions (Flohn, 1974 *a*) remain unaltered when the assumption of a large Antarctic surge is replaced by a cluster of volcanic events or by a sudden decrease of the solar constant. In such a case as the initiation of a new glacial phase – an event which has

3.7 The long-term climatic record

occurred at least eight times during the last million years – imagination has to be controlled by physical reasoning and, finally, mathematical modelling (see Chapter 5).

Prevention of a new ice age?

If the formation of a permanent snowbed constitutes the initial stage of a glaciation, one may ask how far this process could possibly be modified by a concerted action program.

As an example, such a program, under the assumptions outlined above, could be based on repeated blackdusting of the snow surface during the season of high global radiation, from mid-April to mid-September. If dusting reduces the surface albedo of fresh snow (about 80%) by 20% only, this will double the radiation energy available for melting. In contrast to a program for the large Arctic drift-ice (about 10^7 km²), here only several areas of 10^5 km² and nearby well-equipped air bases need be considered.

Let us assume that each efficient dusting needs 10^3 soot particles each of 1 mg m⁻²: this is equivalent to 1 tonne km⁻². One airplane carrying 10 tonne could dust 10 km² per flight. Assuming two flights daily, 1,000 airplanes could dust 2×10^4 km², or 10^5 km² in 5 days. Since one has to take into account efficient snowfall, say, six times a month, this operation has to be repeated once every 5 days.

The necessary supply of soot (or similar) particles would be 10^5 tonne each 5-day period or 3×10^6 tonne for a 150-day campaign. The logistical problems are heavy, but not completely unrealistic. If the operations are sufficient to suppress the initial stages, the costs of repeated campaigns are not too prohibitive when compared with visits to the moon. A few years after the beginning, the snow cover would otherwise become so thick that the available solar radiation would be insufficient for melting; rather soon the feedback processes would become irreversible.

Recently Adam (1973, 1975) has stated: 'We may well be able to start an ice age, but we probably cannot stop one once it begins.' Considering the overall tendency of anthropogenic effects to warm the atmosphere (Flohn, 1973, 1974b; Kellogg et al., 1975) it is more appropriate to state: even if we are unable to start an ice age – under some conditions we perhaps can stop it once it begins.

4. Patterns of shorter-term change and variability

4.1 Climatic variability and extremes
W. J. GIBBS, J. V. MAHER & M. J. COUGHLAN

Despite the apparent simplicity of the traditional concept of climate, there are difficulties in arriving at generally acceptable definitions of climate, climatic variability, and climatic extremes. Much of the difficulty arises from considerations of time scale and the inherent variability of the atmosphere even in the absence of external influences.

In this section we examine some basic concepts and methods of analysis relevant to the study of climatic variability and extremes over time scales of a few hundred years or less.

Concept of climate

There is a variety of meaning attached to the term climate. *The Meteorological Glossary* (Her Majesty's Stationery Office, 1972) defines climate as follows:

> The climate of a locality is the synthesis of the day-to-day values of the meteorological elements that affect the locality. Synthesis here implies more than simple averaging. Various methods are used to represent climate, e.g. both average and extreme values, frequencies of values within stated ranges, frequencies of weather types with associated values of elements. The main climatic elements are precipitation, temperature, humidity, sunshine, wind velocity, and such phenomena as fog, frost, thunder, gale; cloudiness, grass minimum temperature, and soil temperature at various depths may also be included.
>
> Climatic data are usually expressed in terms of individual calendar month or season and are determined over a period (usually about 30 years) long enough to ensure that representative values for the month or season are obtained.

Another notion of climate is that in the report *Understanding Climatic Change* (US Committee for GARP, 1975) which discusses the question of definitions and advances the concept of 'climatic state' as follows:

> Climatic state. This is defined as the average (together with the variability and other statistics) of the complete set of atmospheric, hydro-

135

spheric and cryospheric variables over a specified period of time in a specified domain of the earth–atmosphere system. The time interval is understood to be considerably longer than the life span of individual synoptic weather systems (of the order of several days) and longer than the theoretical time limit over which the behaviour of the atmosphere can be locally predicted (of the order of several weeks). We may thus speak, for example, of monthly, seasonal, yearly, or decadal climatic states.

The report also states:

Except when otherwise indicated, the use of the word 'climate' in this report is to be considered an abbreviation for climatic state.

Under this definition, if the statistical measures (mean, variance, skewness, kurtosis, etc.) of meteorological elements for one month were significantly different from those of the same month of the previous year, it would be considered that the climates of the two months were different. To say, however, that there had been a change in climate on this basis would invoke quite a different concept of climatic change from that contained in other texts where the term is used to indicate changes of a more radical nature, e.g. the onset of an ice age.

A rather different approach to the concept of climate and climatic change is taken by Landsberg (1975) who suggests that the term 'climatic change' be used only for variations in terrestrial climate that are caused:

(i) extra-terrestrially by change in solar emission of an aperiodic nature, or if periodic in nature, of 10^4 years or longer duration;

(ii) by change in insolation received by the earth because of long period changes in orbital elements.

He also suggests, that all other climatic variations should be designated 'climatic fluctuations'.

It is evident that whether variation of climatic elements from month to month, year to year, decade to decade and century to century is regarded as part of the nature of climate itself or whether it is to be interpreted as climatic change is more a matter of individual preference than objective assessment.

Our attitude is that, in the absence of climatic change as defined by Landsberg, the climate of a particular locality is, for practical purposes, best described by a synthesis of:

(i) the frequency distributions of all meteorological elements at that place over as many years as are necessary to obtain a set of statistics sufficient to delineate the frequency distribution with the precision required;

(ii) a description of the variation in time (systematic or otherwise) of those elements;

136

(iii) information about serial correlations within the time series of each element, and in the occurrence of particular meteorological phenomena;

(iv) information about correlations between the time series of the different elements.

In relation to (i) above, the necessary length of record may differ from one climatic element to another. With weather as variable as that in Australia, rainfall records for 150 years and temperature records for 50 years or more may be considered necessary adequately to determine and describe the climate.

Climatic variability

In our description of climate we have implied that variability is an inherent characteristic of climate itself. If we were to include diurnal and seasonal variability, these two by themselves would account for a dominant proportion of the overall variability within climate. In practice, however, their effects may be readily removed to allow investigation of variability arising from other causes.

The recognition of non-stationarity in climate has led to discussion over the use or worth of climatic 'normals'. Indeed the fluctuations of some 30-year normals have been such that they could be used as indices of climatic 'change' – or variability – rather than as bases for the fixed climate of an area or locality (Jagannathan *et al.*, 1967). It seems that the fluctuations of 30-year normals may often be due to the effects of sampling which result from the choice of 30 years as the period used for calculation of the normal.

Another problem is change in the variability or variance of a climatic element with time. Certainly the lengths of available and reliable records in Australia can give only a coarse assessment of stationarity of variance. Gani (1975), using monthly values of district rainfall in southeast Australia, concluded that there is no evidence for a significant change in the variance over 60 years of rainfall data.

The principal factors inducing spatial variability of climate are latitude, elevation, local topography, proximity of the sea and the position and size of land masses relative to oceans and other continents. The spatial variability arising from these factors produces variability in time through the motions of the atmosphere and oceans and the interactions between them. Short-term extra-terrestrial influences, the motion of the earth and the increasing effects of man's activities also produce variability of climate in time. Thus the variability of climatic elements in time may be regarded as deterministic or stochastic – with some element of persistence (Mitchell, 1966) – or a combination of the two depending largely upon the time scale involved.

4.1 Shorter-term variations

In Australia a great proportion of the variability of the climate may be related to the occurrence and intensity of synoptic systems such as tropical cyclones, east coast lows, extra-tropical cyclones and blocking anticyclones. They operate as individual systems but their net effect may be represented by indices such as the mean latitude of the anticyclonic belt (Pittock, 1973 a) or 'cyclonicity' and 'anticyclonicity' (Karelsky, 1956).

Analysis of climatic variability

The increasing realisation that variability must be accepted as an inherent characteristic of climate (e.g. World Meteorological Organization, 1976) has focused attention on the need to describe and document climatic variability on a wide range of time scales. Such information will act as an aid to economic planning, and as a pre-requisite to an understanding of the climatic system and to the development of reliable climatic models. However, its collection and interpretation involve consideration of the nature and reliability of the available data base, and the selection of appropriate techniques such as frequency distribution analysis, time series analysis and multivariant methods such as principal component analysis.

Nature of the data

In general, climatic elements are observed at fixed clock times or represent accumulated totals or extremes over a fixed period. The regular diurnal and annual cycles may be filtered out by selecting data at fixed hours of the day or time of the year respectively. Alternatively those cycles may be described by some form of harmonic analysis with appropriate confidence limits based on the distribution of the elements at the fixed times. In order to summarise the enormous quantities of data observed, it is usual to calculate averages over a fixed period, e.g. pentads (5-day periods), months, years, etc. Care must be exercised in handling such averages because of the existence of increasing persistence or autocorrelation as the time between successive meteorological observations is decreased. Figure 4.1.1 shows how the number of independent days in a monthly average decreases as the persistence expressed in terms of the 'lag one' autocorrelation coefficient increases.

Homogeneity of records

The problem of reliability or homogeneity of climatological data is one of the most difficult obstacles in any quantitative estimate of climatic variability. The earliest instrumental recordings of climate in the Austral-

138

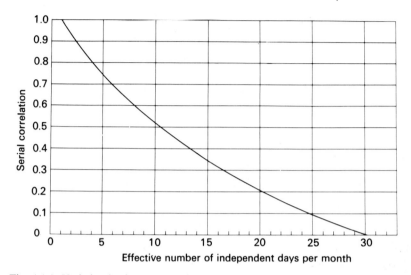

Fig. 4.1.1. Variation in the number of independent days in a month with increasing serial correlation.

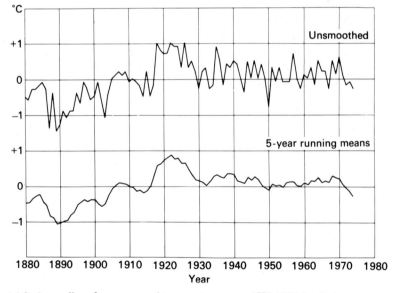

Fig. 4.1.2. Anomalies of mean annual temperature range 1880–1974 for Sydney.

ian region can be dated back to the time of Matthew Flinders, about 1800 (Gentilli, 1967). During the first half of the nineteenth century, meteorological observations were generally sporadic and confined almost exclusively to the four or five major centres of population. The emergence of the Australian States during the latter half of the century led to an improvement in the situation with the appointment of State Government Meteorologists, e.g. Wragge in Queensland, Russell in New South Wales and Neuymayer in Victoria, However, there was little agreement between the States on a uniform set of procedures for measuring and recording the various meteorological elements. For example, there were at least five or six different types of thermometer screen in use throughout the country prior to 1907 when the responsibility for the meteorological services came under Federal control and the Stevenson screen became the standard.

Any study of climatic variability or change covering the period of the instrumental record must take into account inhomogeneities introduced by changes in site, exposure, etc., of the instruments. Temperature and rainfall, the elements most frequently used in climatic studies, are particularly susceptible.

Kemp & Armstrong (1972) have investigated the record of Sydney temperature in some detail. Figure 4.1.2 shows the record of mean annual temperature range anomaly from 1880 to 1974; temperature range $(T_{max} - T_{min})$ can be particularly sensitive to changes in instrument exposure. The most significant feature in this record is the sudden jump around 1918. This jump is associated with a site change in the thermometer screen of a mere 135 m in October 1917. The common practice of taking moving averages to smooth the data will mask such abrupt changes, making them appear more like short-term trends.

The effects of increasing urbanisation on climatic records should also be noted (Dronia, 1967; Mitchell, 1953). The effect, most noticeable in minimum temperatures, is particularly evident in the capital cities of Australia. Mean annual minimum temperatures in the five eastern state capitals have increased by an average of about 1.4 °C (see Section 4.5) since the mid 1940s. This increase has not been matched by records from provincial areas over the same time.

Frequency distributions and indices of variability

Assuming that one does have a single climatological record which is homogeneous and free from trends over the period under investigation, the data may be considered as an assemblage of elements from which a frequency distribution may be prepared. Values for various statistics may be derived to characterise the frequency distribution and serve as a basis for climatic probability assessments. Among these statistics are a number of measures of variability.

The simplest measure of variability is the range, the difference between the largest and smallest values in the set of data. The range, however, depends in a systematic fashion on the number of observations and is not a stable or particularly useful measure of variability. More useful are the mean deviation

$$\text{MD} = \frac{\Sigma(x-\bar{x})}{N} \qquad (4.1.1)$$

and the standard deviation

$$\text{SD} = \left(\frac{\Sigma(x-\bar{x})^2}{N}\right)^{\frac{1}{2}}, \qquad (4.1.2)$$

where x denotes the values of the element under consideration, $\bar{x}$ is their mean value and N the number of observations. Another commonly-used statistic is the variance which is the square of the standard deviation.

Loewe (1948) used a statistic, the Average Variability Index, due to Maurer (1938) to measure variability of annual rainfall over Australia. Maurer's index is so devised as to enable variability to be compared over regions of widely differing mean precipitation.

It is often useful to 'fit' one of a number of theoretical frequency distributions (e.g. the binomial, Gaussian, or Poisson distribution) to the climatic data set. In some cases it is necessary to transform the data to make it conform to a theoretical distribution – the Gaussian or 'normal' probability density function being the most common. In the case of rainfall the transformation may be of the type:

$$X = x^n, \text{ where } 0 < n < 1, \qquad (4.1.3)$$

or
$$X = \log (x+a), \, a = \text{const} > 0. \qquad (4.1.4)$$

In an analysis of the 62 years of record for Australian rainfall districts, Pierrehumbert (1975) has shown that monthly rainfall is rarely normally distributed, n in Equation (4.1.3) most commonly lying between 0.25 and 0.67 for the best transformation, depending on the month and locality. Even with annual rainfall, the assumption of normality was justified in less than one-third of the 107 districts investigated.

This general lack of normality of rainfall distributions over large areas of Australia has led to the use also of non-parametric probability estimates based on the cumulated frequency curve. Cumulated frequency curves or tables are particularly useful when comparing rainfall frequencies at different stations or different months or other intervals at the same station. Gibbs & Maher (1967) used this approach, adopting the decile and decile ranges of the cumulated frequency curve, to study Australian droughts.

4.1 Shorter-term variations

The decile information may also be used to provide a measure of variability. Gaffney (1975), for example, computes an index representing the variability of annual rainfall.

$$\text{Variability Index} = \frac{90 \text{ percentile} - 10 \text{ percentile}}{50 \text{ percentile}}.$$

This index is thus a measure of the spread of the distribution, standardised by division by the median.

Examples of variability

Figure 4.1.3 from Gaffney (1975) illustrates the spatial variation of his variability index for annual rainfall over Australia. It is in general agreement with Loewe's (1948) results. Of particular interest is the area of high variability extending from the dry interior to the northwest coast around Port Hedland, a reflection of the occasional influence of tropical cyclone rains in this region.

Numerous specific examples could be quoted of rainfall variability throughout Australia in time and space, linked with drought and flood, but only two outstanding cases will be given, both of which are linked with the occurrence or non-occurrence of tropical cyclones in the northwest of the continent. At Onslow (21° 43′ S, 114° 57′ E, elevation 4 m) annual totals range from 13 mm to 711 mm and, in the four consecutive years 1921–4, they were 565, 69, 681 and 55 mm respectively. At Whim Creek (20° 50′ S, 117° 50′ E, elevation 32 m), where 747 mm has been recorded in one day, only 4 mm were received in the whole of 1924.

Variability of temperature in Australia has also been the subject of numerous studies, the results of some of which have appeared in Australian Bureau of Meteorology publications.

Time-series analysis

The study of climatic data as chronologically ordered time series has expanded rapidly over the past 20 years with the development of improved techniques and the increase in computational power. That there may be periodic or quasi-periodic effects in the atmosphere, apart from the obvious diurnal and annual cycles, has long been a topic of study and speculation (Kidson, 1925). In many scientific papers on the subject, so-called convincing evidence has been put forward for the existence of a multitude of cycles. Much of the disrepute into which many of the studies fell was due either to a misapplication of mathematical techniques or to a misinterpretation of results. This applied particularly to the use of harmonic or Fourier-type analysis. Cycles which may have mathe-

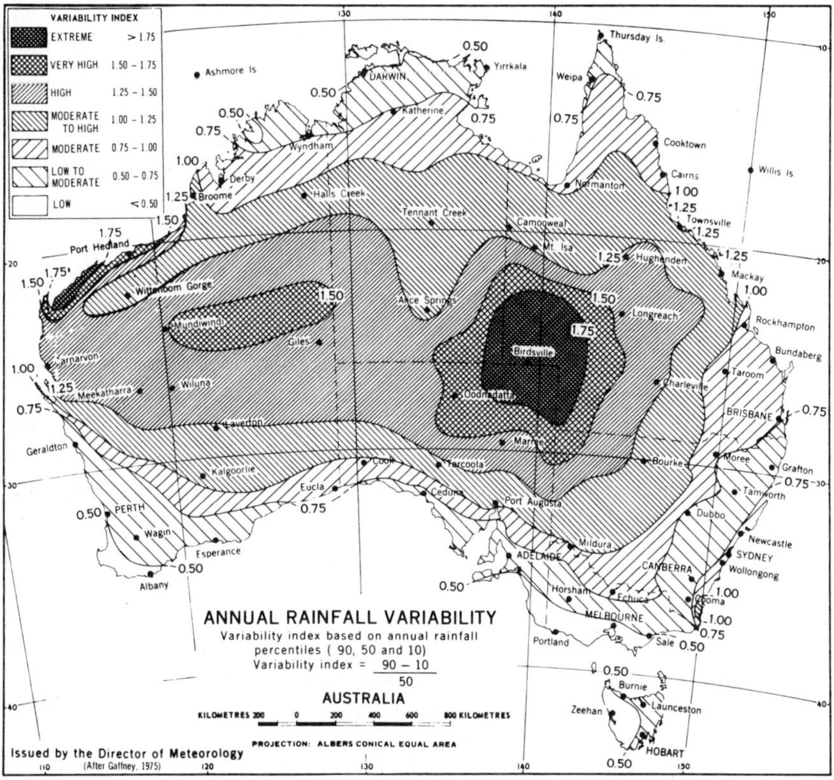

Fig. 4.1.3. Annual rainfall variability over Australia after Gaffney (1975).

matical significance in the reconstruction of a time series cannot be said, *per se*, to have any physical significance. It has been recommended (Mitchell, 1964) that such techniques be used only when a known cycle, viz. the annual or diurnal, is to be described.

Probably the most widely used technique in current use for investigating time series of climatic data is that of power spectrum analysis. This technique is based on the premise that a series is composed of an infinite number of oscillations, spanning a continuous distribution of wavelengths. The spectrum yields a measure of the distribution of variance in a time series over a continuous domain of all possible wavelengths – each arbitrarily close to the next – ranging from an infinite wavelength (linear trend) to the shortest wavelength which can be resolved by any scheme of harmonic analysis (equal to twice the interval between successive observations in the series).

The existence of autocorrelation in many meteorological time series, particularly when the time between successive observations is short, leads

4.1 Shorter-term variations

to the consideration of a particular form of spectrum where a higher proportion of the variance is contained in the lower frequencies (a 'red-noise' null hypothesis) than would be expected if the series were completely random ('white-noise' null hypothesis).

O'Mahony (1961) applied the technique of power spectrum analysis to a number of rainfall series in Australia and found evidence of significant periodicities from two to three years and approximately seven years, the latter probably being a harmonic of the former.

The main shortcoming of power spectrum analysis is the absence of information on the phase characteristics of any fluctuations. The application of suitably designed filters, of which the simple moving average is the most common example, enables one to investigate the phase behaviour of any suspected periodicity. Such filters may be either low-pass, high-pass or band-pass in form. It should be stressed, however, that the application of any filter to climatological time series and the interpretation of the output should be based on the careful formulation of specific physical and statistical hypotheses.

Principal component analysis

A technique variously called empirical orthogonal function analysis, eigenvector analysis or principal component analysis has recently found increasing use in the study of climatic data. This technique has been applied most frequently in climatic studies as a space/time analysis of variance and seeks to apportion the overall variance in the most economical way to a group of patterns, arranging them in rank order of importance. In practice, only a few of the total number of possible patterns – equal to the number of data points under investigation – are needed to explain a large proportion of the variance. Some of these major patterns may be recognised as having an underlying physical explanation. A more detailed description of the technique may be found in Barry & Perry (1973). Pittock (1975) and Wright (1971) have used the technique in studying Australian rainfall, and Veitch (1965) for describing Australian pressure fields. Pittock, in his study of Australian district annual rainfall, found that the first two patterns (Fig. 4.3.1), related to the effects of the 'Southern Oscillation' and the latitude of the subtropical high-pressure belt respectively, accounted for about 54% of the variance. (See also Section 4.3.)

Climatic extremes

As with variability, the notion of climatic extremes involves consideration of time scales. The occurrence of climatic extremes is of great importance

144

in matters such as design of structures where one might be concerned with the strongest expected wind gust, with a duration of a few seconds, or in the design of a water storage dam where the longest expected dry spells with durations of months or even years could be of great importance. Further, climatic extremes may involve consideration of a combination of climatic elements as in 'effective temperature' or 'cooling power'.

Statistics of extremes

Data which are used to analyse climatic extremes, that is, the largest or smallest values in sets of data, should satisfy three basic assumptions, that:
   (i)   they are a statistical variate;
   (ii)   the statistical distribution is not sample dependent; and
   (iii)   they are independent.
The last two requirements are related to the notions of stationarity and autocorrelation mentioned earlier. A particular example of the serial correlation problem in the study of extremes is that the maximum event in one year can usually be considered as independent of maxima in other years but not necessarily independent of the second highest in the same year. Assuming, however, that the three basic assumptions above are satisfied, historical extremal distributions can then be analysed to determine future probabilities of occurrence.

Because of the limited period for which reliable meteorological data are available, it is very difficult to determine the probability of extreme or near-extreme values or, alternatively, the magnitude of an event having a very low probability. To be reasonably certain of an estimate, the length of the period of record should be at least double the 'return period' (the reciprocal of the probability of occurrence, or the probability of exceeding the particular value).

An important consideration, which is not peculiar to extremes, is whether only one population is represented in the data; there is the possibility that two or more populations, each with its own statistical distribution, contribute to the data. Such mixed population distributions can be a result of different meteorological processes; for example, rainfall maxima may be due to tropical cyclones, orographic lifting, frontal processes, large-scale uplift, monsoonal activity, etc. Similar considerations apply to extreme wind gusts, temperature, sunshine hours, etc. Ashkanasy & Weeks (1975) have attempted to fit annual maximum catchment rainfall for the Pioneer River Catchment around Pleystowe in northern Queensland by distinguishing between cyclonic events with strong easterly air flow and all other events. In most cases, however, there are insufficient data to postulate double or more complex distributions. Empirical distributions obtained by analysing numerous recording stations in a specific

4.1 Shorter-term variations

area may be used if the length of record is reasonable. Canterford & Pierrehumbert (1975) have analysed annual maximum rainfall for Queensland in this manner, and differences are apparent between inland and coastal stations for the rarer of the extreme events.

The distributions developed especially for extreme events are known as Fisher–Tippett Type I (or Gumbel), Type II and Type III (Weibull) distributions. Type III has an upper limit to the extreme values. Type I, not being bounded above or below, can be considered as the limit of Type III, while Type II has a lower bound. Annual maximum rainfall and flood discharges appear to belong to Type II although this implies no upper limit to these quantities which is opposed to what is expected on physical grounds.

Extreme temperatures, pressures and maximum wind speeds are usually found to be well described by the Type III distribution with its upper bound. It has been found that these populations retain their Type III characteristics when the upper 20% of events is used, giving estimates of the probability of future occurrences similar to those obtained from all annual maxima.

Extreme climatic events

Australia is a large continent subject to climatic extremes of many types. Probably the most important are floods and droughts.

Gibbs (1975) describes 1973 and 1974 as years of extremely heavy rainfall over much of the Australian continent with most of the areas of excess north of the Tropic and also in the south-east. These were years of widespread flooding, some of which exceeded the highest levels previously recorded.

Studies by Kraus (1955 b), Foley (1957) and, more recently, Pittock (1975) indicate that the early 1940s, on the other hand, was a period of minimum rainfall over much of eastern Australia; such a climatic extreme is of a type somewhat distinct from the episodic heatwave or flood.

Clearly, extreme low rainfalls are most important when the period extends to months or years and more so when the area covered is extensive.

Despite the very large areas sometimes covered by drought, Gibbs & Maher (1967) show that the greatest proportion of the Australian continent under serious rainfall deficiency in any calendar year was 46% in 1961. It is unlikely that the whole of the continent would be affected in any one year.

Recorded world extreme values of meteorological elements have been published by various authors. Riordan (1970) provides an excellent coverage and includes a discussion of the physical processes underlying the occurrence of each type of extreme.

146

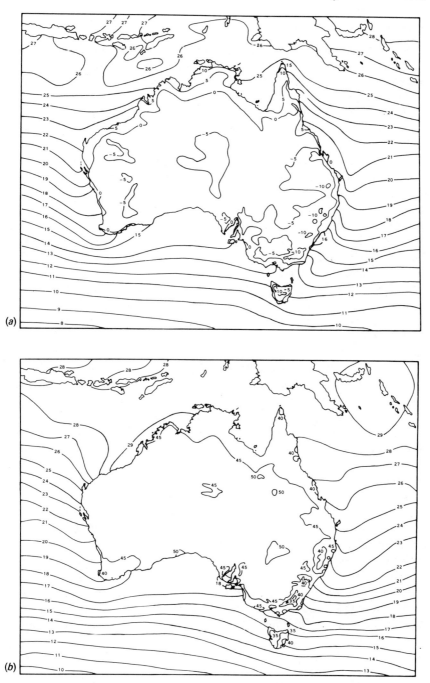

Fig. 4.1.4. Isotherms (°C) of extreme minimum temperature and of July sea-surface temperature (a) and of extreme maximum temperature and January sea-surface temperature (b).

147

4.1 Shorter-term variations

Extreme low temperatures

The effects of latitude, elevation and other factors on extreme minimum temperatures are evident in Fig. 4.1.4.(a) which shows isopleths of recorded extremes for the Australian continent along with mean sea-surface temperatures for the month of July.

McCormick (1958) found that extreme low temperatures result from 'the simultaneous occurrence of an optimum combination of several meteorological elements; absence of solar radiation, clear skies and calm air are the most essential requirements, with the ultimate fall in temperature dependent on the duration of these conditions'. The world's lowest recorded temperature is $-88.3\,°C$ at Vostok, Antarctica (78° 27' S, 106° 52' E, elevation 3,412 m) on 24 August 1960.

Australia's lowest temperature $(-22.2\,°C)$ was observed at Charlotte Pass (36° 26' S, 148° 20' E, elevation 1,759 m) on 14 June 1945 and on 22 July 1947. Charlotte Pass is situated in the Australian Alps and, in this area in winter, the conditions set down by McCormick are met more closely than elsewhere over the continent. Optimum conditions there require a cold outbreak, followed by a sudden change to calm, clear conditions extending from late afternoon through the long winter night with favourable local terrain assisting cold air drainage into low-lying areas.

Under similar conditions, at Melbourne (37° 49' S, 144° 58' E, elevation 35 m), where nocturnal radiative cooling is the dominant factor, the lowest recorded temperature is $-2.8\,°C$ on 21 July 1869 and at Hobart (42° 53' S, 147° 20' E, elevation 36 m), $-2.4\,°C$ on 11 July 1895.

The lowest temperatures experienced by day during cold outbreaks when a cold stream moves rapidly north from the Southern Ocean under favourable synoptic conditions, are somewhat higher. In Melbourne and Hobart on such a day, lowest recorded noon temperatures are $4.2\,°C$ on 21 July 1956 and $1.2\,°C$ on 4 August 1974 respectively.

In New York (40° 43' N, 74° 00' W, elevation 96 m) which is at about the same latitude as Hobart, temperatures as low as $-25.6\,°C$ have been recorded. The presence of a land bridge from a very cold continent on the poleward side facilitates outbreaks of extremely cold air when suitable synoptic systems provide the necessary trajectories for cold air. This happens in Europe, Asia and North America but in the Australian region there is no land bridge and air streams must pass over the comparatively warm waters of the Southern Ocean.

148

Extreme high temperatures

The effects of latitude, elevation and other factors on extreme maximum temperatures in Australia are evident in Fig. 4.1.4(*b*). Lamb (1958) lists the principal influences for very high temperatures as being strong heating on dry desert sand or rock, clear atmosphere, foehn effects, long sojourn or long passage of air over the warmest surface available and subsidence to inhibit both vertical convection and local circulations such as sea breezes. Hoffmann (1963) has postulated the highest possible temperature as slightly over 55 °C. The highest accepted record is 58 °C at El Azizia in North Africa on 13 September 1922 but there have been other claims of higher temperatures which may or may not have been valid. In Australia, the highest recorded temperature is 53.1 °C at Cloncurry, Queensland (20° 43' S, 140° 30' E, elevation 193 m) on 16 January 1889. These extremes occurred under very dry conditions. On the other hand, Priestley (1966) showed that there is a 'rather sharply defined upper limit to which screen air temperature will rise above a well watered underlying surface of sufficient extent'; this is about 33 °C.

Over the Australian continent the latitudinal effect on mean temperatures amounts to about 6 °C for 10 degrees of latitude. It is interesting to note that this effect, although evident in mean maximum temperatures between latitudes 20° and 45°, does not apply to high temperature extremes. Extremes at Melbourne (45.6 °C) and Adelaide (47.6 °C) are higher than those at Perth (44.6 °C), Sydney (45.3 °C), Brisbane (43.2 °C) and Darwin (40.5 °C). Advection of hot dry air from the interior is a very important factor. On the Queensland coast, for example, synoptic situations which will bring hot dry air from a westerly direction occur only rarely. Further south where hot dry streams occur generally as west to north winds in advance of a front associated with a depression, the condition is quite common. In Melbourne and Adelaide it is normal for such streams to bring temperatures near 40 °C at least once each summer.

The same synoptic controls generate considerable variability when air streams from the hot interior are replaced in summer by cool maritime air masses which quite frequently bring temperature falls of 10–20 °C in coastal areas. Far from the coast, in the hot dry interior, the 'heat pool' is well developed, variability is slight and at Marble Bar in the north-west of Western Australia it is on record that the temperatures exceeded 37.8 °C on 160 consecutive days during the summer of 1923–4.

4.2 Shorter-term variations

Extremes of point rainfall

Certain areas of the world are well known for their very heavy rainfall and there are accepted meteorological reasons for these rains. Persistent copious rains throughout a season or year are associated in general with orographic lifting of persistent moisture-laden streams, heavy rains for periods of days with cyclonic storms, and the short-period extreme falls with intense thunderstorms.

Extreme rainfall measurements for the world are available in Riordan (1970) and Australian figures are available in official publications of the Australian Bureau of Meteorology (1961, 1968).

Court & Salmela's (1963) formula for 'improbable extremes' (rainfall in t minutes equal to $2t^{1/2}$ inches)* provides a rough guide and when plotted on the somewhat dated but well-known diagram showing the world's greatest observed point rainfalls against duration on log–log paper (Fig. 4.1.5) the $2t^{1/2}$ values exceed but approximate the highest measured values for all durations tested.

4.2 Climatic fluctuations during the periods of historical and instrumental record
R. G. BARRY†

The primary focus of recent concern about world food supplies has been the nature and causes of climatic variability on the inter-annual to century $(1-10^2$ year) time scale. Climatic records during the twentieth century have graphically demonstrated the fallacy of the concept of a climatic 'normal' state and the need to consider climate as an environmental *variable*. However, a realistic perspective of these recent fluctuations must span at least the last half-millennium if we are to characterise adequately the rates and magnitudes of the changes and their possible causes (Bryson, 1974 *a*). This section outlines the available data base, our present views of the nature of the fluctuations and our understanding of their relation to the general circulation.

The data base

The sources of evidence for short-term climatic variability are of three types: instrumental records, historical evidence of climate-related phenomena and proxy records of environmental variables sensitive to climate.

* 1 inch = 2.54 cm.
† This section was written during the tenure of a Visiting Fellowship at the Department of Biogeography and Geomorphology, Australian National University. Suggestions and comments from R. S. Bradley, G. Singh and B. G. Thom are gratefully acknowledged. My visit to Australia was made possible by the award of Faculty Fellowship leave from the University of Colorado and also National Science Foundation grant OIP-7509736.

150

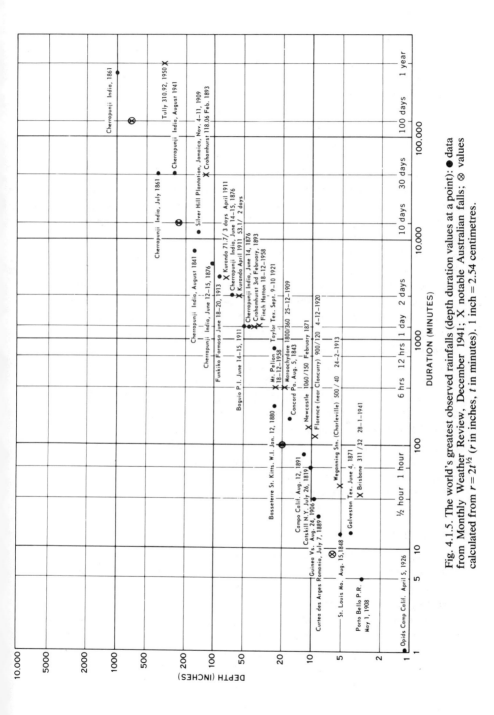

Fig. 4.1.5. The world's greatest observed rainfalls (depth duration values at a point): ● data from Monthly Weather Review, December 1941; X notable Australian falls; ⊗ values calculated from $r = 2t^{1/2}$ (r in inches, t in minutes). 1 inch = 2.54 centimetres.

151

4.2 Shorter-term variations

Instrumental data

Instrumental measurements of the principal climatic elements (pressure, temperature and precipitation) are fairly widely available for the last 100 years although the coverage is uneven, especially in the Southern Hemisphere. Upper-air data span only about 30 years and continuous, consistent monitoring of possibly causally-related parameters – such as solar radiation, atmospheric carbon dioxide and dust content, and snow cover – is largely of more recent origin. Currently, intensive efforts are being made to monitor from satellites global cloud cover, extent of snow cover and sea-ice, and radiation budget terms.

The 100-year period of intensive and relatively accurate measurement coincides with the intensification of inadvertent climatic modification due to man's activities (Landsberg, 1970 a). Accordingly, great importance is attached to assessing the range of 'natural' climatic variability before the advent of major urbanisation and industrialisation. The problem of trying to exclude such effects in assessments of long-term trends has been mentioned in Section 4.1 and is discussed further in Sections 4.3 and 4.5.

Western Europe has the longest and most complete coverage of meteorological observations (Lamb & Johnson, 1966; Manley, 1974). It is perhaps fortunate, therefore, that the North Atlantic sector is recognised to be one which responds strongly to small climatic fluctuations. Much effort is now being applied to extracting and archiving all available information from early weather records and sources such as ships' logs (Jenne, 1975; Oliver & Kington, 1970) in order to clarify our view of spatial patterns of climatic change. A neglected source for the western United States has recently been analysed by Bradley (1976). Records of precipitation kept at the early military forts and other stations were obtained for 135 stations which operated for at least five years during the period 1850–90 in the western states (excluding California where another 120 station records were noted).

Historical records

The wide range of historical documentation of climate-related phenomena (Le Roy Ladurie, 1971, for example) requires only brief mention here. The evidence includes weather diaries, records of droughts, floods, harvests, taxation revenue, glacier extent and sea-ice conditions. For the vast southern oceans, records of iceberg sightings are one source of information that has recently been examined by Burrows (1976). The way in which such varied evidence can be evaluated and co-ordinated is well illustrated by the Proceedings of the Aspen Conference on the climates of the eleventh and sixteenth centuries (Bryson & Julian, 1962).

Proxy data

There is a wide variety of environmental evidence that can yield climatic information but, for the time interval with which we are concerned, only two sources will be discussed. These are dendroclimatology and glaciology. The methodology used in the interpretation of proxy data is discussed below.

Interpretation of proxy data
Dendroclimatology

The study of wood-growth layers in plants (commonly referred to as tree rings) originated with the work of A. E. Douglass in the early years of the century, but its development into a more precise tool for analysing past climate derives primarily from studies since the early 1960s at the Laboratory of Tree-Ring Research, Tucson (Fritts, 1971). Samples of ring widths are taken from trees growing on limiting and therefore sensitive sites.

The extraction of climatic information from the tree-ring record depends on the formulation of an appropriate biological model relating growth to climatic conditions preceding or during the growing season. Fritts (1971) illustrates such models for conifers growing on semi-arid and warm sites where ring widths vary with the intensity and duration of drought, as well as for trees growing in high-altitude or high-latitude sites where the temperature of the growing season is the dominant control. Anomalies of growth over western North America, indicative of moist/cool or dry/warm conditions have been mapped on a decadal basis from 1521 to 1940. More elaborate reconstructions of climate over major sectors of the Northern Hemisphere have been developed using multivariate statistical techniques. Ring-width chronologies have been related to climate data for the period 1931–62 over the western United States by a principal components and multiple regression analysis, thus obtaining a *transfer function* relating the two data sets. Past climatic conditions can then be estimated from earlier tree-ring records. The methodology can be extended to large-scale climatic controls as a result of the spatial patterns of climatic and therefore growth anomalies. Eigenvectors of seasonal pressure anomaly fields for the twentieth century have been matched with eigenvectors of tree-growth anomaly patterns by canonical correlation (Fritts *et al.*, 1971) with useful levels of explained variance over the Pacific and North America. The ring chronologies over the western United States for 1700–1899 have then been used to develop estimates of corresponding seasonal pressure anomaly fields over half of the Northern Hemisphere. Subsequently, Blasing (1975) has grouped the pressure anomaly patterns into types and categorised

153

these in terms of climatic anomaly patterns, thereby providing an estimate of the relative frequency of different climatic types between 1700 and 1899. Verification based on late-nineteenth-century records indicates a very satisfactory agreement between anomalies of the climatic elements and the reconstructed circulation anomalies.

It is important to emphasise that dendroclimatology affords the unique possibility of reconstructing in this way seasonal conditions for *individual* years over at least several centuries. For individual ring-width chronologies, the record extends beyond 8,000 years BP.

Glaciology

The contribution of glaciology and glacial geomorphology to climatic reconstruction complements that of dendroclimatology in that it relates primarily to high-latitude and high-altitude sites. Glacial-geomorphological work in eastern Baffin Island by Miller (1973), for example, has shown the occurrence of several neoglacial advances which terminated just prior to 780, 350 and 65 years BP (dated by lichenometric methods; see Webber & Andrews, 1973). The final phase, spanning the 1600s to *c.* 1900, resulted in the most extensive ice cover of the last 5,000 years. However, such glacial advances (or retreats) have still to be interpreted in climatic terms. Conventionally, glacial advances have been taken to signify colder summers but in high latitudes increased snowfall, which is usually associated with warmer winters, is an equally likely cause. The degree of response is strongly determined by the vertical mass-elevation profile of the glacier.

A different glacial-geomorphological approach has been developed at the Institute of Arctic and Alpine Research. Aerial photographs and Landsat (ERTS-1) imagery have been used to map former snow and ice fields in the eastern Canadian Arctic (Andrews *et al.*, 1975; Barry *et al.*, 1975). These features are predominantly lichen-free and appear light-toned on the imagery, with abrupt borders abutting dark grey-black areas which comprise surfaces having a lichen cover of at least 80% (Fig. 4.2.1). Dates on dead moss and lichen in the lichen-free areas show that they were killed between 500 and 200 years BP, during the 'Little Ice Age'.

The extent of former permanent snow fields, by comparison with the present day, can be used to extract climatic information by means of an energy–mass balance model for a snow cover, developed by L. D. Williams (1974, 1975). This computes the possible combinations of climatic conditions that could produce the observed snowline changes.

Within the last decade, a significant new palaeoclimatic tool has been developed by Dansgaard and his associates (Dansgaard *et al.*, 1975). The oxygen isotopic content of an ice core can be used to determine the

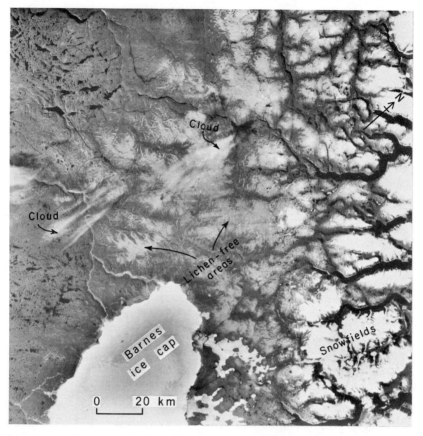

Fig. 4.2.1. North-central Baffin Island as viewed by Earth Resources Technology Satellite I (27 August 1973, multispectral scanner band 5). The image shows light-toned lichen-free areas north of the Barnes ice cap and present-day snow fields to the east.

average surface conditions when the ice was formed. High values of $\delta^{18}O$ indicate warm conditions, low values cold conditions, but the precise relationship is complicated by the exact isotopic composition of the original precipitation, the seasonal occurrence of the snowfall, and the effects of ice flow on the core. At Camp Century, Greenland, the core apparently shows seasonal variations for the last 8,000 years. Recently, several additional shallow cores have been collected and the interpretations of these are discussed below.

155

4.2 Shorter-term variations

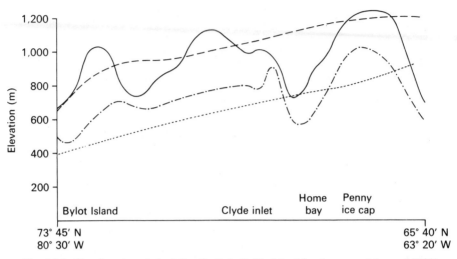

Fig. 4.2.2. The elevation of glaciation limits in Baffin Island for the present day and 'Little Ice Age' (from lichen trimlines) compared with computed equilibrium line altitudes (ELA) for 1963–72 and hypothetical 'Little Ice Age' conditions (based on L. D. Williams, in press): —— present glaciation limit (Andrews & Miller, 1972); – – – – computed 1963–72 ELA; – · – · – 'Glaciation limit' for lichen trimlines; - - - - computed ELA with 5% solar radiation decrease and 1 °C temperature decrease from 1963–72 data.

Climatic characteristics of the last half-millennium
Major fluctuations

The climate of northwestern Europe during the past 1,000 years has been well described by Lamb (1966a) and Le Roy Ladurie (1971). The major features are an interval about AD 1000–1200 with temperatures somewhat higher than those in the twentieth century, followed by a decline to a minimum, known as the 'Little Ice Age', between about AD 1500 and 1800. The temperature and precipitation changes between these two intervals in England amounted, for century averages, to approximately 1.3 °C and 10% respectively (Lamb, 1965b). The cooling was probably more pronounced in higher latitudes, where it was associated with marked differences in sea-ice conditions. The decline leading to the 'Little Ice Age' saw the extinction of the Viking Settlement in Greenland. The lowering of the snowline by some 200–400 m during the 'Little Ice Age' in Baffin Island, compared with modern conditions, can be accounted for by a summer cooling of about 1.5 °C, or a doubling of winter snowfall; see Fig. 4.2.2 (based on L. D. Williams, in press). Following the 'Little Ice Age' there was a marked recovery in the early twentieth century, especially in the higher latitudes of the Northern Hemisphere.

As records have been compiled from other parts of the world, it has

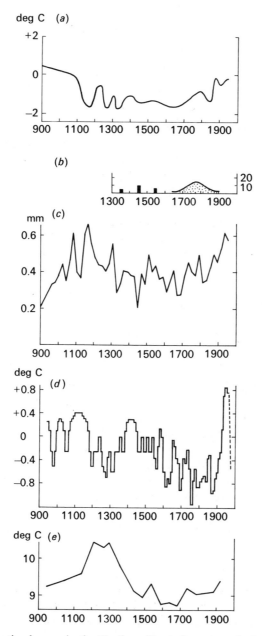

Fig. 4.2.3. Climatic changes in the Northern Hemisphere over the last millennium: (*a*) Temperatures in China (departures from present day, °C) after S. Wen-Hsiung (1974). (*b*) Prolonged rains (frequency/50 years) in Iwate, Japan after Yamomoto (1972). (*c*) Ring-width of bristlecone pine, White Mountains, California after La Marche (1974). (*d*) Mean annual temperatures (departures from average, °C) in Iceland after Bryson (1974*a*). (*e*) Mean annual temperatures in central England from Lamb (1965*b*).

become apparent that rather similar fluctuations have occurred elsewhere. Figure 4.2.3 illustrates the reconstructed temperature trends for China (Wen-Hsiung, 1974) and Japan (Yamamoto, 1970, 1972) compared with England (Lamb, 1965 b) and Iceland (Bryson, 1974 a, after Bergthorsson). Inspection of the graphs indicates that while the changes in China and Japan are broadly in phase with one another, the 'Little Ice Age' appears to have begun in the tenth century in east Asia and temperatures there reached a minimum in the twelfth century. The cooling apparently spread westward, reaching European Russia in the mid-fourteenth century and central Europe in the mid-fifteenth century (J. G., 1973, citing Chu Ko-Chen). Such displacements and longitudinal relationships are suggestive of adjustments in the tropospheric wave structure, a topic which is discussed further below.

Few records in the Southern Hemisphere span more than a century and the coverage is very uneven. For Chile, Taulis (1934) compiled a precipitation record for Santiago and new information for this area is being extracted from dendroclimatic reconstructions. La Marche (1975) has identified a long dry period between AD 1250 and 1450, and a long wet period from AD 1450 to 1600.

In the Southern Alps of New Zealand, moraines indicate glacial advances beginning in the mid-thirteenth century at Mt Cook and the early fifteenth century in the Cameron Valley, with further advances in the latter area c. AD 1650, 1740, 1850 and 1890 (Burrows, 1973, 1975). The chronology is broadly similar with the Northern Hemisphere events (Denton & Karlén, 1973). Orheim (1972) also concluded that summer temperatures on Deception Island (63° S, 62° W), estimated from ice stratigraphy, have paralleled those of the Northern Hemisphere since the early 1800s.

In Antarctica there was little change in the long-term accumulation at the South Pole between the two periods 1500–1750 and 1760–1957, based on snow stratigraphy (Giovinetto & Schwerdtfeger, 1966). However, there was a 40% increase in accumulation about 1820 (from 5.4 g cm^{-2} yr^{-1} for 1760–1825 to 7.5 g cm^{-2} yr^{-1} for 1892–1957).

The instrumental period

The last 100–150 years, which is well covered by instrumental records, includes some remarkable climatic fluctuations. In the Northern Hemisphere, as reported in surveys by Lysgaard (1963), Mitchell (1963), Dzerdzeevski (1969) and Lamb (1966 a), these are:

(i) a period of weak zonal circulation and cold winters in the late nineteenth century;

(ii) increasing vigour of the circulation with considerable warmth culminating in the 1920–30s; and

(iii) a sharp decrease in circulation intensity, and cooling in many areas, starting about 1950.

Bergthorsson's temperature reconstruction for Iceland (as updated by Bryson (1974 a)), underlines the abruptness of the twentieth-century warming and cooling and its apparent uniqueness in the context of the last 1,000 years. Manifestly, the standard climate 'normals' for 1901–30 and 1931–60 are not representative of the historical era. These fluctuations were most pronounced in higher latitudes in the Atlantic–European sector (Kirch, 1966), but even in the British Isles the changes were sufficient to be of significance in terms of growing season characteristics and winter snowfall. The same conclusions are reached by Wahl (1968) for the eastern United States. The records there show a warming from the 1830–40s to the 1930s, especially for September–October, with indications of a recent return to the earlier cooler conditions. For the western United States, Wahl & Lawson (1970) show that the mid-nineteenth century was generally warmer than the period 1931–60. They also state that both the eastern and western United States were generally wetter in the 1850s and 1860s than during 1900–50. On the basis of more station records for the western states, however, Bradley (1976) finds that, at least from 1865 to 1890, summer and autumn months were drier than the 1951–60 averages, while winter and spring were wetter. He notes also that, during the 1960s, summer precipitation amounts have been unusually high and winter amounts low over most of the Rocky Mountain states.

The cooling in high latitudes appears to have begun in the Eurasian sector and then shifted westward (Lamb, 1972 b). This is also apparent in Canada for annual temperatures since 1940 (Thomas, 1975). Bradley (1973) reports summer cooling in the eastern Canadian Arctic in the 1950s with a more pronounced decrease in the 1960s, coincident with increased persistence of pack-ice in Davis Strait (Dunbar, 1973). At the same time, winters in eastern Baffin Island became milder, in marked contrast to the pronounced cooling in the Eurasian sector of the Arctic (Rodewald, 1972). Northern Hemisphere snow and ice cover increased dramatically in 1971 (Kukla & Kukla, 1974) but records for 1966–75 show a subsequent decrease over Eurasia and the hemisphere (Wiesnet & Matson, 1975). The cooling over most of the Northern Hemisphere in the 1960s appears to have continued in middle latitudes, but reversed in high latitudes (Painting, 1977). In Europe, however, winters have been mild since 1970/71 as a result of changes in the North Atlantic (Dickson *et al.*, 1975). Dronia (1974) reported a general cooling throughout the lower troposphere in the Northern Hemisphere between 1953 and 1972. This apparently affected

4.2 Shorter-term variations

both hemispheres for 1958–63 (with a 0.4–0.5 °C cooling of the 700–300 mb layer), but since 1963 the cooling has been negligible in the Northern Hemisphere and balanced by slight warming in the Southern Hemisphere (Angell & Korshover, 1975).

Contrary to the earlier conclusions of Mitchell (1963), which were based on a limited number of stations south of 40° S, the pattern of recent fluctuations in the Southern Hemisphere is dissimilar to that of the Northern Hemisphere. According to Lamb (1967 b), the higher southern latitudes did not share fully in the cold interval of the late nineteenth century, nor the subsequent warming. However, he considered that the trend in circulation intensity, with the westerlies in the sectors south of Australia–New Zealand and South America strengthening from the 1860s to 1920s (see also de Lisle (1957)), resembled that of the Northern Hemisphere. The differences in middle latitudes are more striking. Salinger & Gunn (1975) demonstrate a temperature minimum in New Zealand during 1900–30 and subsequent warming. The results of Deacon (1953) and Tucker (1975) for Australia show a similar pattern, although Tucker notes that the spatial coherence of the changes is subcontinental in scale. The changes in precipitation in Australia are discussed below.

In low latitudes, the most obvious changes over the last 100 years involve precipitation. In many parts of the tropics, but excluding monsoon Asia (Rao & Jagannathan, 1963), there was a general decrease of annual totals of the order of 30% about 1900 (Kraus, 1955 a, 1958; Lamb, 1967 b). During the first half of this century the equatorial rain belt tended to be narrower, and the arid zone broader, than during the latter half of the nineteenth century or after about the 1940–50s. The same pattern of change occurred along the *east* coasts of Australia and North America to about latitude 40° (Kraus, 1954, 1955 b), with a recovery in rainfall totals of about 10% after 1940 (Kraus, 1963). In tropical Australia, however, the 'tropical pattern' recognised by Kraus is by no means general (Stewart, 1973, 1975). The regional precipitation changes can be related to changes in location and intensity of the subtropical anticyclones as demonstrated by Stewart (1975) and Pittock (1975). Gentilli (1971) shows a general increase in the extent of arid area in Australia between 1881–1910 and 1911–40, which reached a maximum extent in the 1930s according to Ambe (1967) and thereafter diminished, although in central Australia annual totals showed a general downtrend between 1911 and 1970 (Tucker, 1975). Seasonal precipitation totals may show quite different trends according to the responsible rain-producing systems (Kraus, 1954; Stewart, 1973; Wright, 1974 b).

The evidence for northern Chile (Lloyd, 1973) is not in phase with that for Australia. Precipitation records in the arid zone (26° S) show some increase from 1880 to 1905 and a major peak about 1940, followed by a

160

decline to 1970. Central stations, such as La Serena (30° S) resemble central Australia, with decreasing precipitation from the beginning of this century.

Rainfall amounts in tropical Africa (the Sahel) have been a major item of recent concern. Winstanley (1973 *a*) reports that the summer monsoon rains from Africa to India decreased by more than 50 % from 1957 to 1970. In west Africa, 1968–72 were the driest five years since 1950 receiving only 80 % of the 1941–73 average, whereas 1950–4 were the wettest five years with 120 % (Tanaka *et al.*, 1975). These extremes illustrate the common difficulty of discerning general trends from short-term records. On the subtropical margins of the arid zone in Africa, the Middle East and India, the winter–spring rains decreased from 1951–2 to the early 1960s and then recovered (Winstanley, 1973 *a*, *b*). However, Flohn (1971 *a*) found *no* evidence of general trends at ten stations in southern Tunisia, although deviations of 10-year averages from the long-term mean reached ±14 %. Throughout tropical Africa, recurrent droughts and wet periods seem to be characteristic (Lumb, 1965; Dalby & Harrison-Church, 1973; Bunting *et al.*, 1976; Rodhe & Virji, 1976).

Statistical properties

The statistical properties of climatic variability were discussed in Section 4.1. The characteristic time scales of climatic variability have generally tended to be described loosely, since the variance spectrum of climatic indicators contains no pronounced peaks with period shorter than 10^4 years. On the shorter time scale, Monin & Vulis (1971) show that in general climatic series display only persistence (a 'red-noise' spectrum), apart from the well-known quasi-biennial oscillation (see also Section 4.5).

Recently, Kutzbach & Bryson (1974) examined fluctuations on the scale of $10^0–10^4$ years over the North Atlantic sector using instrumental, historical and palaeobotanical data. They conclude that the spectrum demonstrates strong persistence for periods $> 10^3$ years, randomness for decadal or shorter periods, and moderate persistence at intermediate time scales (Fig. 4.2.4). The historical data for central England and Iceland show considerable variance near 25–30 years, but this is absent from the central England temperature series. From comparisons with the latter series, Kutzbach and Bryson suggest that the historical records over-emphasise fluctuations with periods less than 30 years and underemphasise those with periods greater than 30 years. Periodicities, which may nevertheless be transient, occur in other series. Wagner (1971) found a significant 21-year peak in spectra of MSL pressure in semi-arid areas of the Northern Hemisphere; Tyson (1971) a 20-year peak in rainfall over northern and eastern South Africa. Tyson *et al.* (1975) also identify

161

4.2 Shorter-term variations

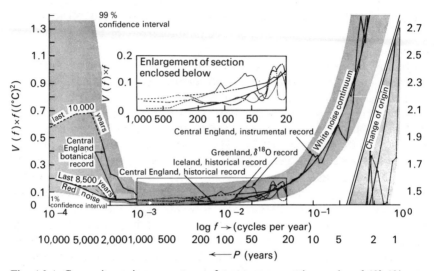

Fig. 4.2.4. Composite variance spectrum of temperature on time scales of 10^0–10^4 years derived from instrumental, historical, $\delta^{18}O$ and botanical records. The ordinate is $V(f)$ times f [units, (°C)²], the abscissa a logarithmic frequency scale. The four lowest frequency spectral estimates of each individual spectrum are connected by dashed lines to indicate that they are unreliable in a statistical sense. Stippling indicates a generalised version of 1 and 99% confidence limits. The insert in the middle of the figure is an enlarged version of the intermediate frequency section. Greenland spectral estimates have been halved. After Kutzbach & Bryson (1974).

oscillations of 16–20 years and 10–12 years with distinctive regional patterns in South Africa.

For the important intermediate range, the Greenland $\delta^{18}O$ curve for AD 1200 to the present has been analysed in detail by Johnsen et al. (1970). Using Fourier analysis they identified periodicities of 78 and 181 years in the record. Other indications of a periodicity averaging about 77 years, as reported by Maksimov (1952) for various historical records, should also be noted.

Despite the evidence for periodic fluctuations, there are also strong indications that many changes are abrupt. This is apparent in the proxy temperature record for Iceland just prior to AD 1200 and about AD 1600. Moreover, the new climatic states then show persistence for about the two centuries. These features are suggestive of Markovian-type step-function changes, as proposed by Bryson & Wendland (1967), rather than quasi-periodic oscillations. This interpretation is strengthened by a similar conclusion for the abrupt climatic deterioration in the White Mountains, based on the death of a large number of trees at the upper treeline, just before AD 1600 (La Marche, 1973).

Circulation features

The material at present available for assessing the circulation characteristics of fluctuations of 100–200 years' duration is essentially limited to
the European-Atlantic sector where, as we have seen, global changes are
probably amplified. Lamb's (1966 a) studies suggest that for this sector
there is an approximately 200-year fluctuation in the strength of the
westerlies, with peaks in the early 1300s, 1500s, 1700s and 1900s. The
appearance of a *c.* 180-year periodicity in the $\delta^{18}O$ record at Camp
Century (Johnsen *et al.*, 1970) may reflect such circulation fluctuations.

A key question is whether variations in circulation intensity in one sector
represent an overall change in the circumpolar vortex* or whether they
indicate only a shift in tropospheric wave pattern. There is now clear
evidence that since 1900 most of the fluctuations of pressure and temperature in the Northern Hemisphere have been related to short-term
displacements of the wave pattern or changes in wave amplitude (van Loon
& J. Williams, 1976 a, b). Only the patterns of change during 1915–29 were
suggestive of fluctuations in the intensity of the basic westerly flow. On
this time scale, therefore, the temperature fluctuations are a response
primarily to differing degrees of advection.† Decadal changes in precipitation over eastern Australia also appear to be associated with the
amplitude of the wave pattern (see Section 4.3).

On the 100- to 200-year time scale there is further evidence of the
primacy of longitudinal displacements. Reference has been made to the
apparent westward spread from east Asia of the 'Little Ice Age' cooling
episode. A similar conclusion has been reached by Dansgaard *et al.* (1975)
from comparison of European data with the isotopic analysis of a 1,420-
year ice core record from Crête, Greenland (37° W). Fluctuations on a
time scale of 60–200 years at Crête are in phase with Iceland and England
whereas longer-term changes are not. Crête experienced minimum temperatures between AD 1150 and 1400 rather than AD 1450–1700, for
example. Lamb (1963) earlier noted the spread of colder winters and wetter
summers westward across Europe from about AD 1250 to 1500 and also
the return eastward of these conditions from AD 1700 to 1900. He
attributed the temperature changes to the respective influences of the
cold upper trough over eastern Europe and the warm ridge over the eastern
Atlantic. Gray (1975), however, identifies a warming spreading westward
from Japan in the sixteenth century to Europe in the eighteenth with a

---

* The zonal index in the Atlantic sector is significantly positively correlated with the
  hemispheric zonal index, whereas that of the Asian sector is negatively correlated.
† A conflicting result is apparent in a diagram presented by Sawyer (1974) which shows a
  15° westward displacement of the 500 mb trough in winter over eastern North America
  between the 1870s and 1950, and a 12° eastward movement of the Asian trough at 75° E
  between 1905 and 1965.

mean speed of 0.61° longitude per year. She attributes it to possible variations in eccentricity of the circumpolar westerly vortex.

Next, let us examine the evidence for overall changes in circulation. Winstanley (1973 *b*) and Winstanley *et al.* (1974) indicate that the Northern Hemisphere circulation weakened sharply during 1968–73, continuing the trend noted since the 1950s (Lamb, 1966 *b*), and they suggest that there was a one-degree equatorward expansion of the circumpolar vortex, although contrary evidence has been forthcoming. Tanaka *et al.* (1975) argue that there was no global expansion of the circulation, although there was a two-degree southward shift of the Saharan subtropical high between 1961–5 and 1969–73. In contrast, Miles & Folland (1974) report a poleward displacement of the mean-sea-level zone of maximum westerlies and the subtropical anticyclone (between 10° and 60° W) since 1955, in summer and for the year as a whole. They also found an absence of trends in the latitude of these features between 1900 and 1970.

On the time scale of 100–200 years, Lamb (1966 *a*) considers that the 'Medieval Warm Epoch' saw a contraction of the northern polar vortex and the 'Little Ice Age' an expansion. A parallel contraction and expansion in the Southern Hemisphere is suggested by La Marche (1975). The Santiago dendroclimatic record may imply displacement of the storm tracks southward from AD 1250 to 1450 and northward between AD 1450 and 1600. In contrast, analysis of circulation indices since the mid-eighteenth century indicates, according to Lamb (1967 *b*), that the wind belts have been displaced northward, or southward, simultaneously in the two hemispheres, rather than simultaneously poleward, or equatorward. This is also suggested by Sanchez & Kutzbach (1974) for changes in the 1960s over the Americas.

In the Southern Hemisphere, Trenberth (1975 *b*) finds no evidence for a general expansion of the southern polar vortex, although there is evidence for displacements of features in particular sectors. The relationship between such displacements and the response of climatic elements has been considered by Bryson (1973 *a*) and Pittock (1973 *b*). Pittock indicates from a study of year-to-year differences that a one to two degree latitudinal shift of the mean position of the major 'centres of action' of the global circulation may be accompanied by a 0.5 °C change in global mean surface temperature but a 10–20 % change in regional precipitation.

The characteristics of tropospheric wave structure and circulation are well known from dynamical theory (see e.g. Palmen & Newton (1969), Chapter 6). Wavelength (L), or wavenumber ($n = 2\pi/L$), are related through the Rossby equation to the zonal wind speed (U) at the midtropospheric level of non-divergence and also to the wave velocity (C).

$$U - C = \beta (L/2\pi)^2,$$

where $\beta = \partial f / \partial y$, the variation of the Coriolis parameter (f) with latitude (y). For a given wavelength and latitude it can be shown that the waves travel faster (a zonal pattern) if the wind speed is greater; slower (a meridional pattern) if the wind speed is less. For monthly mean patterns the degree of wave anchoring is strongly influenced by surface temperatures (Namias, 1972). A readjustment in wave number or spacing may therefore occur in response to a redistribution of the major heat sources/ sinks, as happens on a seasonal basis. Latitudinal displacements of the circulation also have a major effect on the wavelength. Equatorward displacement of the maximum westerlies tends to allow more meridional circulations to develop in higher middle latitudes, although it may also permit an additional zone of cyclonic activity to exist. Lamb (1963) argues that, since 1790, the longitudinal spacing of the upper waves over the Atlantic sector has increased/decreased in parallel with changes in circulation intensity. In addition, however, there was probably a poleward shift of the zone of maximum westerlies between the weak circulation epoch of the early nineteenth century and the much stronger circulation a century later. Changes in circulation intensity, wave spacing, and extent of the polar vortex are evidently complexly interrelated characteristics of the global circulation and simple models of circulation behaviour are of limited value.

The problem of linking changes in climatic conditions with circulation changes has received only limited attention (e.g. Kraus, 1956; Lamb, 1974 b). Dzerdzeevski (1963) has argued that changes in climate result from changes in the frequency of circulation types and support for this idea is provided by a principal component analysis of time series of the Dzerdzeevski hemispheric circulation types (Kalnicky, 1974). It is shown that zonal patterns (represented by particular eigenvectors) decreased after 1950, whereas meridional ones increased by comparison with 1900–50. This matches the observed cooling trend in various parts of the Northern Hemisphere. Cooling is most pronounced in regions affected by northerly flow components such as easternNorth America (Wahl, 1968; Wahl & Lawson, 1970).

Detailed examination for particular regions shows that the postulated relationships are sometimes elusive, however. For example, the twentieth-century warming and subsequent cooling in the British Isles has been attributed to the increase and subsequent decrease in frequency of westerly airflow (Lamb, 1965 a), but analysis of temperature changes at four stations between 1925–35 and 1957–67 in relation to frequency of airflow types shows that the situation is complicated by 'within type' changes (Perry & Barry, 1973). The recent deficit of summer rains in the Sahel has been accounted for in terms of an equatorward expansion of the circumpolar vortex (Winstanley, 1973 b; Bryson, 1973 a) but, as

165

already discussed, this is doubtful. Lamb (1974 b) attributes the drought to a weaker subtropical anticyclone and more prominent northeasterly wind components.

Concluding remarks

Definitive knowledge of climatic fluctuations during the period of historical and instrumental record has advanced remarkably in the last few years. Most major sources of observational and historical weather data have been examined in at least a preliminary manner, although advanced statistical treatment often remains to be pursued. Additional sources of proxy climatic information exist in dendroclimatic series and other environmental indicators, but only the dendroclimatic records can provide seasonal and annual resolution. In high latitudes snow and ice cores are of comparable value, although subject to greater time smoothing. Glacial-geomorphological evidence can yield additional spatial coverage relating to particular climatic events. In all these cases, appropriate transfer functions between the environmental indicator and climatic parameters must be developed.

Links between circulation changes and changes in climatic conditions are hard to establish. Apparent stochastic events, such as fluctuations in snow cover and sea-surface temperature may account for some pronounced short-term climatic anomalies. Mitchell (1966) and Yevjevich (1971) have shown how atmospheric stochasticity may be transformed by the effects of oceanic heat storage into persistent anomalies of temperature and precipitation. More empirical case studies are required to quantify the nature of such interactions.

The principal area of uncertainty at the present time concerns the basic characteristics of atmospheric circulation fluctuations and especially their regional manifestations such as the Southern Oscillation and quasi-biennial oscillation. Statistical studies show that persistence is a key feature of most time series. Periodicities, other than the quasi-biennial oscillation, are relatively rare and most of them are probably transient. It is not certain whether fluctuations of about 80 and 180 years are genuine, but even so the variance which they account for may be rather small. There is considerable evidence of the importance of variations in wave pattern for climatic anomaly patterns of decadal duration and over 100–200 years. However, the complex behaviour of wave amplitude and spacing, circulation intensity and the nature of latitudinal displacements of the global wind systems is inadequately known. Programmes such as the Global Atmospheric Research Programme (US Committee for GARP, 1975) should contribute much to providing more definitive answers to these questions.

4.3 Patterns of variability in relation to the general circulation
A. B. PITTOCK

The broad global pattern of climate is well known. The dominant latitudinal variation of the climatic elements and its modification by topography and the distribution of land and sea were related to the general circulation of the atmosphere and oceans in Chapter 2. In this section we are concerned not so much with the time-averaged picture but with the variability in time of the main circulation features and the impact of such variability on climatic patterns over a range of space scales.

The intensity of the mean circulation patterns and the amplitudes and locations of the major standing waves vary from season to season, year to year and decade to decade and account for much of the observed variability of climate on these time scales. Presumably such shifts in the atmospheric circulation also play an important role in much longer-term climatic changes although many other processes such as changing snow and ice cover, differing sea levels and even eustatic and tectonic changes begin to operate on the longer time scales and would be expected to induce circulation characteristics significantly different from those of the present day.

Main controls

The dominant atmospheric circulation features whose variability leads to distinctive patterns of climatic variation are the mean meridional circulation itself and the major standing waves associated with the distribution of the continents and oceans. The largest land masses in low and middle latitudes induce seasonal inflows and outflows known as monsoonal circulations, of which by far the most pronounced is that over the Indian/southeast Asian region. In fact the longitudinal contrast between the relatively warm tropical subcontinental area of southeast Asia and the relatively cool tropical eastern Pacific Ocean induces a large-scale standing wave pattern which is a dominating influence on Australian climate.

At higher latitudes, the land–sea distribution and topography of the North American, European and Asian land masses leads to a pronounced standing three-wave pattern in the Northern Hemisphere circulation (see e.g. Stark, 1965). This is much less pronounced in the Southern Hemisphere where the smaller land masses of the south-temperate zones lead to a weak three-wave pattern. This gives way in the southern polar zone to mainly a single wave or elliptical pattern (van Loon & Jenne, 1972) often presumed to be linked to the eccentricity about the pole of the Antarctic land/ice mass, but which is probably largely influenced by the dominance of the Indian/southeast Asian/Australia–New Guinea land mass in the

167

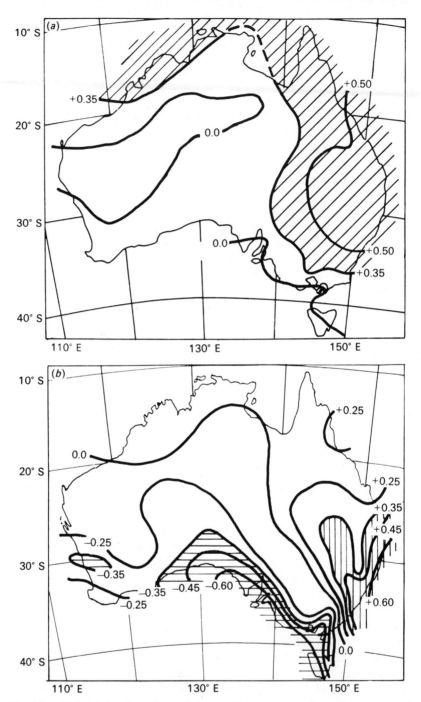

Fig. 4.3.1. (a) Isopleths of equal correlation coefficient between the annual mean Southern Oscillation Index and district mean annual rainfall for the years 1941–70 inclusive. The correlation coefficient $R = 0.35$ is significant at the 95% confidence level and $R = 0.45$ at the 99% level.

(b) Isopleths of equal correlation coefficient between the annual mean latitude of the surface high-pressure belt over eastern Australia and district mean annual rainfall for the years 1941–70 inclusive.

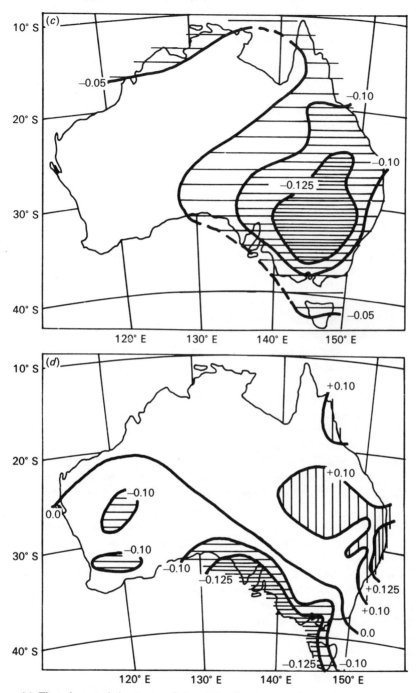

(c) First characteristic pattern of Australian district mean annual rainfall for the years 1941–70 inclusive. This pattern accounts for 36% of the total variance and its amplitude is correlated with the Southern Oscillation Index at $R = -0.53$.

(d) Second characteristic pattern of Australian district mean annual rainfall for the years 1941–70 inclusive. This pattern accounts for 18% of the total variance and its amplitude correlates with the annual mean latitude of the surface high-pressure belt over eastern Australia at $R = +0.83$.

4.3   Shorter-term variations

tropics as evidenced by the linking band of clouds streaming south-eastwards across the South Pacific Ocean (Streten, 1973).

Various studies over the past 50 years have identified a number of these large-scale time-varying circulation features whose behaviour may be regarded as exerting a measure of control over patterns of regional and local climate.

Walker Circulation

Patterns of year-to-year variation of climatic elements around the world were extensively studied by G. T. Walker and his colleagues (e.g. Walker & Bliss (1932, 1937)). One of their principal findings was the existence of strong negative correlations between surface pressure anomalies in the southeast Asian region and the eastern Pacific, with associated changes in rainfall and other variables. Many other workers have studied this pattern of opposed pressure anomalies in the low latitudes, which Walker called the 'Southern Oscillation' but which is perhaps better referred to as the 'Walker Circulation', as suggested by Bjerknes (1969), since it is not strictly periodic but rather has a characteristic time scale of 2–5 years and involves a mean east–west circulation (Troup, 1965; Flohn, 1971 b; Krishnamurti et al., 1973).

As initially described by Walker, the Southern Oscillation was charac-terised by an index based on values of a number of meteorological elements (mainly pressure) at certain key stations in the equatorial belt. Subsequent authors have devised various other indices of the oscillation and have mapped its sphere of influence by examining the distribution over the globe of correlation coefficients between the selected 'Southern Oscillation Index' and a variety of meteorological elements. Berlage (1961), for example, mapped the world distribution of correlation coeffi-cients between annual surface pressure anomalies and an index consisting simply of the pressure anomaly at Batavia/Jakarta and showed that the pattern of variation extends into the high latitudes of both hemispheres.

Pittock (1975) showed that the Southern Oscillation, as described by an index based on pressures at Tahiti and Darwin has a dominant influence on the variation of annual rainfall in eastern Australia (Fig. 4.3.1(a)). The nature and mechanism of the Southern Oscillation and the Walker Circulation are discussed further in Section 4.4.

Standing wave pattern in the Northern Hemisphere

At higher latitudes in the Northern Hemisphere, variations in the amplitude and position of the standing three-wave pattern have been demonstrated by Namias (1970 b) and invoked by him to account for climatic variations

170

on the scale of decades in North America and Europe. Wahl (1968), Kalnicky (1974) and others have invoked similar variations to account for climatic changes on somewhat longer time scales, while Dansgaard *et al.* (1975) suggest such variations in the standing wave patterns are evident in ice-core data over the last one or two millennia, including the so-called 'Little Ice Age' of the fifteenth to nineteenth centuries. Reported increases in the total ozone content of the Northern Hemisphere in the 1960s (London & Kelley, 1974), based largely on continental observing stations, may at least in part be due to the same effect (Pittock, 1974 *b*) rather than to a real hemispheric mean increase.

Latitudinal shift of circulation features

Shifts in the mean latitudes of the climatic zones have been invoked as explanations of climatic variations by many authorities including Kidson (1925), Kraus (1955 *a*), Flohn (1964), Winstanley (1973 *a*, *b*) and Bryson (1974 *a*). Studies in the Australian region have demonstrated significant correlations between year-to-year variations in rainfall along the south and east coasts and the latitude of the subtropical high-pressure belt (Pittock, 1973 *a*), as is illustrated in Fig. 4.3.1(*b*). Strong correlations between the latitudes of the high-pressure belt off the west coasts of North and South America and rainfall in those areas have also been demonstrated (Pittock, 1971 and 1977) and the North American case is illustrated in Fig. 4.3.2.

Characteristic patterns of variation

The recent application of principal component (or characteristic pattern) analysis to time series of meteorological fields has given considerable insight into the dominant modes of atmospheric variation. The technique was discussed briefly in Section 4.1. It has been applied to Southern Hemisphere data by Veitch (1965), Wright (1974 *a*, *b*), Kidson (1975 *a*, *b*), Pittock (1975) and Trenberth (1975 *a*, *b*).

Australian rainfall

Pittock (1975) analysed data from 105 Australian rainfall districts from 1913 to 1974 inclusive. These showed a dominant pattern (Fig. 4.3.1(*c*)) consisting of a large positive or negative anomaly centred over inland southeast Australia and extending northwards over eastern and tropical areas. This pattern had an amplitude which correlated significantly from year to year with an index of the Southern Oscillation, the sign being such that when the surface pressure over the eastern Pacific is anomalously high

171

4.3 Shorter-term variations

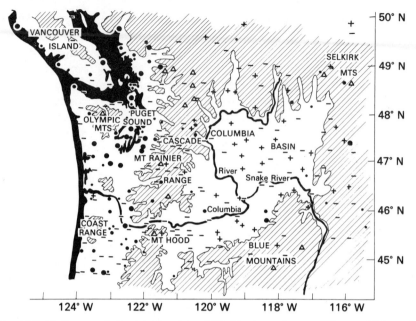

Fig. 4.3.2. Station correlations, in the state of Washington and parts of Oregon, Idaho and British Columbia, between annual rainfall and corresponding annual latitude of surface high pressure along the west coast of North America. Data are for the years 1941–70. Only stations having at least twenty years of data have been plotted. From Pittock (1977).

Topography: ☐ 0–1,000 m altitude; ▨ over 1,000 m altitude; △ major mountain peaks.

Significance   • 2% ⎤   − negative, not significant
level of correlations   • 5% ⎬ negative   + positive, not significant
                · 10% ⎦

relative to that over the southeast Asian region, more rain falls over eastern Australia.

The second most important pattern according to this analysis (Fig. 4.3.1(*d*)) is one showing anomalies of opposite sign over the eastern and southern coasts of Australia, and its amplitude is highly correlated with the latitude of the subtropical high-pressure belt over eastern Australia.

The statistical association of these anomaly patterns with variations in the Southern Oscillation Index and the latitude of the high-pressure belt, respectively, are both physically reasonable and are supported by the mapping of direct correlations between rainfall and these indices of the general atmospheric circulation (Figs. 4.3.1(*a*) and (*b*)).

Nevertheless, the above analysis using 105 very unevenly distributed rainfall districts biases the results towards southeast Australia where the districts are smaller and more numerous. A second analysis using a grouping of data into 45 larger and more evenly distributed districts still

172

placed the Southern Oscillation-related pattern first, but reduced the pattern associated with the latitude of the high-pressure belt to fourth place. This illustrates the point that, while a given mechanism or pattern of variation might dominate the variance on one space scale, other more localised anomaly patterns may tend to greater prominence if the analysis is biased towards or confined to the smaller areas in which they occur. Another case in point is the analysis by Wright (1974 a, b) of data from southwest Australia, which shows patterns dominant in that area which are not prominent in the broader scale analysis.

One result of these analyses of the Australian data is that it can now be said that, on the continental scale, and particularly in relation to eastern Australia, two very important influences affecting Australian rainfall are the Southern Oscillation or standing wave pattern, and the mean latitude of the high-pressure belt, the latter being closely related to the strength of the Hadley circulation (Pittock, 1973 a).

Having identified such patterns, which are the result of the interaction of variations in the general circulation with the general topography of the region, the reasons for particular spatial correlations between, say, rainfall at various stations (e.g. see Priestley, 1963; Anderson, 1970) becomes clear. These are the result of common or opposing influences of given circulation anomalies on the local rainfall in these areas.

Global pressure, temperature and rainfall

Kidson (1975 a, b) applied the technique of principal component analysis to monthly means of surface pressure, temperature and rainfall defined on grids extending over both hemispheres and the tropical belt. Figure 4.3.3 shows the first component of normalised monthly surface pressure departures over a global grid for the period 1951–60.

Kidson's analysis suggests that the most important large-scale monthly departures of pressure, temperature and rainfall throughout the tropics are associated with Walker's Southern Oscillation and confirms that the patterns of variation associated with the Southern Oscillation extend to the high latitudes of both hemispheres. Kidson observes also that 'while both pressure and temperature show coherent departures on a hemispheric scale rainfall components are regional in character . . .'.

Impact on climatic theory
Interpretation of climatic variation

Recognition of these patterns of influence of circulation anomalies on local climate enables some light to be shed on the question of the statistical significance of climatic variations and the nature of the climatological

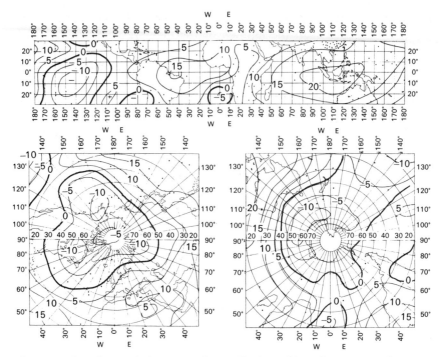

Fig. 4.3.3. First characteristic pattern of normalised monthly surface-pressure departures defined over a global grid after Kidson (1975 a) using data for the years 1951–60 inclusive. This pattern is correlated with variations in the Southern Oscillation or Walker Circulation. Note the difference between the Northern and Southern Hemisphere patterns.

time series, because it reveals the very limited number of pieces of independent information contained in a large spatial array of climatic data.

For instance, applying the simple Student t-test to the difference between 1941–74 and 1913–40 mean rainfalls for the various Australian rainfall districts, we find some ten or more districts which show apparently significant differences at the 95% confidence level. These districts are, however, concentrated in southeast Australia in the centre of the first characteristic pattern of rainfall variations, so that, in fact, the time variation in each of these districts closely reflects that of the Southern Oscillation. We are therefore dealing not with ten independent variables but with a single variable parameter of the general circulation.

Thus the fact that ten or so districts reflect such a variation does not add to its statistical significance on the continental scale. Any random or non-random selection and testing of data from these districts may well show apparently significant local changes, even though the variations in the general circulation may in fact be explicable in terms of some random process, plus persistence (see Curry, 1962; Mitchell, 1966) or some much longer-term causality (Yevjevich, 1971). What is essentially a random-walk stochastic process on the largest space scale may well be locally enhanced by topography into an apparently significant variation on a smaller space scale.

Application to longer time scales

Having recognised the characteristic patterns of year-to-year variations in climate, and some associated mechanisms of the general circulation, it is useful to ask whether the same patterns, and by inference the same mechanisms, operate to produce variations on longer time scales.

In the case of Australia (Pittock, 1975) it turns out that the change which occurred in rainfall over southeast Australia between the two roughly 30-year periods before and after the mid 1940s corresponds rather well to the pattern associated with year-to-year variations in the Southern Oscillation. Indeed the pattern of change between the periods 1881–1910 and 1911–40 as found by Gentilli (1971) is remarkably similar also (although of opposite sign).

This suggests that the pattern associated with the Southern Oscillation is indeed important on longer time scales, and that this should be taken into account in considering climatic changes throughout the Holocene and perhaps earlier in the Quaternary. Changes in sea level would, of course, have more drastic effects on the regional land–sea contrasts and ocean currents, adding plausibility to the argument by Quinn (1971) that the Southern Oscillation during the Quaternary ice ages would be locked in the positive mode, which would tend to offset the aridity in northern Australia and New Guinea which would otherwise result from lower sea-surface temperatures. Particularly wet phases might arise when the Timor and Arafura Seas and the Gulf of Carpentaria were shallower and warmer than at present.

Significant shifts in the mean latitude of the subtropical high-pressure belt would likewise be reflected in rainfall changes of opposite sign on the eastern and southern coasts of Australia. This suggests caution in using results from particular sites in these areas as representative of general trends and gives priority to the examination of 'conjugate' sites in order to determine the importance of this second pattern on longer time scales.

Implications for climate modelling

Numerical modelling of climate is currently being tackled in a number of ways (see Chapter 5). One is the running for long periods of numerical weather prediction models, which explicitly predict the behaviour of individual high- and low-pressure systems, and the averaging or other

4.3   Shorter-term variations

statistical treatment of the results over seasons or years. Such an approach is very expensive in terms of computer time and strenuous efforts are being devoted to simplifying these models and parameterising the individual weather systems by treating them statistically as eddies. Other approaches seek to start from basic energy-balance considerations to model mean states directly.

Characteristic pattern analysis suggests an approach based on modelling only those processes which directly determine the amplitudes or behaviour of the dominant few patterns which account for most of the variance. If, for instance, we could directly model the behaviour of the Southern Oscillation and the latitude of the subtropical high-pressure belt, i.e. two basic variables, we could account for more than 50% of the total variance in Australian rainfall anomalies (Pittock, 1974 a).

Local anomalies and anthropogenic effects

A great deal of evidence has been accumulating from satellite cloud photographs, meso-scale synoptic analyses, and other means, which demonstrate the large local spatial variations in climate which can be induced by topography, e.g. simple orographic effects (Lee, 1911; Coote & Cornish, 1958; Schermerhorn, 1967; Chuan & Lockwood, 1974), valley effects and topographic convergence (Wilson & Atwater, 1972; Longley, 1974; Huff *et al.*, 1975), lake effects (Changnon & Jones, 1972) and even lake-breeze convergence (Harman & Hehr, 1972). These effects are, in general, dependent on weather type and prevailing wind direction, and hence vary with season and with longer-term climatic variations.

The lake-breeze effect on precipitation found by Harman & Hehr (1972) for instance, depends on lake-breeze convergence operating in synoptic situations normally found in summer over the Upper Peninsula of Michigan. Harman & Elton (1971) have suggested that just such a mechanism might operate to produce the much discussed La Porte anomaly which is most clearly manifest in summer thunderstorm activity (Changnon, 1968). The more conventional explanation is either observer error (Holzman, 1971) or pollution effects (Changnon, 1973 a and b). The latter explanation is supported by observed trends of La Porte precipitation with time which broadly parallel those of nearby (upwind) steel production. These trends, however, could be the result of a trend in the general circulation of the atmosphere such as that found by Wahl (1968) and Eichenlaub (1971) to have been operative over North America, which has either moved the mean position of the summer lake-breeze front nearer to La Porte or increased the instability or moisture supply associated with it.

The carry-over of assumptions from La Porte to the large Metromex experiment designed to measure urban effects on climate at St Louis and

176

elsewhere (see e.g. Huff & Changnon, 1972) must raise serious questions as to the magnitudes and causes of other alleged urban effects. There is no lake-breeze effect at St Louis, but in common with many large cities, St Louis is built in a valley with upslopes downwind.

Wilson & Atwater (1972) concluded from their study of topographic effects on rainfall over Connecticut that 'large variations in rainfall can frequently occur over short distances over terrain of minimal relief ($\sim$ 150 m) with a highly repetitive pattern from storm to storm'. Indeed Huff *et al.* (1975) have themselves come to a similar conclusion, finding that 15% more warm season precipitation falls on hills in southern Illinois than on farm flatlands only 120 m lower in elevation both north and south of the hills. This is roughly the magnitude of the supposed urban effects on precipitation, so it is clear that strenuous efforts should now be made to separate out the two effects.

Increases in precipitation anomalies with time are, of course, strong circumstantial evidence for anthropogenic effects, but as we have seen, changes in the general circulation may even lead to selectively enhanced changes in precipitation. A case in point is the apparent anomalous increases in rainfall in parts of Washington State between 1929 and 1966 which Hobbs *et al.* (1970) ascribed to cloud condensation nuclei produced by pulp and paper mills and various other industries.

The possibility of orographic enhancement of secular trends in rainfall in the Washington State area has been checked (Pittock, 1977) by determining the annual mean latitude of the maximum surface high pressure down the west coast of North America from 1941 through 1970 and correlating annual rainfalls for individual stations against this. The resulting distribution of correlation coefficients is shown in Fig. 4.3.2. Bearing in mind that Hobbs *et al.* (1970) include stream flow data, we can see that significant negative correlations occur in most of the areas found by Hobbs *et al.* to have had anomalously large increases in precipitation. A trend analysis on the latitude of the high-pressure ridge shows that it indeed moved about one degree per decade equatorward over the years 1941–70, which is about the right magnitude and direction to account for most of the observed changes in precipitation. Inspection of a topographic map of the area shows that this result is not surprising, since most of the areas of industrial activity identified by Hobbs *et al.* are in situations where convergence and uplift are likely to be pronounced or at least strongly affected by changes in prevailing wind direction.

In the case of the Washington State study, Hobbs *et al.* produce undoubted evidence that pollution has increased the concentration of cloud condensation nuclei. The point at issue, however, is how much this, rather than natural climatic variability in space and time, has contributed to the observed pattern of rainfall changes. On the more general question of urban effects, there is somewhat conflicting evidence from studies of variations of rainfall with days of the week (e.g. Changnon, 1970, 1973 *b*: Lawrence, 1971; Shulman & Brotak, 1973), and from particular case studies (e.g. Atkinson, 1971), that there may be real urban effects on rainfall. Again, the need is to separate out the effects of human activity from the patterns of meso-scale climate and its variability, which would seem to be a major contributor to many supposedly urban effects.

Other applications

An understanding of the larger-scale patterns of climatic variability has important application to contingency planning for climatic fluctuations or change. This is discussed more fully in Section 7.1.

Another possible application is to the evaluation of weather modification experiments, where one could use historical regression relations between target areas and one or more parameters of the general circulation to 'predict' an 'expected' rainfall for the experimental period in the hypothetical absence of seeding. The usefulness of this approach could be further improved by adopting the same procedure with a 'control' area having a similar relationship between precipitation and the general circulation, and calculating a double ratio of actual to expected precipitation in the target area compared to that in the control area.

Finally, in relation to local weather forecasting it is worth noting that the typical patterns of climatic variation associated with particular anomalies in the general circulation, contain useful information right down to the meso-scale influences of topography which enhance and localise anomalies. Understanding these will thus put us well on the way to being able to interpolate and interpret on the meso-scale the broad-scale information contained in the large grid-size general circulation numerical weather forecasting model outputs.

Such interpolations and interpretations are commonly made today by the skilled forecaster who is familiar with the forecast area. However, this skill presently takes considerable time to acquire either from personal experience or from that handed down by others. Sets of detailed meso-scale maps of characteristic anomaly patterns associated with various general circulation anomalies could provide a useful short-cut to this knowledge.

Concluding remarks

It was with a rare touch of scientific frankness that Schmidt-Ten Hoopen & Schmidt (1951) stated 'We also compared the precipitation of West Java with the winter temperatures of De Bilt (Holland)...There seems to be no connection.' Nevertheless the literature abounds in cases where such seemingly unlikely 'teleconnections' are established, and we must conclude that the observed patterns of climatic fluctuations from year to year and place to place are to some extent organised on a global scale.

Such teleconnections are the natural consequence of fluctuations in the global-scale characteristics of the general circulation, most notably those due to fluctuation in the polewards transport of momentum and sensible and latent heat from the tropics. Dominant amongst these fluctuations are those related to changes in the standing-wave patterns induced by land–sea and topographical contrasts.

The local expressions of these large-scale patterns of fluctuation are strongly influenced by topography. Thus certain areas are particularly sensitive to changes in the general circulation and often appear as anomalous. This has led some workers to ascribe local spatial and temporal climatic variations to anthropogenic or other causes when such 'anomalies' are at least in part natural local manifestations of broader-scale phenomena.

The insights to be gained from a full appreciation of the nature of the interactions between the general circulation and climate on smaller space scales are of potential value in many areas related to the theoretical and practical nature and consequences of climatic change. A number of these practical applications have been suggested above.

It is worth stressing however, that the effect of the topography is to amplify quite small changes in the general circulation so that much larger magnitude effects are often observed in particular locations. This applies particularly to precipitation in mountainous areas, and thus to watersheds and runoffs. The differential nature of the water balance relationship adds to the critical nature of these effects. It is therefore not surprising that the most dramatic fluctuations in observed climatic variables are those affecting water supply and river flow. The social and economic consequences of these fluctuations loom larger as populations grow and existing water resources are increasingly utilised.

The implications of the existence of spatial patterns for the organisation and interpretation of palaeoclimatic data should be emphasised. These involve the choice of priority sites for examination in order to test particular climatic reconstructions, the careful interpretation of highly localised palaeoclimatic data such as much in the Australian record, and greater consideration of the role of standing-wave patterns in the reconstruction of palaeoclimates in the Southern Hemisphere.

In the Australasian context the effects of the varying exposure of the Sunda and Sahul shelves north of Australia and the consequent variations in latent and sensible heat inputs would seem to be crucial. Carefully dated absolute chronologies of sea-level changes, linked with similarly dated evidence from Australia of wet and arid periods, are vital to this task.

4.4   The Southern Oscillation
P. B. WRIGHT

The Southern Oscillation is a fluctuation of the atmospheric circulation with irregular period. It may be summarised as a variation in the strength of the 'Walker Circulation', a circulation in the plane of the equator involving ascent over the Indonesian sector and subsidence over the eastern Pacific. When the Walker Circulation is strong, sea-surface temperatures in the equatorial eastern Pacific are low, rainfall and cloudiness are low in the same area but high over Indonesia and Australia, and there are associated anomalies of weather and circulation in many parts of the world. When the Walker Circulation is weak, the reverse anomalies occur. The strength of the Walker Circulation fluctuates irregularly but with a characteristic period of two to five years, and either of the extreme states, once established, usually persists for a year or more. A useful review is given by Boer & Kyle (1974). The importance of the Southern Oscillation is illustrated by Kidson's (1975 b) finding that it accounted for 9.9% of the variance of standardised monthly pressure anomalies over the globe as a whole during 1951–60, and 18.4% of that in the tropics. For seasonal anomalies the percentage would be even higher.

If the mechanisms of the Southern Oscillation could be understood, predictions of seasonal weather and circulation anomalies in many parts of the world might be possible up to two years ahead. There are three general aspects requiring explanation:

(i) the existence of spatial correlations ('teleconnections') over large parts of the globe;

(ii) the reason for the persistence of either of the extreme states, once established; and

(iii) the mechanism for the changeover from one state to the other.

The teleconnections can mostly be explained as resulting from variations in the Walker cell (involving, for example, variations in heat and moisture supplied to the upper troposphere in the Indonesian sector) and in the Pacific Hadley cell (which is weaker when the equatorial sea is colder, i.e. when the Walker cell is strong). Rowntree (1972) has simulated many of them numerically. The cause of the persistence is also reasonably well established, following the work of Bjerknes (1969). It is a good example of a feedback mechanism. The cold sea surface in the equatorial eastern Pacific favours a strong Walker Circulation; a feature of the latter is strong easterly winds in the equatorial eastern Pacific; these winds produce Ekman divergence and strong upwelling along the equator, so maintaining the cold sea. On the other hand, the cause of the changeover has not been elucidated, although it seems certain that this also involves feedback between atmosphere and ocean.

180

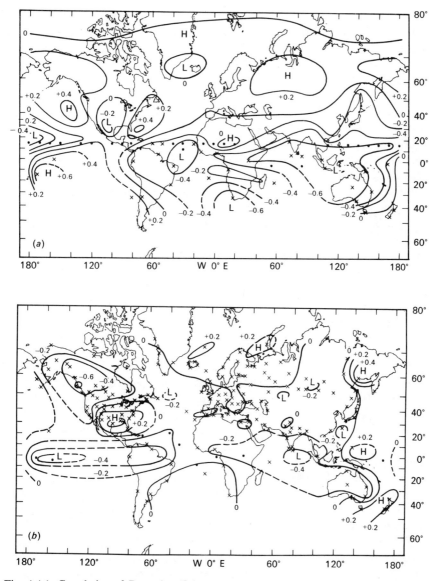

Fig. 4.4.1. Correlation of December–February pressure (*a*) and air temperature (*b*) with simultaneous Southern Oscillation Index: × indicates stations or grid points with at least 30 years' data; • indicates stations with less than 30 years' data. Complete coverage was available north of 20° N in the case of pressure.

181

4.4   Shorter-term variations

Mapping the Southern Oscillation

The Southern Oscillation may be mapped on a global basis by correlating pressure, temperature and other parameters with an appropriate index of the Oscillation. The one used here is a weighted mean of pressures at Cape Town, Bombay, Djakarta, Darwin, Adelaide, Apia, Honolulu and Santiago. Full information is given by Wright (1975). The season is taken as the time unit with maps prepared for each season separately, because the pattern of the Southern Oscillation is known to differ through the year, substantially in some regions.

Figure 4.4.1 illustrates the effect of the Southern Oscillation on pressure and temperature over much of the world in December to February. When the Walker Circulation is stronger than usual (high index), pressure is relatively low over the Indian Ocean, Indonesia and Australia, and high over most of the eastern Pacific and southwest North Atlantic, while temperature is low over most of the Tropics, Australia and Canada, and high over the United States. Effects over Australia are among the strongest. As already indicated in Section 4.3, Australia also exhibits some important rainfall correlations. The actual values differ from season to season and with the choice of index; for example, Adelaide rainfall for the June–August period is positively correlated (correlation coefficient +0.48) with the Southern Oscillation Index used here.

Lag relationships

A thorough mapping of lag relationships has not been undertaken before, although Walker noted a few in his studies during the 1930s, including a tendency to low sea and air temperatures in the Indian Ocean region after a period of high index. Some lag relationships have been invoked as the basis for speculative hypotheses about the cause of the changeover. Most such hypotheses have suggested a movement of oceanic waters (or at least, advection of surface temperature anomalies) in the Pacific, and are quite plausible, but will be difficult to verify without much oceanic data. The limited sea-temperature evidence examined refutes the suggestion of a gradual movement of anomalies over a period of many months.

Figure 4.4.2 illustrates a few lag relationships of the Southern Oscillation Index with air temperature and sea-surface temperatures. Clearly there are lag effects in several parts of the world. It is unlikely that they are *all* involved in the feedback which leads to the changeover. The problem resolves itself into determining which lag effects are involved, and which of several plausible mechanisms is the true one. It may be pointed out that such a widespread occurrence of lag effects is fully to be expected. The persistence of an atmospheric anomaly, such as a strong Walker

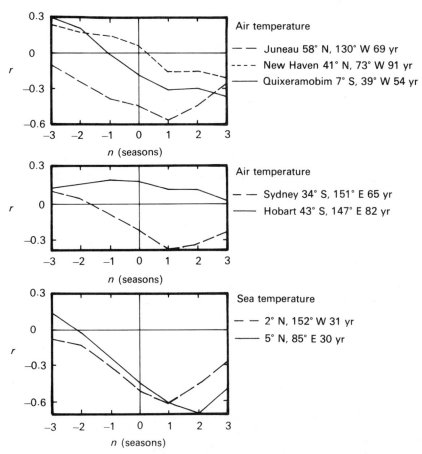

Fig. 4.4.2. Correlation, *r*, of various parameters in December–February with the Southern Oscillation Index '*n*' seasons previously. For example, the value plotted against '3' is the correlation of December–February parameter with the previous March–May index. The 5% significance levels are approximately ±0.21 for 90 years' data, ±0.28 for 54 years' and ±0.37 for 30 years'.

Circulation, for several months will cause anomalies to be set up in the state of the surface, and some of these may persist after the atmospheric anomaly has ceased. These surface anomalies will then affect the atmosphere, resulting in some lag effects in the circulation also. If one of these effects involves the Walker Circulation, there exists a potential mechanism for the changeover. Thus there exists sufficient possibility for a feedback mechanism involving only the atmosphere and the ocean surface, without any need to postulate advection of anomalies. Moreover it will be shown below that there is sufficient scope even if consideration is restricted to the equatorial zone.

183

Mechanism of the Oscillation

The Walker Circulation is driven primarily by the surface temperature difference between the Indonesian sector (warm) and the east Pacific (cold). Now one of the more notable lag relationships is in the Indian Ocean; during a period of high index, this region gradually cools. The main factor in this is probably that the relative cloudiness of this sector during high index reduces the input of solar radiation. The cooling implies a reduction in the 'driving force', and so the Walker Circulation weakens; but this weakens the Pacific upwelling, the temperature rises there, reducing the 'driving force' still further, until eventually a low index state is established. This then favours a warming trend in the Indian Ocean, and the second half of the cycle takes place.

This process is quite plausible as the dominant mechanism, and agrees with the data so far investigated. However, it needs two important refinements. The first is to postulate a limit to the amplitude of the swings. The second is to bring in some factor which will allow the Indian Ocean sector to cool for some time before the index starts to fall; because the atmosphere reacts to the surface within a few days, and the upwelling probably responds to the atmosphere within a few weeks at most, so the above mechanism alone would have a period of not more than a few months.

The Walker Circulation may be more accurately regarded not as a simple cell, but as a complex set of vertical motions involving the whole equatorial zone. If so, the strength of the easterlies in the Pacific will be a function not only of the temperatures of the eastern Indian Ocean and the eastern Pacific, but of the complete equatorial temperature profile. In this connection it is interesting to note that the Atlantic sea-surface temperature shows significant lag relationships, different from those of the Indian Ocean. It is then reasonable to suggest that the Southern Oscillation involves an interplay between these three sectors.

A mechanism of this sort is just as plausible as any others that have been suggested, and it is relatively easy to test. Conclusive testing must await the development of suitable numerical models including atmosphere–ocean coupling. However, the coupling need not be elaborate; one needs merely a reduction of temperature along the equator related to the strength of easterly winds (to represent upwelling) and a much slower cooling of the surface under cloudy conditions. Meanwhile, identification of the lag relationships will greatly improve prospects for seasonal forecasting.

4.5 Regional mechanisms and variations

The previous sections of this chapter have dealt with general, statistical and historical aspects of shorter-term climatic change and variability and have examined these in relation to the global atmospheric circulation. It remains to consider finally some regional features of the responsible circulation mechanisms and of recent climatic variations in the Southern Hemisphere. The results of an analysis of variations in the Atlantic–Pacific circulation and their relation to extreme climatic events in the Tropical Americas and Africa are described by Hastenrath in the following paragraphs. Williams and van Loon report on an examination of trends of winter temperature in the Southern Hemisphere over the 15-year period 1955–69. Finally Coughlan summarises a recent examination of evidence for climatic change in Australian meteorological data (rainfall and temperature).

*Variations in the Atlantic–Pacific circulation**
S. HASTENRATH

Among the most spectacular variations of tropical circulation and climate are such extreme atmospheric/hydrospheric regimes as the El Niño events (high ocean temperatures and torrential rainfall) on the Pacific coast of South America, the Sêcas (drought) of northeastern Brazil, and the recent subsaharan drought. Less publicised, in the Central American–Caribbean region, runs of predominantly wet years have commonly alternated with periods of prevailingly dry ones, throughout the twentieth century. Variations in rainfall productiveness have had important consequences for the large-scale water budget, natural vegetation and land use. A portion of the tropical belt extending from the African Continent through the Atlantic and the Americas to the eastern Pacific Ocean has been studied with emphasis on the Central American–Caribbean region and with a view towards a causal understanding of these extreme climatic events in terms of the large-scale circulation.

Observational basis

As a major basis for the study, ship data comprising more than seven million individual observations for the period 1911–72 were obtained on magnetic tape from the US National Climatic Center at Asheville, N.C. Monthly mean surface pressure charts for the period 1899–1972 derived from the historical weather maps of the Northern Hemisphere were

* This study was supported by the National Science Foundation Office for Climate Dynamics.

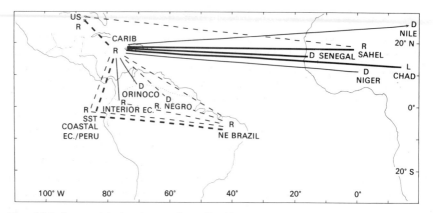

Fig. 4.5.1. Large-scale interconnections. Significant correlations (5% chance probability level) heavy, others thin lines; positive solid, negative broken. R = rainfall, SST = sea-surface temperature, L = lake level, D = river discharge.

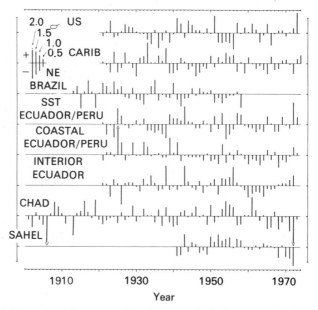

Fig. 4.5.2. Normalised departure of hydrometeorological parameters plotted in terms of standard deviation σ. US: annual precipitation at 26 stations in the US Central Great Plains. CARIB: annual precipitation at 45 stations in the Central American–Caribbean region. NE Brazil: annual rainfall at 40 stations. SST Ecuador/Peru: annual sea-surface temperature departure along the coast of Ecuador and Peru. Coastal Ecuador/Peru: annual rainfall at 9 stations along the Pacific coast of Ecuador and Peru. Interior Ecuador: annual rainfall at 6 stations in the interior of Ecuador. Chad: annual changes in lake level. Sahel: annual rainfall at 20 stations in subsaharan Africa.

186

acquired from the same agency. Climatic records at long-term land stations, and series of lake levels and river discharge compiled from a variety of sources, were a further important part of the data base.

Large-scale rainfall variations

Climatic events in low latitudes are primarily reflected in the rainfall activity. Fortunately, this is at the same time the most significant, most robust, and most commonly available element, but care is needed to ensure areal and temporal representativeness. To this end, regions homogeneous with regard to the occurrence of unusual years were delineated. Only entire rainy seasons were considered. An index of rainfall departure was compiled for each homogeneous region. This procedure was applied to collectives of long-term stations in selected regions as indicated in Figs. 4.5.1 and 4.5.2.

Figure 4.5.2 shows that extreme weather conditions tend to come in sequences of several years that are predominantly either wet or dry. Thus, in the Central American–Caribbean region (CARIB) the early and mid thirties and the mid fifties to early sixties were abundant in rainfall, whereas the early to mid forties and the late sixties to early seventies were comparatively dry. Extreme years in this region may be identified as follows: 1925, 1930, 1939, 1940, 1946, 1947, 1957, 1959, 1965, 1972, as dry; and 1924, 1927, 1932, 1933, 1936, 1938, 1954, 1956, 1960, 1966, as wet.

Figures 4.5.1 and 4.5.2 illustrate the parallelism in the quality of the rainy season in the Central American–Caribbean region and the African monsoon belt, and the inverse behaviour of precipitation in the US Central Great Plains, in particular. Positive linkage with river catchments broadly decreases equatorward, and a weak positive correlation is still indicated with rainfall in the mountainous interior of Ecuador; negative linkage is particularly strong with Ecuador–Peru coastal sea-surface temperatures, but weaker with rainfall. Precipitation along the Ecuadorian–Peruvian littoral parallels coastal sea-surface temperature anomalies, and is essentially unrelated to the interior of Ecuador. Rainfall in northeastern Brazil shows weak negative correlations with the Central American–Caribbean region and the Ecuadorian Andes; it is strongly negatively correlated with sea-surface temperatures off the Ecuador–Peru coast, but not with coastal rainfall. It is recognised that the behaviour revealed by the correlation does not exclude the occasional coincidence of extreme events in the different regions. Thus, Caviedes (1973) noted the occasional simultaneity of Sêcas in northeastern Brazil and El Niño in northwestern Peru. Figures 4.5.1 and 4.5.2 do not suggest that such a pattern is followed consistently, although the sizeable negative correlation between northeastern Brazil rainfall and Pacific sea temperatures is noteworthy.

187

4.5 Shorter-term variations

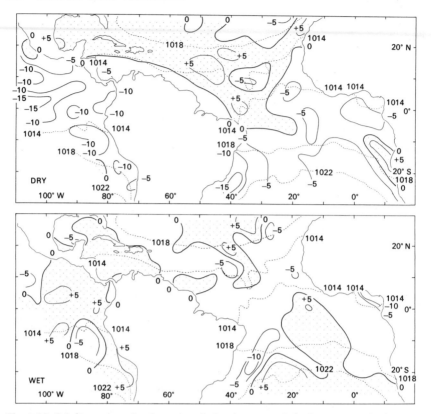

Fig. 4.5.3. July/August sea-level pressure during the composites of ten extremely dry years (top) and ten extremely wet years (bottom) expressed as departures from the 60-year mean pattern (thin broken lines, mb). Areas of positive departure are stippled. Isopleths are at intervals of 0.5 mb.

Atmospheric circulation of drought and flood regimes

Composite pressure maps for the ten driest and ten wettest years in the Central American–Caribbean region are shown in Fig. 4.5.3 for July/August, the middle of the rainy season in the area. The maps portray departures from the 60-year mean patterns, being understood as extreme composite minus long-term mean. For reference the 60-year mean fields are also entered with light signature and only wide isopleth interval. Atmospheric circulation during composites of extreme dry and wet years shows significant but only small departures from the long-term mean conditions. Similar maps were constructed for wind speed and other elements.

During deficient rainy seasons in the region (Fig. 4.5.3, top), the North Atlantic high extends farther southward, meridional pressure gradients

188

are steeper on its equatorward side, and the Trades are stronger, albeit in a band further south. Over the eastern Pacific, the Intertropical Convergence Zone (ITCZ) is also at a more equatorward position, as is borne out by an examination of pressure, wind, and divergence fields. This is paralleled by the varying frequency of weather systems: temporales, storms originating in the ITCZ over the eastern Pacific, have nearly disappeared from Central America since the early 1960s. Sea-surface temperature also has a significant departure pattern, although maps are not shown here. The North Atlantic in the realm of the Canary Current and in a band equatorward of 10° N is colder than in the long-term mean, whereas the Southern Hemispheric and equatorial regions of the Atlantic and all of the eastern Pacific are colder.

During abundant rainy seasons in the Central American–Caribbean region (Fig. 4.5.3, bottom), departure patterns are broadly inverse to those characteristic of the dry composite. The North Atlantic high extends less southward, meridional pressure gradients on its equatorward side are slacker, and the Trades are weaker. The ITCZ over the eastern Pacific stays further away from the equator. The sea-surface temperature pattern is characterised by positive departures in an area extending from the eastern North Atlantic to the western equatorial Atlantic and the Central American seas, and negative anomalies in the equatorial eastern Pacific.

Departures in low-latitude circulation during winters preceding unusual rainy seasons in the Central American–Caribbean region are also significant. During the northern winter preceding a deficient rainy season in the region, a southward extension of the North Atlantic high, a strengthening of the Trades, and an equatorward shift of the ITCZ over the eastern North Pacific as compared to the 60-year mean conditions, are already apparent. Likewise, the equatorward flank of the South Atlantic high has weakened already, precursory to the July/August pattern. Sea-surface temperature is higher than the long-term mean over the eastern South Atlantic and in an area extending from the Caribbean and Gulf of Mexico into the Gulf Stream region. Negative sea-surface temperature departures are found in the low-latitude Atlantic, but also in the equatorial eastern Pacific. While the sea-surface temperature departures have broadly the same sign as during the deficient rainy season to follow, the sign reversal in the equatorial eastern Pacific is noted as an exception. This reversal takes place during the spring transition.

During the northern winter preceding an abundant rainy season in the Central American–Caribbean region, a corresponding build-up or preparation in the pressure and wind patterns is noticed. This holds also for sea-surface temperature in most areas. Thus, the waters of the eastern North- and the equatorial Atlantic are anomalously warm; much of the South Atlantic, the waters off northeastern South America and an area

189

extending from the Gulf of Mexico into the Gulf Stream region are too cold. However, the eastern Pacific has positive sea-surface temperature anomalies at the height of the northern winter preceding an abundant rainy season in the Central American–Caribbean region, a pattern to reverse during the spring transition. In this respect the wet and dry year composites are again complementary.

Departure maps indicate the important role of variations in the sub-tropical highs and the major ocean currents on sea-surface temperatures in the low-latitude Atlantic and Pacific. Thus, equatorward advection of cold waters in the domain of the Canary and Benguela Currents appears favoured by an equatorward expansion of oceanic highs in the respective hemispheres. Thermal control from the South Atlantic reaches well into the equatorial region. In turn, the extension of the South Equatorial Current may contribute to sea-surface temperature anomalies during northern winter further downstream, in the Caribbean, Gulf of Mexico, and the Florida–Gulf-Stream system. Atmospheric forcing with seasonal lag as suggested by Wyrtki (1973 b) and Namias (1973 a) may be active over the eastern Pacific.

As illustrated by Figs. 4.5.1 and 4.5.2, rainfall in the Central American–Caribbean region has a strong correlation with sea-surface temperature and precipitation on the coast of Ecuador and Peru. The maps composited according to extreme dry years in the Central American–Caribbean region can therefore be expected to resemble, albeit not duplicate, the patterns resulting from a stratification according to Ecuador/Peru El Niño years. While the term El Niño implies a warming starting as early as Christmas, sea-surface temperature time-series actually reveal a marked tendency for sudden anomalous warming to be preceded by strong negative departures. This circumstance seems pertinent in appraising the opposing temperature anomaly patterns apparent in the winter and summer charts composited for dry years in the Central American–Caribbean region. Of further interest in this connection is evidence presented by Quinn (1974), to the effect that major El Niño events tend to be preceded by an abrupt change in atmospheric pressure distribution over the eastern South Pacific.

Rainfall in the Central American–Caribbean region is negatively cor-related with northeastern Brazil (Figs. 4.5.1 and 4.5.2). The study of extreme climatic events in northeastern Brazil shows an equatorward extension of the North Atlantic high, and a southward shift of the ITCZ during extreme wet years; at the same time, the equatorward flank of the South Atlantic high weakens.

A particularly strong positive linkage is indicated between rainfall in the Central American–Caribbean region and subsaharan Africa. Composite maps stratified according to extreme dry and wet years in the Sahel belt indeed show departure patterns very similar to those for composites of extreme climatic events in the Central American–Caribbean region.

190

Trends of winter temperature in the Southern Hemisphere
J. WILLIAMS & H. VAN LOON

As part of a study aimed at refining earlier estimates of global temperature trends, an analysis was made of the relationship between changes in temperature over different regions and changes in the atmospheric circulation during the same period. Initially, regression lines were computed using overlapping 15-year sets of winter temperature and pressure data from North America for the period 1900–69. The geographical distribution of the slopes of the regression lines of temperature in each 15-year period (1900–14, 1905–19, . . . , 1955–69) showed that: in no period was the winter temperature going down (or up) over the whole of North America; since 1950 the predominant trend had been downward but there were still areas with an upward trend. Comparison with pressure-trend distributions showed that temperature trends were indeed related to circulation changes, so the analysis of the trends was extended to the whole Northern Hemisphere (van Loon & Williams, 1976a, b). For the Southern Hemisphere an analysis of temperature and pressure trends was undertaken for the period 1955–69 and the results are summarised below.

Data and data treatment

The Southern Hemisphere temperature and sea-level pressure data were station data from the World Weather Records. For each station regression lines were calculated for both temperature and pressure. The slopes of the regression lines were plotted on maps so that the geographical distribution of the trends could be examined.

Temperature trends at two Antarctic stations

Winter (June–July–August) temperature data for the period 1955–69 and the regression lines through them are shown in Fig. 4.5.4 for the South African National Antarctic Expedition (SANAE) station and for McMurdo. These two stations in Antarctica are almost 180° in longitude apart. It is therefore interesting to note that within this 15-year period they have opposite trends. The slope of the regression line at SANAE is −0.26 °C/yr, while that of McMurdo is +0.25 °C/yr.

Burdecki (1969) in describing the climate of SANAE has documented the change in station location of 20 km in 1962. The small change of geographical position and height above sea level are not obvious in the temperature data. The periods 1957–60 and 1962–9 both exhibit downward trends (−0.68 °C/yr and −0.48 °C/yr respectively).

Figure 4.5.4 illustrates the fact that during a single 15-year period the temperature does not have the same trend at all stations. The two trends

191

4.5 Shorter-term variations

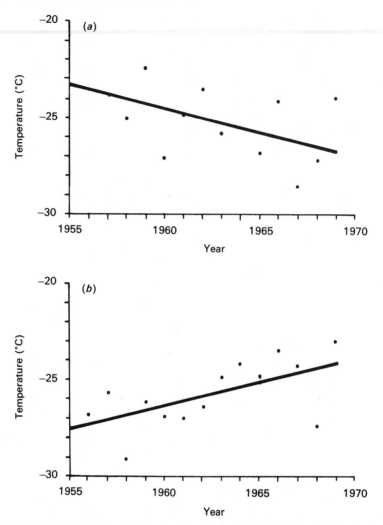

Fig. 4.5.4. Winter (June–August) temperature data (°C) for 1955–69 and the regression lines through them for (a) SANAE (70.5° S, 2.9° W) and (b) McMurdo (77.8° S, 166.6° E).

are the largest found in the Southern Hemisphere in this period and are found on opposite sides of Antarctica.

Temperature and pressure trends over the Southern Hemisphere

Figure 4.5.5(a) shows the geographical distribution of temperature trends in the Southern Hemisphere winter for the period 1955–69. Since data are unavailable over wide ocean areas, a complete hemispheric map is not

192

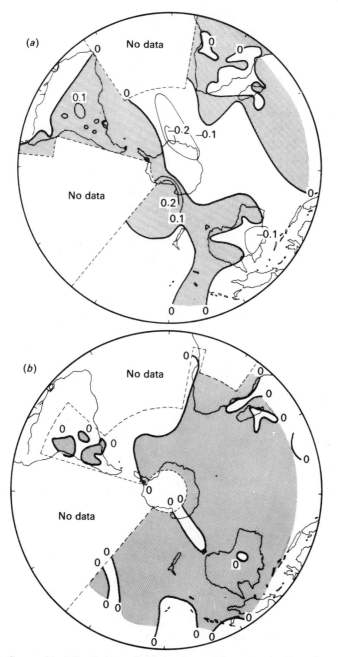

Fig. 4.5.5. Geographical distribution of: (*a*) the slope of the regression line of temperature (°C/yr), (*b*) the trend of sea-level pressure, in the Southern Hemisphere winter for the period 1955–69. Stippled areas indicate upward trend.

193

possible. The distribution of sea-level pressure trends compiled from available station data is shown in Fig. 4.5.5(*b*). Areas with no data coverage are more extensive in the case of pressure observations.

The distribution of temperature trends (Fig. 4.5.5(*a*)) suggests that south of 40° S much of the change occurs within wave number one. Vowinckel & van Loon (1957) found that the temperature change between 1920–34 and 1945–55 over the Antarctic Ocean was positive from 40° W eastwards to 180°, and negative elsewhere. This pattern also suggests that much of the change is occurring in wave number one. Van Loon & Jenne (1972) show that, at 500 mb between 45° S and 65° S, wave number one accounts for 75–95% of the variance of geopotential height. Thus it is possible that the temperature changes to a large extent are occurring in the climatologically dominant wave, as was found also for the Northern Hemisphere.

For the geographical distribution of pressure trends, the limited amount of data makes it difficult to contour the map. Over the Indian Ocean and Australasian region, pressure is rising during the 1955–69 period. Over Australia the trends are largest over the south east (+0.25 mb/yr). In the Indian Ocean, the trends are greater than 0.20 mb/yr between 35° E and 80° E and 35–50° S. Over the parts of South America and the South Atlantic for which data are available, the pressure is generally falling or the trend is slightly positive. The lack of data prevents any conclusions being drawn about the role of advection in the temperature changes in the way that was possible in the Northern Hemisphere (van Loon & Williams, 1976*a*, b).

Changes in Australian rainfall and temperatures
M. J. COUGHLAN

Continuous instrumental records of climatological elements in the Australasian region span in general 100 years or less. Reliable records of 60 years or more, while comparatively large in number for rainfall, are very few for temperature. A study has been made of Australian rainfall and temperature data, paying particular attention to the reliability and homogeneity of the series, to assess the evidence for recent climatic change and variability in the Australian region.

Rainfall trends

The study of rainfall trends was based on a set of 107 district monthly rainfalls from 1913 to the present. Certain districts, after preliminary analysis, were found to have had changes in station lists which resulted in obvious and spurious trends. These districts were omitted.

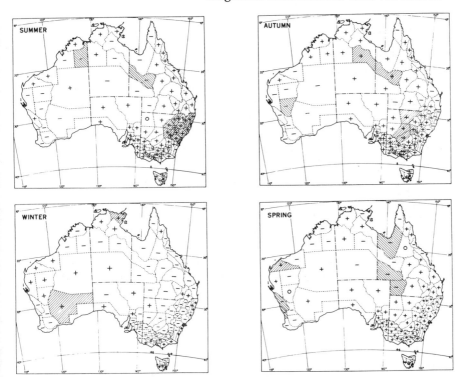

Fig. 4.5.6. Geographical distribution of the sign of the Kendall rank statistic applied to seasonal rainfall from 1913/14 to 1973/4. Hatching indicates districts with significance (95% level) in either rank statistic or linear trend.

Figure 4.5.6 shows the sign of Kendall's rank statistic, which is an indication of the sign of the trend – not necessarily linear – for seasonal totals. Those districts which demonstrated significant linear regression or significant rank statistic (95% confidence level) are shown as hatched. It may be seen immediately that trends can vary from season to season in any one district. The spatial coherence in areas where districts are small, strengthens confidence in the data, i.e. the changes are real, for example, over northeast New South Wales in summer. In areas where districts are large, however, confidence is lower, for example, inland southwestern Australia in winter. It should be noted further that this spatial coherence or correlation between districts, which appears to vary with season, makes it difficult to assign a reliable number of degrees of freedom for tests of statistical significance. Perhaps the most striking seasonal variation in trend is that over much of eastern Australia between winter and the other seasons. The upward trends in spring, summer and autumn rainfall in southeastern Australia would appear to be mainly the result of the quasi-abrupt change in rainfall during the mid 1940s, discussed in Section 4.2.

4.5   Shorter-term variations

Rainfall periodicities

Power-spectrum analysis on the Australian district rainfall data set indicated that significant periodicities, i.e. in the statistical sense, were to be found. The quasi-periodicity most frequently found was that around 2.2 years, and can be cited as an example of a manifestation of the well-known quasi-biennial oscillation (Landsberg et al, 1963; Trenberth, 1975 a). A second quasi-periodicity centered around 3.1 years also featured prominently in the analysis. The periodicity of around 6 to 7 years, quoted in some studies (O'Mahony, 1961), was not strongly in evidence in this analysis. It is possible, however, that it may have appeared as the shorter period harmonic at around 3.1 years.

There is no convincing evidence of the 11-year and 22-year sunspot cycles. There were certainly peaks covering these periodicities in the rainfall spectra for certain areas – generally in the northern half of the country. The limited length of the data series, however, precludes the attachment of significance to any particular periodicity within the three spectral bands spanning 9 to 27 years. Willett (1974) has suggested, however, that the 11-year sunspot cycle is most readily observed in lower latitudes.

Temperature trends

There have been few systematic attempts to investigate recent temperature trends in Australian climatic records. Deacon (1953), comparing the two periods 1880–1910 and 1911–50, noted that mean daily maximums in the interior of the country were appreciably lower during the latter period; this being a consequence of the latitude of the subtropical high-pressure belt. The extent to which Deacon's study was affected by inhomogeneities, principally those due to changes (generally prior to 1911) in thermometer screen type, is uncertain. Schwerdtfeger (1958) has expressed doubts on this matter. There have also been few attempts to determine systematically the homogeneity of Australian temperature records; two notable exceptions are those of Kemp & Armstrong (1972) for Sydney, and Cornish & Evans (1964) for Adelaide.

A comparison of the five Australian state capital cities – Adelaide, Brisbane, Sydney, Melbourne and Hobart with a group of nine provincial centres situated around the same, roughly triangular, area in southeast Australia was undertaken to assess the effects of increasing urbanisation on temperature trends. Figures 4.5.7(a, b) show the coarse of average annual temperature anomalies from 1911 to 1974 for each of the groups. The most obvious difference between the two groups is in the course of

196

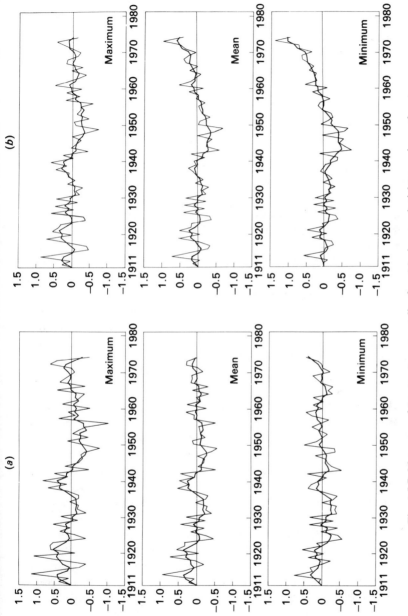

Fig. 4.5.7. Mean annual temperature anomalies for: (a) nine provincial towns in southeast Australia; (b) the five capital cities Adelaide, Brisbane, Sydney, Melbourne and Hobart. The thick line is the 5-year running mean.

197

4.5 Shorter term variations

Table 4.5.1. *Correlation coefficient between average annual temperature anomalies of five state capital cities and nine provincial centres in southeast Australia*

All are significant at the 1% chance probability level except * which is marginally significant at 5%

Period	Maximums	Means	Minimums
1911–74	0.91	0.81	0.58
1941–74	0.86	0.72	0.33*

average annual minimum anomaly. Whereas there has been an increase of about 0.3 °C for the provincial centres since the mid 1940s, the corresponding increase for the capital cities has been around 1.4 °C. The course of mean temperature for each group, calculated as a simple arithmetic average of maximum and minimum temperatures, also reflects this difference, although to a lesser extent. The courses of maximum temperature, on the other hand, follow each other closely. Table 4.5.1 gives the correlation coefficients between the two groups for the whole period and for a shorter period, 1941–74.

It would seem then that urbanisation effects are most evident in minimum temperature records – as a result of the surface heating being confined to the shallow stratified nocturnal surface layer. A consequence of this would appear to be that maximum temperature alone might be a more reliable indicator of larger-scale climatic variability than some combination of maximum and minimum temperature.

If one were to accept that the anomaly curves of annual maximum temperature in Figs. 4.5.7(a, b) are roughly representative of trends in southeast Australia, it would appear that there has been an overall rise of this temperature in this area of about 0.4 °C from the late 1940s to the early 1970s. This would be in accord with rises observed in average annual mean temperature in New Zealand (Salinger & Gunn, 1975; Salinger, 1976; Trenberth, 1976), and the results of an analysis of tree-ring records in Tasmania (Pearman et al., 1976).

In the remainder of Australia, the situation is less clear. Tucker (1975) has inferred that temperature trends, in accord with rainfall trends, are probably subcontinental in scale. In Western Australia there is no overall pattern of either rising or falling trend, nor is there any clear latitudinal pattern to be seen from an admittedly sparse network of seven stations investigated. Increasing temperatures would seem to be the case in central and north-central Australia, on the basis of 30-year trends at Alice Springs, Darwin and Camooweal. On the other hand, Innisfail in northeast

Queensland shows a significant fall in annual maximum temperature from 20-year and 30-year trends. The nature of the 30-year trend in the Innisfail record suggests possible inhomogeneities. Until further work is done, however, not only on this station record but on all the others as well, such statements must remain tentative as indeed must the above results.

5. Models of climatic change

5.1 Global cycles and climate
R. W. FAIRBRIDGE

Introduction: the scale of cycles

In the earth sciences today there is an urgent need to integrate the dynamic modelling and the historical approaches to geophysical problems, for usually the enthusiasts for one approach regrettably exclude the other with which they are less familiar. It is proposed, here, to explore the historical record back to the origin of the planet, and to draw in some of the evidences of the dynamic school in order to suggest ways in which the climatic prediction problem may be profitably pursued. The fact that there are two groups of scientists (with contrasting background and training) approaching one and the same topic means that there must be some misunderstanding. Take steady state, for example: an excellent first approach to global dynamics, but constantly disturbed by short- and long-range factors. There are many agencies on this earth which are non-uniform and non-steady state, so that any hypothesis of short-term climatic change on a steady-state model is a matter of conceptual or modelling convenience, which may not bear a close relationship to the total facts. As another example, consider the increase of entropy in natural processes. This is a fundamental law of physics: yet the natural scientist finds that through history both the sum-total of biologic complexity and lithologic variety in the earth's crust become progressively greater. Thus the earth must be viewed as an extremely dynamic body evolving progressively, and becoming more complex as it changes. Entropy in the history of the earth, that is, in geology and in biology, is recognised by the geomorphologists as negative.

Cyclic or repetitive events in nature are on many scales, from the rhythmic pulse of the human heartbeat to the age of the universe. In the open systems of the earth and universe all cycles tend to interact, so that the picture is complicated by 'noise', and by the complex intermodulations of the various cycles. For example, cyclic thermal runaway in mantle convection provides periodic pulses in heat flow to the crust from within the earth. Thermal expansion of the crust is followed by cooling and contraction over an 85 m.y. cycle. Now, each periodic thermal runaway will cause an expansion of the mid-ocean ridge, an acceleration of the rate of sea floor spreading by an order of magnitude above the present, and a resulting world-wide change of sea level of the order of 500 m. This will profoundly change the oceanicity: continentality ratio of the globe and hence the climate.

There are many geophysical cycles, and we must consider them all, from the shortest to the largest. There are the diurnal cycles, the yearly cycles, the 2- and 5-year cycles, longer cycles that correspond to the sun spot 11- and 22-year periods, astronomic cycles of 550, 1,100, 1,650 years and so on. One of the more naive assumptions of twentieth-century meteorology and geomorphology has been that the patterns were essentially fixed standards. Extremely complex cycles of solar activity and celestial mechanics are constantly at play, sometimes out of phase, sometimes in phase (though rarely with several multiple cycles in phase), but never constant or static.

Perhaps the best analogue for the climatologist to study is the level of the ocean and its response to energy input from wind which generates a whole series of waves, of differing characters, amplitude and wavelength. Interactions occur along coastlines, on the ocean bottom and with density stratification. This analysis could also include the effects of tides (extra-terrestrial gravitative energy), the seasons (also celestial mechanics), the sunspot cycle, the Milankovitch cycles or the cycles of sea-floor spreading. But although the present state of mean sea level is difficult enough to establish, it can be measured at established points with a reasonable degree of accuracy and predicted for periods of the order of 10 years for purposes of coastal engineering, geodesy, satellite navigation and so on.

Consider a contemporary shoreline. It is marked by a beach, an offshore belt where waves break, and a littoral belt marked by high-tide accumulations, coastal dunes or cliffs. This geomorphic association is marked by details, such as borings by intertidal organisms or by the texture of sand grains identifiable only under the electron microscope, that are so characteristic that similar associations can be identified in the geological record. Another strandline is commonly identified about 5 m above the present one, and on a low coast possibly a kilometre inland. By absolute dating techniques, that strandline is found to have been formed about 100,000 years ago, and furthermore this picture is valid in many far separated parts of the world. Despite the fact that many parts of the earth's crust are rather unstable (and the old strandlines are deformed in those places), the relative uniformity of this former seashore around the world persuades us that it is a eustatic phenomenon – that is to say, it was produced by a world-wide change in the level of the ocean.

Detailed study of the changing level of the sea has shown some remarkable results. For example, in the past 20,000 years or so, not only has

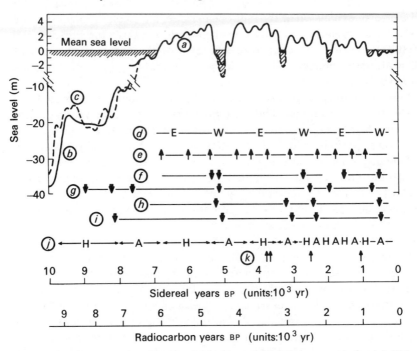

Fig. 5.1.1. Trace of eustatic sea level over the last 10,000 years, as indicated (*a*) by an example from Brazil (Fairbridge, 1976), based on radiocarbon-dated beach deposits, coral reefs and Indian middens, for the last 7,000 years, and for the previous 3,000 years by data (*b*) from Scandinavia (Mörner, 1969), and (*c*) elsewhere (Fairbridge, 1961). Note that time is given in both sidereal and radiocarbon years. Comparisons are presented (from top to bottom) with (*d*) the magnetic declination cycle (approximately 2,850 years) with its E–W inflections; (*e*) the $^{18}O/^{16}O$ isotope ratio peaks from Greenland (Dansgaard *et al.*, 1971) indicative of maximum warming; the glacial readvances (*f*) from New Zealand (Burrows, 1973, 1975), (*g*) from Greenland (Ten Brink & Weidick, 1974), (*h*) from Alaska and (*i*) from Lapland (Denton & Karlén, 1973); (*j*) the Sahara climates (Geyh & Jäkel, 1974) where H denotes humid, and A denotes arid; and (*k*) maximum flood levels of the Nile (Fairbridge, 1962, and pharonic inscriptions). Note that imperfections in the datings exist that might present false impressions of one region 'leading' another.

there been a large oscillation from glacial to interglacial of the order of 130 m, but also the record shows secondary oscillations in the past 7,000 years – see Fig. 5.1.1. It is clear that these variations are real. In particular, at about 4,300 radiocarbon years BP (during the documented historical period of Egypt) there was a sea-level drop that exceeded 5 m. Such a sea-level drop corresponds to an extraordinarily abrupt climatic change especially since the sea level returned to its previous height over 250 years. There was no 'ice-age', but there was a catastrophic change. This 5 m of sea water involves more ice than is contained in all the mountain glaciers of both hemispheres put together so that certainly Greenland and perhaps even Antarctica would have to be drawn on to provide the quantity of water to make this oscillation. There was also a fairly big oscillation in Roman times, of some 2 m, which had a profound effect on glacier advances in the Alps and on the weather in the Mediterranean region. It produced, as a matter of fact, a very pleasant cool equable climate which historians would recognise immediately as being favourable for fostering intellectual activities.

The questions that the climatologist must ask are: (i) to what extent are these sea-level events tied to climatic controls? and (ii) to what extent are they cyclic, and therefore predictable?

Other parameters besides the changing sea level may be studied: winds and pressure patterns, for example. Atmospheric pressure of the past can be established with some surety (see Lamb *et al.*, 1966; Loewe, 1974, and Chappell in the next section), and ancient wind systems may be demonstrated by structural analysis of former sand dunes, ripples and sediment textures. In this way the approximate centres of former anticyclonic cells may be identified together with the regular storm tracks. In desert regions, a very characteristic distribution of longitudinal dunes helps to identify the transition between westerly and Trade Winds (today, about 25°). The dunes in the northern Sahara, for example, advance from WNW towards the ESE, then gradually swing to S and then to SW or WSW, describing a gigantic 'hair-pin' swirl. The mean bisectrix of this swirl is sometimes called the 'wheel-round latitude', and if established for an ancient geological period, firmly demonstrates the tropical zone of that time (Fairbridge, 1967).

Let us now review the history of global environments, as we read the story today.

The dawn of history (4,500–3,500 m.y. ago)

According to the most popular model of the origins of our planets, the 'quick-boil' concept would be expected to leave a crystallised surface deposit on the earth, which would be the beginnings of the planetary crust as we know it today. Other models, however, postulate melting in stages as a response to the critical temperature/pressure relationships, so that the crust would be of primary material intruded progressively by rising plumes of lighter molten ingredients from the furnaces below. But in either case the oldest rocks contain conglomerates, stones rounded by rolling in streams or along the seashore, which indicate that the early earth, 4,000 m.y. ago, already had a hydrosphere with water in its liquid state – not ice or vapour. Thus the earth had an average temperature in the range of 25–50 °C. What of the atmosphere? Some of the oldest conglomerates include pebbles of the minerals pyrites and pitchblende. These consist of compounds that are *unstable* in a moist, oxygen-rich air so their presence

5.1 Models of climatic change

indicates a reducing atmosphere, i.e. one free from oxygen. The geo-chemists have made experiments to demonstrate that the *first life* on this planet could only have evolved in such a reducing atmosphere and that this consisted largely of hydrogen, helium, carbon dioxide, ammonia and methane (see e.g. Landsberg, 1953; Sneath, 1970). Life was Revolution I in Earth history.

First life and oxygen

Traces of bacteria have been found in South Africa and in other places, but larger life traces are the 'stromatolites' found living today as far apart as Florida and Western Australia. These structures of calcium carbonate or silica are usually found in Precambrian sedimentary rocks younger than 2×10^9 years and are formed by the action of primitive marine algae which require environments in the range $25 \pm 15\,°C$ in shallow ocean water, neither too saline nor too fresh. The mineral types in the sediments of the day suggest a barren landscape without much life, subject to rainstorms, flood, river erosion and sedimentation, but with volcanic rocks more prevalent than today. The earliest land areas did not include high mountains and broad plateaus as today, but rather consisted of large numbers of small nuclear islands within a broad, shallow world-encircling ocean. There were no great ocean deeps. Climates were maritime and continentality effects were minimal, although day/night contrasts were greater than today.

About $2,900 \pm 200$ m.y. ago, with the evolution of photosynthetic algae, a carbon-dioxide-rich atmosphere began to receive oxygen, the 'waste produce' of photosynthesis. This was Revolution II in Earth history. Initially all the oxygen produced was immediately captured because there were many reduced (oxygen-poor) chemicals in the early soil and water of the infant Earth. Iron, for example, was mainly in the ferrous state and highly soluble, so that the early ocean was rich in dissolved iron. Liberation of oxygen began to produce ferric iron (Fe_2O_3) which precipitated. In the middle Precambrian there was a tremendous world-wide formation of iron-ore deposits, largely in the form of Fe_2O_3. It is significant that *all* of the really enormous iron-ore reserves of the world came from this specific time in Earth history.

Free oxygen evolved only slowly at first and it was not until about 1,600 m.y. ago that the O_2 level reached 1% of its present level. In the meantime, water vapour and carbon dioxide continued to build up the new atmosphere. As oxygen began to appear, strong ultraviolet radiation from the sun tended to dissociate it thus leading, by recombination, to formation of ozone (O_3). The land surface would have been fatal for any life as we know it. But gradually with the water vapour, CO_2, oxygen and, most

204

importantly, the ozone in the higher atmosphere, a so-called 'greenhouse effect' was initiated that did two things: firstly it screened the land surface from potentially lethal radiation; secondly it helped to create milder climates. Thus there was a complete replacement of a primeval atmosphere by a new one that has gradually evolved through time (see e.g. Berkner & Marshall, 1967).

Mountains, trees and coal

The initial crust of the earth, with its low islands of continental nature within broad shallow seas, was rather thin, not very stable, and incapable of supporting high mountains or great plateaus. There were, however, innumerable small volcanoes. An important change occurred around 2,500 m.y. ago. After that time (the 'Proterozoic' interval) troughs began to form along borders of the land masses and ring-like sedimentation sites evolved. Since long-continued sedimentation in such troughs leads eventually to an unstable situation, it is here that the crust tends to fracture and buckle, and thereby produce mountains.

One of the most remarkable features of mountain-building in geologic history is that it is definitely episodic. There have been many long, relatively quiet times during our protracted global history, interspersed with crescendos of activity, marked by volcanicity, crumpling and mountain-building ('orogeny'). The discovery of 'absolute' dating techniques (utilising radioactive disintegration rates), produced statistical proof that there was indeed a long, rhythmic periodicity of earth movements. Once poetically described as the 'mighty heart-beats of the Earth Mother', this periodicity is approximately 200 m.y., which is about the same as the period of the rotation of our galaxy, but no convincing proof of correlation has yet been offered. Nevertheless, it is tempting to speculate that our earth movements are 'triggered' gravitationally by our cosmic motion. There is, parenthetically, evidence also for a 200 m.y. cycle in glaciation, and theories have been advanced to link the galactic rotation rate to this long-term climate oscillation by G. E. Williams (1975).

Earlier than 2,500 m.y. ago, with small, low continental slabs of crust surrounded by the world ocean, the global climatic picture would be maritime, and given such extreme oceanicity, there could be no ice ages, nor even small mountain glaciers. In the geologic record of the Proterozoic, however, there are numerous glacial 'tillites', and examples of these glacial episodes are found today on every single continent. It is clear that the first mountains brought with them the first major glaciers and ice fields. Now, if the orogenic episodes of the Proterozoic are cyclic, is it not probable that the glacial phases are similarly episodic? This seems to be borne out by the field evidence. The tillites are world-wide because under plate tectonics and sea-floor spreading, continents would have drifted to different latitudes. Major changes in palaeolatitude are strongly suggested by studies of palaeomagnetism.

At some unknown time in the Proterozoic the first animal life began to evolve. Clearly a certain minimum oxygen level would be required, but the precise time and circumstances are elusive. All that we know is that by Cambrian times, 570 m.y. ago, a large and sophisticated marine biota of soft-bodied organisms had evolved. Their impressions and trails are widely distributed and are now well studied. They tell us more about the existing environments: oxygen was present with temperatures generally equable (say 15 ± 10 °C) and marine conditions comparable with today.

There is a popular theory that during the Proterozoic the moon was much closer to the earth, which would mean a very short day, a high rate of rotation and immense tides. The sediments of the time should contain evidence of vast submarine sand dunes created by these violent tidal currents: there are none that we have been able to find. In contrast the stromatolites speak for relatively calm shallow-water conditions.

Following shortly after the late Cambrian Ice Age came a long period of calm, mild conditions that saw one of the most exciting and obvious revolutions in Earth history – the first sea-shells, in great variety and complexity, ranging from the mollusc to the trilobite. The almost 'instantaneous' appearance of this complex biota has astonished geologists and biologists for a century or more. The explanation is partly climatic (rising temperatures), partly geochemical (rising alkalinity of the world ocean, making $CaCO_3$ excretion difficult), and partly biological (increasing population-pressure and predation which led to Darwinian selection of those organisms that developed hard carapaces).

The early Palaeozoic Era (Cambrian/Ordovician) closed with another great ice age. At around 450 m.y. ago, that is, 200 m.y. after the late Precambrian refrigeration, continental ice conditions returned – and on a great scale, lasting probably at least 10 m.y. At this time the African continent came, by pole motion and continental drifting, to lie in the South Polar region. Following upon the late Ordovician 'Sahara' Glaciation the world saw a return to almost universally mild-to-warm conditions. On land the great feature of the age was the appearance of the first big non-aquatic plants and trees some 400 m.y. ago. The first forests naturally grew in the dampest places – the low swampy delta areas close to the lakes or sea coasts. This has always been called the Carboniferous Period, because of the widespread coal in the Northern Hemisphere. But at this same time (about 340 m.y. ago), in the Southern Hemisphere another great ice age descended. While the Northern Hemisphere lands had split up again after multiple collisions, the southern lands (known as 'Gondwanaland') had remained welded together as a single continent since the Cambrian time.

In the intervening 200 m.y. the South Pole had shifted progressively from West Africa, through Brazil to South Africa, thence through Antarctica to Australia. All these 'Gondwana' continents (with India) were then one. Glaciation appears to have been only scattered and relatively minor at first, being limited to certain mountain ranges. But, in a cyclic way, with ever-increasing waves, the glacial phases grew greater and greater, until by the late Carboniferous and early Permian time giant ice sheets were expanding over land areas ranging from South America to Australia, and including even India.

With glacial cycles large oscillations of sea level took place, drowning the great tropical forests and their population, and burying them beneath blankets of muddy sediment. Thus were born the great coal deposits that we exploit today. Geochemical studies of the isotopes of sulphur can now prove that during both Carboniferous and Permian Periods there was a negative swing in the oxygen supply. Being a 'continental' time, sea levels were quite generally low and outside the glaciated regions and apart from equatorial or coastal coal swamps the land climates tended to be extreme (hot summers, cold winters) and dry like the Gobi Desert today, but there were appreciable evaporite-forming lakes – mainly ephemeral or seasonal. In this way oxygen compounds were being buried.

The Mesozoic (c. 225–65 m.y. ago)

With the end of the Permian Period came also the end of the Palaeozoic Era. The northern continents suffered another severe series of collisions: Africa and America collided as documented by the Appalachian–Mauritanide revolutions: northern and southern Europe collided along a belt extending from southern Ireland to the Carpathians: eastern Europe and Siberia collided along the Urals and have remained welded together ever since. So again we see 'continentality' as the dominant climatic feature of the time. Deserts and desert lakes began forming already in the late Permian and persisted throughout the Triassic. It was against this backdrop that the first of the giant reptiles, the dinosaurs, evolved and eventually came to dominate both lands and seas of the entire era (225–65 m.y. ago). Their bones and footprints are often perfectly preserved in the muddy deposits of the former desert lake margins. This reptilian paradise may have had a climatic explanation. There were almost no high mountains, but extensive plains, and at no time, on no continent, was there any evidence of major glaciation throughout this 160 m.y. interval.

In the last quarter of the Mesozoic, the Upper Cretaceous Period, a remarkable 'flowering' of small floating marine organisms took place. These are the 'nanoplankton', unicellular plants (coccolithophoridae) and animals (foraminifera), the shelly parts largely constructed of calcium carbonate. Previously most of marine life was bottom-dwelling or free swimming. The result of this 'flowering' was a rain of dead shells onto the deep sea floor – to form what we all know as *chalk*, and now found all over the world, from Dover to Texas, from the Ukraine to Australia. Of special significance to the climatologist is the semi-permanent withdrawal of carbon dioxide (as $CaCO_3$) from the atmosphere. So our atmosphere lost CO_2 to gain still more oxygen. This would lead to a slight reduction of the 'greenhouse effect', favouring a cooling trend that was perceptible throughout the next era.

The Cenozoic (c. 65 m.y. ago to the present)

As often remarked, the ending of the Palaeozoic and of the Mesozoic were marked by great extinctions, but the mechanism of this phenomenon is clouded in deep mystery. From the evidence of sediments and soils we are satisfied that no great ice age or violent climatic revolution marked each transition. Plate tectonics may furnish a clue. There was a crescendo of sea-floor spreading activity near the end of each era that must have led to an accelerated heat flow along the global spreading axes, and this would cause crustal expansion, leading to a universal rise of sea level. Around 80 m.y. ago, sea level reached over 500 m above its present level, greatly reducing the world's available land areas, and in the late Cretaceous Period only 15% of the earth's surface was dry land. The population crowding must have been Malthusian. Furthermore, the lack of natural catastrophes (such as ice ages) during the Mesozoic did nothing to trim the population growth. There is on the other hand, abundant evidence from the fossils that predators (e.g. *Tyrannosaurus rex*) developed on an immense scale – a phenomenon suggestive of food shortage and desperate competition. A feature of the reptile metabolism is that they thrive in a carbon-dioxide-rich atmosphere, whereas mammals do not. The biogenic oxygen build-up of the global atmosphere was presumably increasing steadily throughout the Mesozoic and towards its close the environments were becoming more attractive to the reptiles' mammalian competitors. At the same time there was removal of carbon dioxide in the deep-sea chalks. Sulphur-isotope studies show that a serious oxygen deficit existed during the Mesozoic and this was reversed at the start of the Cenozoic.

But there is one serious flaw in these arguments in that they only extinguish a few classes of organisms. Much more satisfying, and more likely to be correct, would be a geophysical explanation that would treat all organisms alike. It is known that the earth's magnetic field from time to time becomes weaker and may reverse itself. It is not likely that the faunas of the time would be so disoriented that they could not survive, but it is likely that climate may be coupled to this fluctuation. In fact Reid *et al.* (1976) have suggested that at times of magnetic reversals, the shielding effect of the earth's magnetic field disappears, and high-energy particles from the sun enter the atmosphere in abundance. The mutations produced by these, and by enhanced cosmic rays, may have harmful effects on fauna (Uffen, 1963), but certainly not enough to cause wholesale extinction of species. On the other hand, the solar particles, as Reid *et al.* show, will interact with the nitrogen of the high atmosphere and will eventually lead to the formation of nitric oxide in large quantities. Recent studies on the effect of nitric oxide on the ozone layer of the stratosphere, undertaken to determine the possible effects of supersonic transports, show that the ozone is catalytically destroyed by the nitric oxide. But, as was pointed out above, it is the ozone layer which shields life on earth from the energetic, and inimical, ultra-violet solar radiation. As well, the partial destruction of the ozone will affect the heat absorbed by the atmosphere and, given intense proton bombardment from the sun during periods of weak magnetism or magnetic reversal, the earth's climate may also change. It is likely then, that many species which had adapted to the stable conditions preceding a magnetic reversal may have been unable to cope with the increased radiation and changed climate.

The climatic history of the globe during the Cenozoic Era is very illuminating, for it shows us how an ice age is initiated. At the beginning of this era we have evidence of world-wide semi-tropical conditions – that is, warm and moist. The size of the continents was initially very small, and the oceanicity factor was dominant. Continental drifting and sea-floor spreading during the Cenozoic progressively led to a series of intercontinental collisions from Spain to the Himalayas. What had been, in Mesozoic times, a wide equatorial seaway, the 'Tethys', or ancestral Mediterranean, that joined the North Atlantic to the South Pacific, now became blocked. By about 30 m.y. ago no further east–west migration was possible. By the late Miocene (12 m.y. ago) the Mediterranean became limited to a salt lake (see Hsü, 1972). Other barriers were also rising: in the East Indies the uplift of island chains almost connected Australia to southeast Asia; in Panama, volcanic chains eventually joined North America to its southern partner.

The continentality of the earth increased, and so the passage of warm, moist air was blocked both longitudinally and latitudinally. The mountains, moreover, would force the predominant horizontal airflow upwards and so would extract moisture (and latent heat) from the atmosphere. Dry, desert regions would tend to be formed, and the albedo of the earth would change. In short, orogeny would disrupt the energy balance of the earth (see e.g. Claiborne, 1970). Since the earth's heat balance is mainly related to heat loss, this blockage in the global temperature regulating system led to an ice age. Minor climatic cycles were superimposed on the major trends so that episodically the glaciated regions grew larger. As long ago as 30 m.y. there were scattered signs. These grew more and more extensive – from Antarctica to Alaska – until around 2 m.y. ago the present ice age (the 'Quaternary Period') is deemed to have begun in earnest. Icelandic faunas began to appear in the Mediterranean. In ever-greater cycles of about 100,000 years each – the great glacial stages closed down over the globe. We are now in one of the mild 'interglacial' phases. But how long will it last?

Since 1940 the mean world temperature has been dropping, whereas the mean magnetic intensity has been rising. However, there are also regionally reversed trends: in some parts of North America and the Southern Hemisphere there is a warming climatic trend during the same period, paralleled by a falling magnetic intensity. Wollin *et al.* (1974) in discussing this phenomenon conclude that the evidence from their studies indicates a link between higher magnetic intensity and colder climates. This is in accord with the work of Reid *et al.* (1976) discussed above, who showed that a weak magnetic field should allow a greater heating rate at the ground compared to the present-day value. The observed evidence of Wollin *et al.* thus supports the conclusion that the magnetic field of the earth modulates climate through its ability to act as a shield against corpuscular radiation. One reason for the regional anomalies in geo-magnetism mentioned above is the eccentric distribution of the magnetic dipole field of the earth and its secular (westward) drift. Since the global climate average is likely to be steered by effects that coincide with the world's largest land mass (Eurasia) and not with the greatest oceanic regions (Pacific and Southern Hemisphere), it is the strengthening magnetic field and dropping temperatures over Eurasia today that are most significant.

World climatic patterns are much affected by feedback phenomena. Firstly, there is the Eurasia albedo effect where increased snow or ice cover will increase the albedo of the ground, which in turn will reflect more solar radiation back to space, thus positively enhancing the colder temperatures. Secondly, during a glacial phase the transfer of moisture from the oceans to continental glaciers lowers the ocean level, the glacio-eustatic effect, and global 'continentality' increases about 12%. Thirdly, the northern seas lose moisture, become shallower, and cool and freeze more easily. The sea-ice is white, however, and this contributes still further to the albedo heat loss. Fourthly, icebergs are carried by ocean currents into lower latitudes, and will cause more widespread cooling. Lastly, a fifth consequence of global oceanic cooling should be to lower the mean evaporation rate; thus glaciations should coincide with global aridity. During the last 5,000 years there has been a progressing drying up of the Sahara landscapes. There are numerous geological proofs of global cooling of about 2.5 °C in this period which appear to follow the Milankovitch pattern – see the following section by Chappell.

It is important to realise that these climatic changes do not proceed steadily. Rather they advance with strong oscillations – sometimes warmer, sometimes colder – but ultimately reaching a threshold limit, when the mighty feedback equations begin to play their role. Then the next climatic deterioration will be rapid and absolute.

Conclusion

It is concluded that in long-term processes connected with geology and biology the crust of the earth and its populations become progressively more complex. The following evolutionary trends are basic:

the *atmosphere* has been progressively generated by emanations from the earth;

the *hydrosphere* has also grown progressively and become more alkaline;

the *continents* have become higher and thicker;

the *ocean basins* have become deeper and more segregated from one another, while continental shelves have become smaller; and

in the *biosphere*, as speciation increases, total populations of each tend to decrease.

Whether these trends are unique and unrepeatable or the result of intermodulary and complex interactions between existing internal global cycles, and external solar and galactic cycles is still not determined, though the evidence favours the latter. It is abundantly clear that the supreme problem in climatic prediction is the resolution of long-term secular trends and multiple short-term cycles.

5.2 Theories of Upper Quaternary ice ages
J. CHAPPELL

The question of causes of repeated glaciation, during Pleistocene times, has acquired contemporary relevance as attention focuses on questions of climatic change and world food production. Whether analysis of Pleistocene climates can aid prediction of the future has been answered in the negative by some (e.g. Hare, 1971), on the grounds that the record is so blurred as to admit alternative explanations. However, records of major variables such as ice volumes and ocean temperatures over the last 140,000 years or so, are now established to such a degree that several previous theories for glaciation can be eliminated, and certain strong determinants of major climatic changes can be identified. Such records are reviewed below and the elimination of certain hypotheses about ice-age

5.2 Models of climatic change

causes is discussed. The strongest factors affecting major climatic changes of Upper Pleistocene times are, so it is argued here, firstly the changes in annual march of radiation caused by orbital perturbations and secondly the surface temperature and sea-ice conditions in the northern Atlantic Ocean.

Comparison of selected geological records

Analysis of Pleistocene climates necessitated correlation of the different types of records from different regions – deep-sea cores analysed palae-ontologically and geochemically, ice cores and cave speleothems (e.g. stalagmites) analysed isotopically, and various terrestrial stratigraphies. Correlation requires adequate chronologic control, often difficult beyond the range of radiocarbon dating. Figure 5.2.1 illustrates this, with 4 records within the range 140,000 to 60,000 years BP. The first shows temperature changes in Mexican Gulf surface waters, identified palaeontologically from rapid sedimentation rate sea-floor cores (Kennett & Huddlestun, 1972a). The second and third curves show $\delta^{18}O$ profiles, respectively from a stalagmite from southern France (Duplessy et al., 1971) and from the basal part of the Camp Century, Greenland, ice core (Dansgaard et al., 1971). Both of these curves have been interpreted as indicating temper-ature of the precipitation. The last curve indicates sea-level changes, identified stratigraphically from flights of raised coral reefs in Barbados, New Guinea and Timor (Bloom et al., 1974; Chappell, 1974a; Chappell & Veeh, 1977). Plotted time scales are as given by the original authors. The dashed lines show possible correlations, including several which have been made by other authors. Not only are there temporal offsets in Fig. 5.2.1, but also there are major form differences between the curves.

Dating uncertainties differ markedly between these records. The chron-ology of the coral reef sea-level record is the best established, with the portion shown being tied to 39 $^{230}Th/^{234}U$ age determinations from Barbados, New Guinea and Timor, and supported by a further 23 dates from the 135,000–120,000 years BP reefs of Hawaii by Ku et al. (1974). Dating of the Greenland ice core, on the other hand, is based on extra-polation of an ice-cap flow model; it is admitted to be tentative for the basal section by its authors, and is likely to be in error by up to 20% (Budd, 1975a). Dating of the Mexican Gulf cores is little better: lacking radiometric ages, Kennett & Huddlestun (1972b) use correlations with Caribbean cores dated by Broecker and co-authors, and uncritically use the '89,500 years BP' abrupt event in the Greenland ice core as support. The French stalagmite is better dated, based directly on 6 $^{230}Th/^{234}U$ determinations with good counting statistics.

Interpretation of these records as indicators of 'global' climate is quite

212

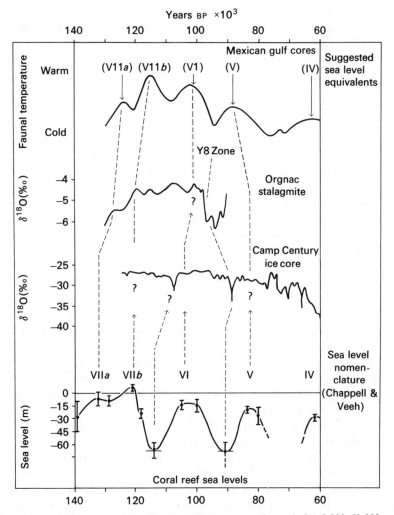

Fig. 5.2.1. Variations of various climatic indicators over the period 140,000–60,000 years BP compared with sea levels (note: error estimates shown on sea-level curve). For references, see text.

doubtful. Once again the ice-core record is vulnerable because the sharp ^{18}O changes in the highly compressed basal layers could have occurred over much longer periods than indicated or 'simply reflect variations associated with large dunes of winter accumulations or large-scale topography and accumulation irregularities' (Budd, 1975b). Climatic interpretation of the stalagmite ^{18}O record is equivocal (Emiliani, 1971). Empirically it is not a simple ice-age gauge because it fails to show any excursion during 120,000–105,000 years BP when sea-level changes indicate

213

an important glacial cycle (accompanied by a major North Atlantic cooling–warming cycle: Sancetta *et al.*, 1972). The stalagmite ^{18}O reflects meteorologic changes in its local region, which may be relatively unaffected during early phases of glaciation, until some threshold between alternative circulation modes is passed. The abrupt change at 97,000 years BP could well indicate a change in circulation mode occasioned by ice cap growth. Similarly, rapid cooling in the Mexican Gulf cores may reflect only changing ocean surface climates in the Gulf region, rather than general global cooling, as was implied by Kennett & Huddlestun (1972*b*). The sea-level curve reflects changing ice volumes, and as such reveals the course of glaciation of the northern continents, but the rates of change between high and low sea levels are not yet accurately identified.

Figure 5.2.1 indicates cautions necessary when estimating the course of Pleistocene ice ages and speculating about causes. Bearing this demonstration in mind, we now examine relationships between certain of the best-established records for the last 250,000 years, shown in Fig. 5.2.2. Curves (*a*), showing orbital perturbation effects, will be discussed later. Curve (*b*) shows sea-level changes over the last 0.25 m.y., based on analysis of the raised coral reefs of Barbados, Timor and New Guinea (the last curve in Fig. 5.2.1 is extracted from this record). Also shown on curve (*b*) are migrations of the southern margin of the Wisconsin ice sheet, to the limit of ^{14}C dating (from Dreimanis & Karrow, 1972). Coherence between the two records corroborates the inference that the sea-level changes are glacio-eustatic. Curves (*c*) show $\delta^{18}O$ variations in surface waters of the equatorial mid-Pacific (core V28–238: Shackleton & Opdyke, 1973) and of the Caribbean (core P6304–8: Emiliani, 1966*b*). From differences between these $\delta^{18}O$ curves and the sea-level curve, Chappell (1974*b*) estimated that the northern ice caps had a mean $\delta^{18}O$ level around $-25\%_0$ from $\sim 120,000$ to $\sim 80,000$ years BP, and about $-37\%_0$ from $\sim 80,000$ to $\sim 10,000$ years BP. Curve (*d*) shows that these estimates compare well with the main trend of the Camp Century ice core. Curves (*e*) show meridional migrations of polar and subtropical water-mass fronts in the eastern North Atlantic, identified from deep-sea core palaeontologies (from McIntyre *et al.*, 1972). Time scales in Fig. 5.2.2 are from the original authors, with the exception of Emiliani's core P6304–8, for which the time scale has been re-evaluated by Broecker & van Donk (1970), and Shackleton & Opdyke (1973).

Dating errors for main peaks in these curves are as follows. The sea-level curve from 135,000 years BP to present is tied to over 70 satisfactory $^{230}Th/^{234}U$ age determinations, and the high sea-level peaks are thought to be accurate within $\pm 4,000$ years (Bloom *et al.*, 1974; Chappell & Veeh, 1977). From 250,000 to 135,000 years BP the age determinations are much fewer and errors are estimated as $\pm 15,000$ years for each peak

214

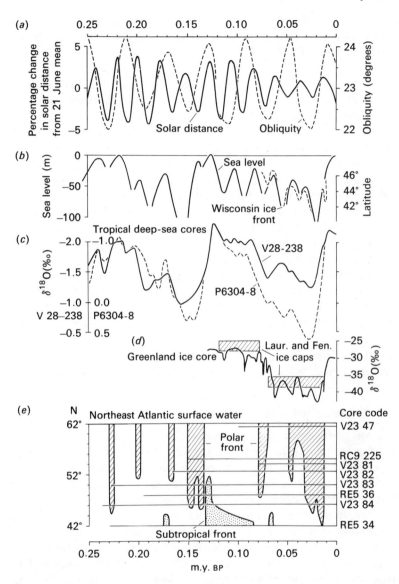

Fig. 5.2.2. (*a*) Variations of mean solar distance for 21 June and of obliquity (from Vernekar, 1972). (*b*) Sea-level variations identified from flights of raised coral reefs in New Guinea and Barbados (Chappell, 1974*a*), and latitude variations of later Wisconsin southern ice front (Dreimanis & Karrow, 1972). (*c*) $\delta^{18}O$ variations in equatorial Pacific core V28–238 (Shackleton & Opdyke, 1973) and Caribbean core P6034–8 (Emiliani, 1966*b*, redrawn on time scale of Shackleton and Opdyke). (*d*) $\delta^{18}O$ variations in Camp Century, Greenland, ice core (Dansgaard *et al.*, 1971) and estimated mean $\delta^{18}O$ levels in combined Laurentide and Fennoscandian ice, derived by comparison of (*b*) and (*c*) (Chappell, 1974*a*). (*e*) latitudinal movements of subtropical and polar water mass fronts in northeastern Atlantic (McIntyre *et al.*, 1972).

215

5.2 Models of climatic change

(Chappell, 1974*a*, *b*). Core V28–238 extends through the Brunhes–Matuyama magnetic boundary (700,000 BP), and Shackleton & Opdyke (1973) argue that uniform sedimentation rate can be assumed, yielding the time scale as shown. The tie between the 120,000 BP high sea level and the $\delta^{18}O$ minimum confirms this chronology. It is estimated that ages given to other peaks are accurate within 10,000 years, on the basis of reviews of sedimentation-rate data by Turekian (1965) and Broecker & van Donk (1970). Curves (*e*) are tied by correlation of the exceptionally warm sea-level peak with the 120,000 years BP high sea level (i.e. the last interglacial climax). Other peaks in curves (*e*) are taken as accurate within 10,000 years. No weight is given to the time scale for the ice-core record, (*d*), beyond about 40,000 years, for the reasons explained by Budd (1975*b*) and outlined above. Finally, the curves of orbital perturbations affects in Fig. 5.2.2 (*a*) (obliquity variations – dashed line; combined eccentricity and precession effect on 21 June solstice distance to sun – solid line) are judged to be accurate within ±2,000 years. The orbital extrapolations by Vernekar (1972) from which these curves are taken, have been further refined (Berger, 1975), but with small effect on age-estimates of peaks within the 250,000 years BP range.

From such records, and relationships between them, certain possible mechanisms for repeated glaciation in Pleistocene times can be explored. In the light of Figs. 5.2.1 and 5.2.2 the status of several contemporary theories of ice ages briefly can be established.

Status of various ice-age theories

Explanation of repeated glaciation during Quaternary times has been the object of many different theories. Leaving aside hypotheses invoking extra-terrestrial mechanisms such as variations of solar constant (e.g. Öpik, 1965), these can be grouped into two broad groups: those which attribute a major influence to orbital perturbation effects and those which do not. Prominent amongst the latter are suggested bistable or 'flip-flop' global climatic systems, invoking either alternations between open and frozen Arctic surface waters as triggering glacial–interglacial alternations (Donn & Shaw, 1966; Donn & Ewing, 1968), or periodic surges of the Antarctic ice cap causing substantial expansions of circum-Antarctic ice which in turn induce extensive global cooling (Wilson, 1964; Hollin, 1972; Flohn, 1974*a*), or alternations of state in ocean circulation (Weyl, 1968; Newell, 1974*a*). Yet other suggested mechanisms invoke factors such as episodic explosive vulcanism (e.g. Budyko, 1969), or effects of the changing magnetic field of the earth (Wollin *et al.*, 1971). These are discussed first.

216

Alternately open and frozen Arctic

Direct palaeontologic evidence for the condition of the Arctic surface comes from Arctic ocean-floor cores. Foraminifera productivity is very much higher in open waters than beneath floating ice, and because all reported Arctic cores show uniformly low foraminifera densities at least over the last 70,000 years (Van Donk & Mathieu, 1969; Hunkins *et al.*, 1971), the last three ice-cap fluctuations cannot be attributed to partial disappearance of Arctic ice. Concerning the critical period immediately before termination of the last interglacial maximum, Arctic deep-sea cores appear to indicate cover comparable to the present (Hunkins *et al.*, 1971). This theory thus appears to be eliminated.

Periodic major surges of Antarctic ice

The suggestion that northern continental glaciation would follow any substantial increase of circum-Antarctic floating ice has recently been given theoretical support by Flohn (1974*a*, and Section 3.7 above). Types of evidence which could show whether a major Antarctic ice-cap surge actually caused any Pleistocene glaciation has been discussed by Hollin (1972) and include sharp excursions of $\delta^{18}O$ levels in deep-sea cores, ice cores and other records. Flohn (1974*a*) refers to the sharp changes around 95,000–90,000 years BP in the Mexican Gulf cores, Orgnac stalagmite, and Camp Century core as evidence of 'instant' glaciation following an Antarctic surge. As discussed above, alternative explanations are likely for each of these records, and in any case the first advance of the last glaciation commenced shortly after 120,000 years ago, not at 95,000 years BP. Perhaps most strongly supporting the surge-ice age hypothesis would be evidence showing a sudden rise of sea level, indicating sloughing of considerable Antarctic ice into the sea, immediately followed by slower eustatic fall corresponding to northern glaciation (outlined by Hollin, 1972). No such eustatic pattern has been identified yet, although a possibility is indicated by the rise to the 120,000 years BP high sea-level peak, VII*b* in Fig. 5.2.1. The rate of this rise has yet to be determined, and hence the surge-glaciation hypothesis cannot with finality be ruled out for initiation of the last ice-age period. However, 'instant' glaciation is not yet demonstrated although models suggest that Antarctic surging is a possibility (see Budd & McInnes in Section 5.3).

Alternations of state of ocean circulation

It is certain that ocean circulation patterns, as well as surface temperatures, have varied sympathetically with Pleistocene glaciations. Illustra-

5.2 Models of climatic change

tion is given by curve (e) of Fig. 5.2.2 showing meridional migrations of subtropical and subpolar water-mass boundaries in the northeast Atlantic. Dynamic circulation changes are involved – for example, the North Atlantic Current appears not to have flowed during the Wisconsin maximum, but a separate anticlockwise cold gyre established instead, north of about 45° N (data from Ericson, reported in Weyl, 1968). It is unclear to what extent such dynamic changes represent responses to the altered atmospheric circulation of a glacial epoch, and to those attendant modifications of the salinity–temperature structure stemming from cooling and from transfer of surface waters into ice caps. The idea that the ocean may have two circulation modes – one conditioning glaciation, and the other engendering interglaciation – has several times been suggested (Adam, 1969; Weyl, 1968; Newell, 1974a). The 'saw-tooth' pattern with roughly 0.1 m.y. period, seen commonly in ^{18}O records of deep-sea cores (Broecker & Van Donk, 1970), may be interpreted as indicating perseverance of one mode while glaciation advances to climax (with many shorter oscillations superimposed), followed by rapid change of ocean structure, accompanying deglaciation. However, in the light of present inability to distinguish causes from responses, only the course of change can be chronicled in the hope that insights can be gained from recurring patterns in the past. Certain recorded changes of ocean surface conditions, similar to those shown in Fig. 5.2.2 will be discussed later.

Volcanism and magnetic theories of glaciation

Volcanism as a cause of glaciation is discussed sympathetically by Bray in Section 5.5 below. Changes of geomagnetic intensity have been suggested to affect global temperatures, because the flux of cosmic rays into the upper atmosphere increases as magnetic field strength decreases (Wollin et al., 1971; review in Lamb, 1972a). Additionally, excursions of the principal dipole may affect climate, because the field itself may exert a steering effect on the positions of troughs in the circum-polar circulation (Lamb, 1972a). Substantiation has been claimed from correlations in deep-sea cores between palaeoclimatic indicators and variations of magnetic intensity and dip on the one hand, and with volcanic ash falls on the other (Wollin et al., 1971; Kennett & Watkins, 1970). However, closer examination of the records shows little support for either case (Chappell, 1975). A second objection arises from Fig. 5.2.2 which shows that several cycles of glacial advance/retreat correlate strongly with the precession-dominated solar distance curve. If volcanism and/or geomagnetism were primary causes of ice-cap variations, then orbital factors should correlate with volcanism or geomagnetism. Not only is there no indication of this in reported records, but also there are good geophysical reasons for there

218

being no such connections (Chappell, 1973, 1975). In short, although neither mechanism for global cooling can be dismissed, there is no evidence to show that either controlled Pleistocene glacial variations.

Summarising thus far, the only factor from these various theories for which there is clear evidence of direct association with glaciation is that of sea-surface changes, and cause–effect relationships are still unclear. Contrasting, however, is the clear emergence in Fig. 5.2.2 of a strong association between ice-volume variations (curve (*b*)) and orbital factors, especially during earlier phases of the last glaciation from 120,000 to 60,000 years BP. This points strongly to the Milankovitch hypothesis of glaciation, which now is discussed.

The Milankovitch hypothesis: orbital perturbations and glaciation

Exact interpretation of Fig. 5.2.2 is qualified, however, by the dating errors outlined above: especially, caution must be exercised when phase relationships are being examined.

Comparisons initially must be made within the period of best dating control, viz. the last 135,000 years. Within this period, the most striking correspondence in Fig. 5.2.2 lies between sea level and variations of 21 June solar distance. Sea-level peaks at 120,000, 105,000, 83,000 and 61,000 years BP almost exactly coincide with times when northern summer solstice was at perihelion (see Section 2.1). With a lower degree of confidence (because of dating uncertainty) the same can be said for sea-level peaks shown at 240,000, 220,000, 195,000 and 175,000 years BP. At the very least, this pattern of correspondence bears out the point made by Broecker *et al.* (1968) that the effect of precession on glaciation appears to be at least as strong as the obliquity effect. It can be noted that numerical modelling by the GFDL joint atmosphere–ocean model (Wetherald & Manabe, 1972) shows that enhanced snow cover follows reduced seasonality, which is consistent with ice advance as northern summer moves to aphelion. It must also be noted that there is no clear correspondence between ice volume changes when ice caps are near their maximum size (viz. 55,000–20,000 years BP, and apparently also around 165,000–140,000 years BP) and the orbital factors, at least as presented in Fig. 5.2.1.

The main question is whether insolation changes directly cause ice-cap variation or whether they trigger self-amplifying interactions between ice caps, atmosphere and oceans. Kukla (1975) shows contemporary evidence for a direct relationship between seasonal geographic changes of snow cover and radiation. He demonstrates a close correspondence between autumnal snow advance and southward march of the 200 Ly/day (837 J cm^{-2} day^{-1}) radiation line, especially in central USSR, and concludes that variations of September radiation levels had a major influence on

5.2 Models of climatic change

Pleistocene ice-cap fluctuations. A simple test of this idea is to compare timing of ice minima with both June and September radiation maxima, over the period when ice and radiation curves are most similar – i.e. 130,000–60,000 years BP. Sea-level maxima occurred at 121,000, 105,000, 83,000 and 61,000 years BP while autumnal radiation maxima were at 122,000, 101,000, 77,000 and 55,000 years BP. There seems to be a drift of peak correspondence from a coincidence of autumnal peak with high sea level at the interglacial, to summertime peak – high sea-level coincidence 0.06 m.y. ago. The month of most critical radiation levels apparently advances as glaciation advances. If correct, this suggests some conditioning of the radiation-change effect by the ice sheets themselves. More cannot be made of this apparent phase shift, because of the chronological uncertainties attached to the records.

Whether orbital effects on radiation are sufficient to induce ice-cap changes of the known magnitudes and rates is a long-standing question, argued negatively by many who reckon in terms of mean annual budgets (e.g. Budyko, 1969). The question must be addressed in seasonal terms, however. Consider the crude energetics of the last deglaciation, 17,000–7,000 years BP which is the most rapid glacial change so far properly documented (as explained above, claims for very rapid ice-volume change around 95,000 years BP are highly equivocal). A landbounded ice cap essentially is unstable, and wasting persists once it is initiated by increased ablation, or decreased accumulation, or raised snowline (Weertman, 1961). Rate of retreat depends on available energy which, in a simplified version of the problem, may be estimated for either the ablating margins or the central accumulation area. For the ablating margins an energy balance is difficult to estimate because ice-margin calving in proglacial lakes is an important process of uncertain magnitude (see Andrews, 1973). For the accumulation area, the mass balance change required to shift the ice cap from stable to wasting at maximal rates of the last deglaciation is around 50 cm/yr, or 1.4 mm/day (Chappell, in press). If this were equally distributed between reduced winter snowfall and increased summer ablation, the mean summer heat increase required is 46 J cm^{-2} day^{-1}.

This compares quite closely with the perturbation-induced summer radiation excess during the last deglaciation. Figure 5.2.3 shows insolation departures from present levels. The summer excess 11,000 years ago was 125 J cm^{-2} day^{-1} (30 cal cm^{-2} day^{-1}) at 50–60° N (Vernekar, 1972), yielding about 42 J cm^{-2} day^{-1} net excess at the surface. Hence the deglaciation is energetically consistent with the radiation change, especially if winter snowfall is concomitantly reduced. The earlier ice-volume fluctuations of the last glacial period, 120,000–60,000 years BP, are associated with yet larger swings of insolation levels (Fig. 5.2.3) and thus the near-perfect

220

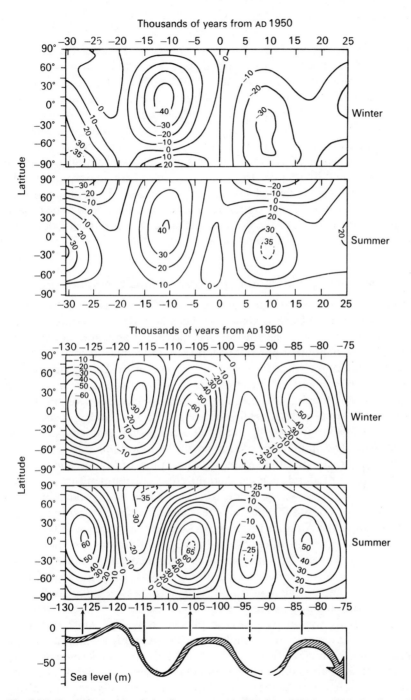

Fig. 5.2.3. Deviations of insolation from present levels, from 90° N to 90° S, for the periods 30,000 years BP to 25,000 years hence, and 130,000–75,000 years BP (from Vernekar, 1972). The lower insolation history is compared with sea-level changes. Units: cal cm^{-2} day^{-1} [1 cal ≡ 4.18$_4$ Joule].

tracking of the 'driving' and 'driven' phenomena during these earlier phases seems consistent with energetics.

A question remains, however, about conditions leading to glacial initiation and termination. This is indicated by comparing the termination, just discussed, with the retreats of 105,000, 84,000 and 61,000 years BP which stopped well short of deglaciation and yet were associated with summer insolation excesses greater than 125 J cm^{-2} day^{-1} (30 cal cm^{-2} day^{-1}) (Fig. 5.2.3). It was suggested above that the last termination required diminution of winter snowfall, relative to the level required for ice-cap growth or stability, as well as enhanced summer radiation. Significant diminution of winter snowfall during later stages of a glacial period appears to be caused by progressive cooling of the northern Atlantic Ocean, in turn induced by glaciation itself, as is now explained.

Glacial initiation, termination and North Atlantic temperatures

The last glaciation began shortly after 120,000 years BP when summer–winter radiation contrast at higher northern latitudes was approaching a minimum because the June solstice was moving to aphelion, eccentricity was high, and obliquity approached a minimum value. Survival of winter snow into the summer was favoured. However, survival of the snow cover is at least as dependent on winter snowfall as on reduced summer radiation (Wetherald & Manabe, 1972). Development of glaciation requires that heavy snow years persist while the ice cap grows, and that meteorological changes induced by the ice itself are not such that ice-cap starvation sets in. The influence here of North Atlantic temperatures can be indicated by reconstructing maps of wintertime upper air and surface pressure, for a moderate ice cap and different sea temperatures.

Figure 5.2.4 (a) shows the reconstruction by Lamb & Woodroffe (1970) of mean January 500–1,000 mb thickness and of surface pressure over the North Atlantic and America for the post-Alleröd cold stage ($\simeq$ 10,000 years BP). The 500–1,000 mb thickness is based on observed and theoretical relationships between this thickness and surface temperature, tied to geologic estimates of surface temperatures 10,500 years ago. Figure 5.2.4 (b) shows a comparative map from Chappell (in press), calculated by the same methods with the same ice cap retained but with North Atlantic temperatures as for January of the present day. The most important difference in assumed 'initial conditions', between Figs. 5.2.4 (a) and (b) is the sea–ice boundary. The most important difference in derived pressure fields is the appearance in (b) of a pronounced low over Newfoundland–Labrador, a condition which today has a noted association with heavy snowfall over the eastern part of the glaciated area (Barry, 1973). As argued by Chappell (in press), the snowfall over Labrador–Hudson Bay

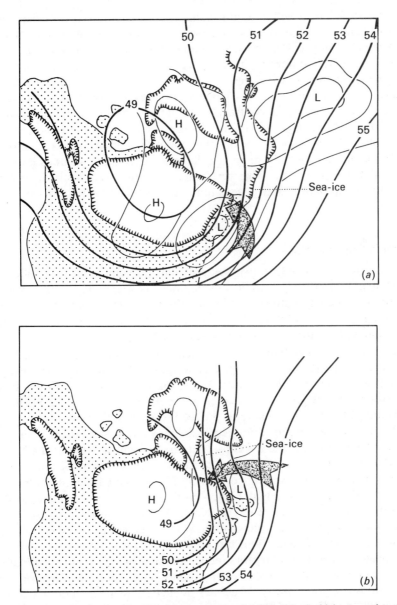

Fig. 5.2.4. North America showing (*a*) reconstructed 1,000–500 mb thickness and surface pressure for 10,500 years BP (January case) with sea–ice boundary shown (Lamb & Woodroffe, 1970). (*b*) Similar reconstruction but with sea–ice boundary in present position (Chappell, in press). Units: 10^2 m.

223

will be substantially greater for Fig. 5.2.4 (*b*) than (*a*) partly because frontal snow bands will extend substantially deeper into the ice cap, and partly because precipitable moisture will be greater. The latter is estimated by Chappell (in press) to be 1.6–2.0 times greater in (*b*) than in (*a*).

Reconstructed synoptic patterns in Fig. 5.2.4 are qualified by several criticisms, stemming from the relative simplicity of the estimation technique (discussed briefly by Chappell, in press). Irrespective of such limitations, however, the contrast between the two reconstructions should stand because absence of sea ice in the northwest Atlantic and south Labrador Basin should swing the 500–1,000 mb-thickness contours meridionally, as shown, enhancing thermal steering of cyclones into the Labrador–Ungava area. This result appears consistent with the North Atlantic temperature and glacial initiation–termination patterns of Fig. 5.2.2.

Concluding discussion

Results of the preceding two sections can be summarised as follows. Glacial initiation and/or rapid development is associated with a warm northern Atlantic, coupled with an orbital pattern yielding minimum seasonal contrasts around 50–65° N – i.e. summer solstice approaching aphelion with relatively high eccentricity and minimum obliquity. Glacial retreat, on the other hand, is associated with approaching maximal seasonal contrast – ideally, summer solstice approaching perihelion with high eccentricity and/or high obliquity. When this orbital condition for retreat is associated with extensive sea-ice and cold conditions in the northern Atlantic, then retreat is likely to proceed to termination.

A little further amplification is possible, in the light of records in Fig. 5.2.2. Once initiated, the ice caps continue growing as long as the North Atlantic is warm and a 'maximal-contrast' orbit pattern does not arise. This is suggested by the advance about 95,000 years BP, for which summer–winter radiation contrast at higher northern latitudes is similar to the present (Fig. 5.2.3). However, as North Atlantic temperatures fall, not only does the possibility increase that 'maximal contrast' radiation will induce complete deglaciation, but also probably the ice-cap growth rate diminishes unless orbital factors reduce summer–winter contrast. Finally, the growing ice caps induce North Atlantic cooling, which in turn preconditions their own decay. Feedback linkages here are several, including direct albedo effects of the ice, through to the effect of eustatic change on salinity flux in the Gulf Stream system (Weyl, 1968). The question of dependence of North Atlantic temperatures and sea-ice cover on salinity–temperature–depth structure of the North Atlantic – which itself is affected by ice-cap growth – should be accessible to numerical modelling.

A concluding prognosis about future ice ages can be sketched. If the foregoing results are applied simple-mindedly, the next glaciation appears remote rather than very near, firstly because orbital projection show no 'initiation condition' of markedly reduced summer–winter radiation contrast at higher northern latitudes in the next 25,000 years (Fig. 5.2.3). Secondly, North Atlantic temperatures are significantly lower now than near the end of the last interglacial, 120,000 years BP. Confidence in this forecast is qualified, however, by uncertainty about whether the simple conjunction of this orbital pattern and North Atlantic temperatures are necessary and sufficient conditions for glaciation. Chappell (in press) argues that the orbital factor appears persistently important through records several times longer than discussed here; whether the same is true of North Atlantic temperatures is unknown. Elucidation of this point, and of the question as to whether other factors not discussed also contributed to glacial initiations in Upper Pleistocene times, await yet further detailed analysis of adequate geologic records.

5.3 Modelling of ice masses: implications for climatic change

*Some introductory considerations**

As pointed out in Chapter 2, the cryosphere is an important active participant in climatic control. It is clear that as continental ice or sea-ice extends equatorward, strong horizontal thermal gradients associated with the ice margin will lead to a concomitant shift in the mean position of the Polar Front Jet and a general intensification of the mid-latitude Westerlies. Poleward transfer of energy will be increased and cyclonic activity enhanced. Fletcher (1969) has discussed the importance of sea-ice as a climatic parameter, and Kachelhoffer (1973) has examined the relation between the sea-ice margin and the Polar Front: while the meteorology of ice ages has been discussed by Albrecht (1962), Lamb *et al.* (1966), Fairbridge (1967) and others.

Computer-aided approaches to these problems have been initiated by J. Williams (1975) and Gates (1976). An important input is clearly the horizontal and vertical extent of the polar ice caps, together with the associated sea-ice. Modelling of the growth (dynamics) of these caps, and their dependence on external (climatic) and internal (thermodynamic) temperature regimes will be treated in this section. The models fall into two main groups: models of very large ice masses (Antarctica, for example), and models of smaller ice masses (Alpine glaciers, for example). These two regimes overlap, inasmuch as portions of Antarctica, for example, may be treated in a quasi-independent manner as autonomous glacier systems.

* Editorial contribution.

225

5.3   Models of climatic change

The relevance and importance of the small glacier models to larger ice fields resides in the treatment of the dynamics of so-called 'surging' glaciers. Modelling of these systems is now at a stage where many observed surging glacier systems can be quite accurately simulated, so that there must be a high degree of reality in the flow law used (Budd, 1975a; Budd & McInnes, 1974). In fact this model has also been used successfully to treat glacier systems which do not exhibit surging (see Kruss, 1976; Smith, 1976).

These glacier models so far have either been applied to temperate glaciers, or it has been assumed that the flow law does not depend critically on the temperature of the ice – a reasonable assumption if the temperature remains fairly uniform throughout the glacier. When treating the very large ice masses, however, the temperature does change drastically – in Antarctica, for example, there are temperature variations exceeding 60 °C – and the thermodynamics of the system must also be treated.

Thermodynamics of the ice masses (which include glacier systems, floating shelf ice and large continental ice sheets) have been studied by analytical and computational means. Robin (1955), Zotikov (1961, 1963), Crary (1961) and Budd (1969) have developed analytical solutions for the vertical temperature distribution within the ice for a number of realistic boundary conditions. Even so, the solution for any general, arbitrary case is not possible, and only numerical solutions can be determined. Computer modelling is even more versatile since few assumptions need be made, and quite complex boundary conditions, varying both in time and space, can be specified. Early computer models by Jenssen & Radok (1961, 1963) treated varying accumulation (or ablation) along the ice flow, changing surface and bedrock topography, climatic effects, and geothermal heat flux. Later studies by Budd *et al.* (1971) extended the models to include phase changes of the ice, variable density, specific heat, thermal conductivity and diffusivity, non-uniform horizontal velocity and internal deformational heating. In general though, while the vertical velocity is a direct result of continuity requirements, the horizontal velocity distribution is usually prescribed.

There has been no attempt made to model the temperature dependence of the flow law of ice, but any fully realistic model of large ice masses must take this into account. Unfortunately, the flow of ice is still not fully understood, and field and laboratory data can be approximated only by empirical rules as yet. Budd (1969) discusses this aspect fully, while Budd & Radok (1971) give a useful summary as well as reviewing the computational aspects of the thermodynamic and dynamic modelling of large ice masses.

Field and laboratory measurements of the flow of ice have enabled the

226

relation between the shear stress and shear strain rate of ice to be graphically represented (e.g. see Budd & Radok, 1971). (The 'shear strain rate' is equivalent to the meteorological concept of wind gradient or shear: shear strain rate, then, may be a deformation field $(\partial u/\partial y + \partial v/\partial x)$, for example, or a divergence field $(\partial u/\partial x + \partial v/\partial y)$. 'Shear stress', on the other hand, may be thought of as a directed force per unit area, in which not only the force is directed, but also the area is treated vectorially. A stress is set up, for example, by a force acting in the plane of the area being considered: pressure is a normal shear and is a special case of the general shear.) The relation between stress and strain in ice is complex, but nonetheless some simple analytic forms have been suggested. For a given temperature, various relations between shear $(\dot{\epsilon})$ and stress (τ) have been accepted, the most widely used being known as 'Glen's Law'.

In principle the flow law for any given situation may be computed by integration of three simultaneous tensor equations which hold for equilibrium conditions, but the boundary conditions are so poorly known that it is not, in fact, possible to perform this integration. In practice, theoretical studies of the dynamics of large ice masses tend to examine each situation separately, make simplifying assumptions, and develop specific laws for those cases. If the situations are chosen with discretion, these specific cases can be of quite some generality.

The most characteristic feature of large-scale ice flow, its variability in time, is linked to the occurrence of sliding on the bedrock, and reaches its most spectacular manifestation in the rapid advances known as 'glacier surges'. The pioneering theoretical study was made by Weertman (1957) who suggested ice may flow past obstacles either by melting upstream and reforming downstream, or by deforming visco-plastically. Lliboutry (1964) extended the possibilities by considering that the ice may lose contact with the rock over small features. However, these theories cannot account for the larger sliding velocities which are observed and have limited utility since many of the parameters are not observable or are poorly understood.

Observation and theory suggest that the velocity of glaciers at a point is not determined by conditions at that point alone, but is dependent on processes up and down stream. Budd & Radok (1971) concluded that for non-surging glaciers, the empirical data suggest that base stress increases with ice thickness and surface slope. But, if at some stage of relatively high meltwater production the base stress is dramatically reduced, the glacier can move very rapidly downstream due to this lubrication effect, and so enter the surge mode. The variation of base stress is represented by the relation
$$\tau_b = \rho g \alpha Z/(1 + \phi \rho g Z \alpha V),$$

where ρ is density, α is surface slope, V is mean horizontal velocity, Z

is the ice thickness, g is the acceleration due to gravity, and ϕ is called the friction lubrication factor.

Modelling surging glaciers and periodic surging of the Antarctic ice sheet
W. BUDD & B. McINNES

The model

The sudden advance or 'surge' of a glacier or ice cap after years of apparent stagnation or retreat remains one of the most fascinating and puzzling phenomena associated with these large ice masses. In such surges the glacier often moves forward many kilometres, over periods of months to years, at speeds one or two orders of magnitude faster than normal. Surges occur in most glaciated regions of the world, and it has been suggested by Wilson (1969) and Hollin (1969) that their occurrence in the Antarctic or Greenland ice sheets could have catastrophic consequences on a global scale.

Numerical models already developed by Budd & Jenssen (1975) treat the glacier flow as a non-sliding phenomenon and are stable in that if the input data consisting of bedrock configuration, accumulation–ablation balance distribution, and properties of the ice, are kept constant then the glacier eventually tends to steady state from any initial configuration.

The essential new feature in a model developed by Budd & McInnes (1974) and Budd (1975a), which includes the class of periodically surging glaciers as a subgroup of all glaciers obeying essentially the same laws, is the separation of the average velocity of a vertical column into a basal sliding component and an average internal deformation component. The local base stress at a point is then lowered according to the rate of production of water lubrication by the frictional energy dissipation of the sliding.

The local base stress increases linearly with surface slope and ice thickness until the average velocity of the vertical column reaches a critical value at which frictional lubrication sets in, when the base stress is suddenly lowered, and a surge is initiated. In this way an expression can be derived for the ice velocity and through the equation of continuity new thicknesses can be derived.

The model has been tested on a number of ordinary glaciers (Stor in Sweden, Hintereis and Aletsch in the European Alps, and Fedchenko in the Pamirs) as well as several surging ice masses for which there are sufficient data (e.g. Medvezhii, Brúarjökull in Iceland, and Otto Glacier in Ellesmere Island). Close matching has been found between the results of the model and the observed surges in respect to the main features such

228

Table 5.3.1. *Characteristics of surging glaciers from the model compared to measured values*

Glacier...	Medvezhii		Brúarjökull	
	Measured	Model	Measured	Model
Surge				
Period (yr)	10–14	11.8	70–100	77
Duration (yr)	0.2	0.8	0.6	0.1
Maximum speed (km/yr)	~ 30	19	45	120
Maximum length (km)	14	15	45	47
Length change (km)	~ 2	4.5	8–10	8
Thickness (m)	~ 180	180	700	500
Thickness change (m)				
Lowering	70	90	—	—
Rising	150	150	—	—

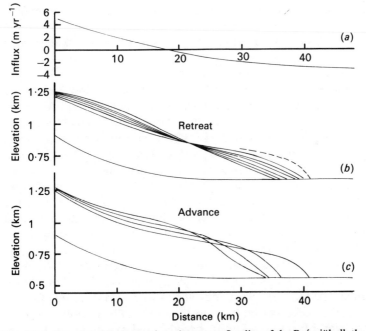

Fig. 5.3.1. For an idealised smooth analogue to a flowline of the Brúarjökull, the influx curve shown in (*a*) represents an accumulation/ablation balance curve modified to take some account of the varying width. A corresponding smoothed bedrock profile is shown in (*b*) and (*c*). The resultant surface profile of the ice mass after about 700 years is shown for a retreat phase at 10-year intervals in (*b*) and for a rapid advance phase in (*c*) at intervals of 0.1, 0.1 and 0.15 years from the start of the advance. The broken line in (*b*) shows the position at the end of the previous advance.

229

5.3 Models of climatic change

as: surge period, surge duration, maximum velocity, length change, maximum thickness and thickness change. Typical results for the Medvezhii Glacier and the Brúarjökull are shown in Table 5.3.1, and that for the Brúarjökull – is pictured in Fig. 5.3.1.

The confirmation of the model provides some encouragement to its application to other ice masses including the Antarctic ice sheet.

Application to an East Antarctic ice sheet flowline

The main differences between the Antarctic ice sheet and the known surging ice masses are its large scale and low temperatures. The effect of scale changes can to some extent be examined by comparison of the Vatnajökull and Medvezhii Glacier. The effect of lower temperatures can be examined by comparison of the temperate surging ice masses with the colder ice masses such as the Otto Glacier. The lower temperatures result in high effective viscosities of the ice which give lower strain rates and longer durations of the surge. For the Antarctic there are a number of major drainage basins, the main ones being those of the Ross, Filchner–Ronne, and Amery ice shelves and the Wilkes Land–Queen Maud Land ice-sheet zones. Similarly to the drainage basins of Vatnajökull, which appear to surge independently, it may be expected that those of the Antarctic ice sheet also act largely independently.

The programme of study adopted here is to consider flowlines of the Wilkes Land ice sheet, the Lambert–Amery basin and the West Antarctic –Ross basin. To begin with, a flowline in Wilkes Land, inland of the Soviet station Mirny from the summit of the East Antarctic ice sheet, was chosen because there was a detailed profile of surface and bedrock elevation data available (e.g. cf. US Geographical Society 1:5,000,000 map).

The net accumulation was obtained from Bull (1971) modified with results from Vinogradov & Lorius (1972). Mean flow parameters for the ice were obtained from the temperature distribution in this region derived by Budd et al. (1971). The model was then run using the same sliding factor as obtained from the known surging ice masses. In this case surging commenced quickly in the coastal zone and as the interior of the ice sheet built up the surging gradually spread inland. Eventually, after about 2×10^5 years a regular multiple periodicity developed with a total period of about 23,000 years. The ice sheet built up to a substantially greater size than the present: over 1,000 m thicker at the edge and extending to the continental shelf and afterwards collapsing to a profile remarkably similar to the present profiles. The time taken to reach the present type shape from the last major advanced state was about 12,000 years, but a minor coastal collapse also occurred some several thousand years ago.

The elevations along the profile as shown in Fig. 5.3.2 undergo major

230

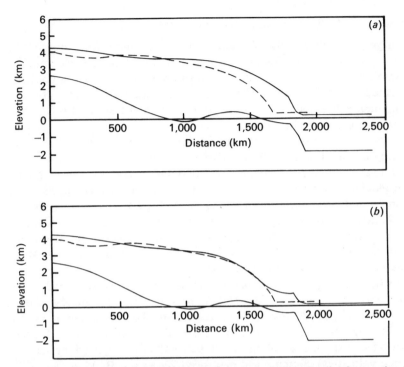

Fig. 5.3.2. Computer-simulated profiles of ice thickness along a flowline from an ice divide (the origin) to the coast (about 1,850 km distance) for two different phases of a surge cycle. Also shown are the bedrock profile (lower curve in both figures) and the present measured profile inland of Mirny (broken line). (*a*) depicts the situation during an advance, and (*b*) shows flowline after the advance, when the ice has 'collapsed' back in during the post-surge state.

changes at different locations at different times. Inland near the summit the changes are small but near the coast they exceed 1,000 m. A sharp up-and-down change near the coast is caused by the passage of a flux wave to the coast. Jenssen (Section 3.3 above) has shown that elevation changes of this type give rise to borehole isotope profiles which have many similarities to the measured deep isotope profiles.

The duration of the major surges was about 250 years, and the surge speed several km/yr. The resultant volume change for the Wilkes Land sector of East Antarctica would be about 2×10^6 km³ which is several times less than the amount considered by Flohn (1974*a*) as required to initiate a global ice age. The other drainage basins, however, particularly those of the large ice shelves, could be expected to have larger volume changes if they surge – a hypothesis which we now examine.

5.3   Models of climatic change

Assessment of other drainage basins in Antarctica

The large ice-shelf basins of Antarctica are characterised by surface elevation profiles along flowlines which start at inland domes, reach a maximum slope, and from there on, towards the coast, have a concave up surface with small slope. The point of maximum slope also typically has maximum basal shear stress as shown by Budd *et al.* (1971). Thus beyond this point the basal shear stress decreases whereas the velocity increases. This also occurs with our model using the double-valued type of sliding law as described above. The high-speed, low-base stress mode of flow is also a common feature of many outlet glaciers of polar regions. These glaciers can be maintained in steady state in the high-speed mode if there is sufficient mass flux coming in, and if the viscosity is not too high. Such a fast-flow steady-state profile with concave-up surface resulted for the Mirny flowline when the viscosity was lowered to that more typical of warmer regions with lower surface elevation, such as West Antarctica or the Lambert Glacier basin.

Thus the concave-up surface is not in itself evidence for non-steady state as suggested by Hughes (1973). The simplest criterion for steady state is whether the measured outward flux rates are equal or not to the total snow mass accumulated in the interior. Since flux rates are less directly measured we examine calculated velocities which are required for balance (balance velocities) compared to measured velocities. Budd *et al.* (1970), showed that around the major ice-shelf basins the balance velocities tend to be considerably higher (by about a factor of two) than the measured velocities. This is also in agreement with many studies of the Antarctic mass balance which indicate a buildup in the interior. This has been most clearly established for the Amery ice shelf basin by Budd *et al.* (1967) and the Ross basin by Giovinetto (1970).

However, the out-of-balance state is not a sufficient condition for surging, and the surge potential of the Antarctic must be assessed by other means. Budd (1975a) has shown that for the temperate ice masses of thicknesses about 700 m, the flux needs to be greater than about 0.2 km^2 yr^{-1} for surging to occur. But, from Budd *et al.* (1971) it is clear that this flux rate is exceeded for balance around the interior of the major ice shelves, and also at some places around the rest of the periphery. On this basis then each of the three major ice shelves has a high potential for surging. Furthermore the flux rates are not sufficient to maintain the grounded ice in the very high speed flow of the fast outlet glaciers of Greenland.

Work is still continuing on the modelling of the West Antarctic ice sheet but even at this stage a number of important results are at hand. Firstly, it is clear that without the high speed sliding resulting from the double valued sliding law, the West Antarctic ice sheet builds up to a dome shape grounded ice mass out as far as the continental shelf, some 500 km beyond the present edge. The volume difference between such an advanced profile and that of the present ice sheet would be of the order of 5×10^6 km^3, which is borderline for the initiation of the ice age considered by Flohn.

This now raises the interesting question of the approximately simultaneous surging of two or more of the large drainage basins. For example, if there were five basins surging independently at about 25,000-year intervals with some random variation from one to another, then if the effects of surges become superimposed if they occur within the same 1,000 years then the probability that two basins surge together would be of the order of once in 100,000 years. For such occurrences it appears that sufficient ice would be produced to give a substantial global cooling. It should be pointed out at this stage that if the West Antarctic ice sheet does develop periodic surging then neither a surge nor a rapid build-up are imminent because the present profile is one typical of the post-surge slow build-up phase.

Assessment of other ice sheets

Finally an examination of the Greenland ice sheet also reveals that a substantial part of the west coast has flux rates sufficient to give surging with the model considered here. In fact since the flux rate depends primarily on the magnitude of the accumulation rates and the size of the accumulation area, relative to the outlet width, it is possible to consider the surging potential of other ice sheets, including those of the Northern Hemisphere during the last ice age. It then becomes apparent that surging is a plausible phenomenon for those ice sheets.

There is still a long way to go in the modelling. The surging considered so far is an inherent ice-sheet mechanism with external conditions constant. With changing climate, and in particular interactive ice sheet–climate changes, the picture is more complex. As yet these have not been studied, nor have such effects as the changes in sea level or the isostatic variation in the bedrock height. It is clear, however, that a surging ice mass in times of changing climate can pass from a surviving state to an extinction state after a surge.

Although still at the beginning phase of modelling of ice sheets, it is expected that the road is now clear for developing more sophisticated models for examining all the complications described above, in order to help unravel the history of climate changes through the ice ages. In the meantime it appears, even at this stage, that periodic surging of the large ice sheets of the past may be expected to have been the rule rather than the exception.

5.4 Models of climatic change

Ice-mass modelling and climate models*

A full space–time simulation of large ice masses combining both the thermodynamics and dynamics of ice flow is possible, using the thermo-dynamic equation in three dimensions, with provision for internal heating due to the deformation of the ice. This involves a feedback between velocity shear (or ice deformation) and internal heating, which in turn changes the velocity field by altering the temperature. A pilot study of the Greenland ice sheet by Jenssen (1977b) encountered severe computational restrictions due to the large number of steps necessary, but nevertheless this limited study is most encouraging. Mass continuity problems at the edge of the ice sheet point to the need for a fine grid to be added near the ice boundary (McInnes, 1976), and work is continuing.

The role of an ice-covered surface in climate is well known, e.g. see in particular Untersteiner (1975), but also Lamb (1966a), Donn & Ewing (1968), Flohn (1974a), Wilson (1964, 1969). Models of changes of ice-surface topography and extent such as are being developed will therefore provide important input to numerical models of climate. It is difficult to argue how the climate will be affected *without* the use of these models, although early attempts have been made using fixed 'ice-age' boundary conditions (J. Williams, 1975; Gates, 1976).

The way in which a fully developed dynamic–thermodynamic ice model could be used is to: (a) run a climate model at the onset of a glaciation to determine accumulation and other parameters over the ice field, (b) run the ice model for a thousand years or so to determine the new ice shape and extent, and (c) recycle from (a). In this way the ice changes – which will be slow on the scale of the climate model, unless surging is allowed – will provide changing boundary conditions for the climate model, and will allow the study of the relation between jet-stream position and strength, polar-front location, strength of the Westerlies, magnitude of poleward transfer of energy, and so forth, to be studied adequately.

5.4 Mathematical modelling of climate

Basic considerations in climate modelling
J. S. A. GREEN

It will be supposed that 'climate' may be described by a set of statistics on the atmospheric circulation, which include mean values at many levels and positions, variances, and correlations. Reasonably well-defined statistics can be found averaging over periods as short as one month, which sets a lower limit to the time scale of climate. Upper limits depend very

* Editorial contribution.

234

much on the processes that are being studied. A 1-year period includes much of the interaction between the atmosphere and the upper layers of the ocean. Similarly, a 10-year period includes interaction with the deeper ocean and even interaction with the land surface – though climatic variation of plant species begins to be important. After 1,000 years changes in the ice cover and at 10^7 years continental drift set the appropriate scale.

It is essential that the *effects* of all the important processes that contribute to the circulation be represented in a model, even though it is not always possible or desirable to represent the details of the processes. The most important mechanisms will be listed here with speculations on the role that they play in the general circulation, and with discussion on how these roles can be modelled and the nature of the results obtained.

The main components and mechanisms

Solar radiation provides the energy source for the atmospheric engine; its distribution and intensity depend largely on the geometry of the earth and its orbit, and on the albedo, which is a function of cloud cover and the kind of surface, including its cover of snow and ice (see Chapter 2). Calculation presents little difficulty once the cloud has been specified though there are still some uncertainties associated with multiple scattering, and absorption in the extended path length traversed by the radiation. Volcanic activity may vary significantly on a geological time scale, and extinction due to volcanic dust may just be significant (see Section 5.5).

Terrestrial radiation provides the energy sink for the atmospheric engine; it depends on the distribution of water vapour, the main absorber at long wavelengths, and on the extent and temperature of clouds. Radiative cooling varies with height in a complicated manner, whereas the upward flux at the top of the atmosphere varies little with position because of a compensatory relation between the density of water vapour and temperature. This may be important: because the upward transfer of energy through the troposphere is dominated by convection, errors and irregularities in the distribution of radiative cooling with height are likely to be smoothed away quickly. Another consideration is that, with the typical cooling rate of about 1 °C/day, it takes a few days for an air parcel to change its potential temperature appreciably, so that only the cooling integrated over such intervals is important.

Because of the complexity of the absorption spectrum, direct calculation of cooling rates is almost as time-consuming as the calculation of the fluid dynamics. If the spectrum is grouped into a small number of bands, calculation is easier, but the likely error of 50% incurred in the local

235

5.4   Models of climatic change

cooling rate (Rodgers, 1975) is worrying. If the errors in the calculated cooling rates were unsystematic they would not be important, but it is likely that they are indeed systematic. To avoid this difficulty, radiation schemes are usually adjusted so that when applied to present-day climatic data at least some of the systematic errors are eliminated. It is true of at least some models, that fixing the radiative cooling at present levels reproduces present climate better than allowing the model the flexibility to adjust its radiation field.

Convection of small scale achieves rapid redistribution of energy in the vertical, principally in a surface boundary layer about 1 km deep. Over land the thermal inertia is so small that the convection serves to redistribute the net (mostly solar) radiative energy through a depth determined mainly by the relative magnitude of the fluxes of latent and sensible energy.

The large thermal inertia of the ocean makes the sea-surface temperature variation from the initial state unimportant for time scales of a week or two. The amount of energy entering the atmosphere then depends only on the depth of the atmosphere affected by small-scale convection. This is a poorly understood function of larger-scale variables like subsidence and wind shear, but current models reasonably use (or make equivalent use of) a characteristic temperature structure for the convective boundary layer, whose depth is then determined by intersection with the larger-scale temperature field.

Cumulonimbus convection, where this is local, represents the conversion of potential energy stored in the boundary layer into kinetic energy on the scale of the cumulonimbus. To the extent that this is more readily dissipated than energy on larger scales the process acts to deny energy to the larger scale and is appropriately modelled as 'deep' cumulus convection. This amounts to ascribing to it a different characteristic temperature profile (which can allow for precipitation cooling down-draught air), and ignoring any contribution to the larger-scale kinetic energy.

Where the cumulonimbus are organised, principally in the tropical convergence zones, but perhaps just as significantly in fronts and in hurricanes, their action is more complicated for they are organised by, and contribute to, the larger scale of motion. There is some evidence that they act so as to enhance the original shear, i.e. to unmix the momentum. There is a fundamental lack of understanding here which may be unimportant for climate models if the overall contribution of cumulonimbus to large-scale kinetic energy is small; it is not so obvious that this is so in the intertropical convergence zones.

The thermodynamic role of such convection is relatively straightforward. Cumulonimbus establish an approximately wet-adiabatic temperature profile in the tropics, which sets the characteristic temperature

236

profile for the remainder of the troposphere through horizontal temperature gradients defined by the large-scale motion.

Baroclinic waves, of wavenumber 3 to 9, effect a large fraction of the horizontal transfer of energy, at least in middle latitudes, and are therefore responsible for determining the horizontal gradients of temperature given the sources and sinks of radiative energy. Through synoptic experience, numerical prediction of the weather and stability analysis, a fair understanding of the mechanism of such systems results. They represent slantwise convection, carrying warm air poleward and upward, and cold air equatorward and downward, deriving their mechanical energy from a reduction in their potential energy.

More-or-less *stationary waves* of wavenumber 1 to 3 are generated topographically and define variations of climate with longitude. These include the winter continental highs and summer continental lows that dominate the surface distribution of pressure and winds. More fundamentally they organise the location of transient eddies by distorting the horizontal temperature gradients. Feedback (by the transient eddies modulating the mean wind, which changes the structure of the stationary patterns and leads to a redistribution of temperature gradient) may be responsible for the short-period fluctuations in climate of order one month, such as the 'index cycle' and 'blocking'. Thermal forcing comes from geographical variation of thermal inertia (land and sea) and albedo (ice or persistent cloud cover), while orographic forcing comes from the more extensive ranges of mountains, especially the Cordillera of the Americas and the Himalayas.

Because these waves are long they penetrate well into the lower stratosphere, where their energy is absorbed or reflected. Several kinds of evidence suggest that the waves may be close to resonance, so that energetically small modification of these upper layers may lead to substantial changes in tropospheric flow. Such a notion is attractive because it could provide a link between the known solar variability at short (ultraviolet) wavelengths and long-period trends in climate: perhaps not all of the weather 'cycles' at the sunspot period which the analysts have delighted in finding are spurious. If this resonance phenomenon is genuine, models of climate will have to include some dynamics (involving radiation and even photochemistry) for the stratosphere.

Of the *zonally averaged circulation*, the east–west component is by far the largest, but represents only a geostrophic response to the latitudinal gradient of temperature. North–south/vertical circulations, though slow, carry large amounts of latent and sensible heat, potential energy and angular momentum. Essentially, however, they convert one type of energy into another and transfer angular momentum in opposite directions at different heights and in each case achieve comparatively little net transfer.

237

5.4 Models of climatic change

For example, in the tropical Hadley cell circulation, latent and sensible energy is liberally converted into potential energy in the ascending branch, while the net poleward transfer of the sum of latent, sensible and potential energy across, say, latitude 15° is comparatively small. Similarly, westerly relative wind is generated in the upper troposphere, and easterly near the surface, but the total momentum generated, integrated with respect to height, is small.

Sea-surface temperature is governed by the net radiation integrated over long periods of time, by ocean currents, and particularly by the depth of the oceanic mixed layer. The ocean has time constants ranging from a day or so for the surface layers to perhaps 30–100 years for the deep ocean and forecasts in these ranges must carry relevant details of the ocean circulation. Poleward transfer of energy by *ocean currents* is some 30% of that carried by the atmosphere and must be taken into account. For periods of a few months the observed transfer at the initial time could be used, if it were known accurately enough, which is probably not so at present, but for longer periods and to augment the available data more specific modelling is required. Ocean currents are driven mainly by the stress of the wind, but momentum transfer between air and water (particularly the role of breaking waves) is poorly understood. Perhaps this does not matter very much because if the atmospheric engine determines the surface stress rather than windspeed, the 'law' connecting the two is not very important. The processes that transfer momentum and energy at the thermocline (determining the depth of the mixed layer and hence surface temperature) are even less well known and understood.

Snow cover clearly tends through high albedo to sustain cold surface conditions. More generally the *albedo of land* is dependent on the dominant plant form, which has a time constant of at least a few years. An example of such feedback has been suggested for warm deserts (Charney, 1975) where diminishing plant cover gives higher albedo, the desert acts as a sink of radiative energy, and so enhances subsidence, bringing dry air to low levels.

Numerical techniques

Most numerical models manipulate data defined at a three-dimensional grid of points in space, forecast the data one time-step ahead and use this as the basis for a new time-step. A fundamental law of such *finite-difference methods* states that no parcel of the fluid should have a speed such that the parcel will move more than one grid unit in the period of the chosen time-step. Thus, if Δt is the time-step, Δx the grid-spacing and c the velocity at which information is propagated we must have: $\Delta t < \Delta x/c$. If this 'computational stability' criterion is violated, then it can be shown

238

(Haltiner, 1971) that small errors in the data will amplify exponentially with time, and the results rapidly become meaningless.

This means that if the spatial resolution is improved by reducing Δx then the time-step Δt must be decreased. This makes substantial increases in resolution prohibitive. More important, the largest value of c (which is determined by those processes that carry information *in the model*) is critical.

Vertically propagating sound waves are the most critical redistributors of information (in the form of a pressure pulse) in the troposphere. Thus for a vertical grid of 1 km the time-step must be less than 3 s. Since sound waves, which are meteorological and climatic 'noise', depend on adiabatic compression and rarefaction, they may be eliminated simply by assuming incompressibility (Thompson, 1961), but this is a fairly drastic restriction. Fortunately, we know also that sound waves quickly rearrange the pressure and density fields to become close to hydrostatic equilibrium. Assuming continuous hydrostatic equilibrium in a model eliminates vertically propagating sound waves with little loss of accuracy.

The next largest value of c, and hence the next restriction in the time-step, comes from the horizontal propagation of sound and of external (or surface) gravity waves: two types of motion that merge into each other depending on the upper boundary condition to the system. In each case a horizontal grid of 200 km demands a time-step of less than 10 minutes, which sets the scale of most high-resolution models.

With some trepidation it is possible to eliminate inertial gravity waves too after noting that they serve mainly to bring the velocity and pressure fields into geostrophic balance. Assuming that such equilibrium exists at each moment, though the pressure and velocity fields are allowed to change, gives the 'quasi-geostrophic' equations where information is carried at the wind speed (~ 30 m/s) and time-steps of 2 hours are possible.

To go to steps of a day or two, some even more balanced state must be postulated. This can be done by taking into account the action of synoptic-scale disturbances on the slowly evolving flow of larger scale, which produces models in which the baroclinic waves are implicit.

A more numerical technique can be used to ease the time-step problem. The terms governing the propagation of slow waves are used explicitly to forecast the motion, but those terms governing the propagation of fast waves are calculated implicitly in retrospect. Such 'semi-implicit' schemes retard the fast waves, thus allowing longer time-steps, but include some of their effects so giving greater accuracy than if they were excluded.

The mathematical theory of finite-difference methods is mainly about the achievement of accuracy, while the prediction of climate is concerned

5.4   Models of climatic change

more with consistency: something closely analogous to the difference between convergent and asymptotic series. One fairly tangible aspect of this distinction concerns *conservation*. Equations expressing the advection of some property θ by velocity $\mathbf{v}$ can be written in the advective- or flux-forms $\mathbf{v}.\nabla\theta$ or div ($\mathbf{v}\theta$) respectively. A finite-difference replacement of the first is the more accurate, but the finite-difference replacement for the second is constrained (by Gauss' theorem) to ensure that contributions to the total spatial integral arising from adjacent boxes vanishes, as it should. The more accurate system can (and does) steadily accumulate the total amount of a quantity (θ) under conditions such that it is supposed only to move around from one place within the system to another. At the present, *schemes that conserve appropriate quantities* (principally entropy in adiabatic flow, energy in non-dissipative flow) have been developed (see e.g. Haltiner, 1971). Such 'conservative' finite-difference schemes tend to produce impossible solutions less readily than non-conservative schemes, but whether this is a good thing is debatable. It may be that the symptom of some more fundamental deficiency in finite-difference methods is being concealed by this device. Indeed there are many unexplored problems associated with the aims and achievement of numerical simulation of climate, ranging from the problems of asymptotic as opposed to convergent analysis to the interpretation of predictability.

One disadvantage of grid-point representation lies in its very versatility. No matter how irregular or meteorologically unlikely the field, the grid points will try to represent it. This is largely avoided by the use of a truncated series of orthogonal functions, such as a *Fourier series*, where the coefficients involved in these functions and their evolution in time is to be found. The big conceptual advantage of such schemes is that each coefficient surveys some aspect of the global circulation whereas a grid-point value knows only of the remainder of the fluid after the numerics has carried information to it. Again, because of this property, rather plausible patterns result using only a few coefficients, and this feature can be exploited even further by using not well-defined mathematical functions but empirical orthogonal functions derived from current meteorological analysis. In this way the number of variables needed to represent the data is an absolute minimum, and the model is constrained so as always to produce meteorologically-plausible fields.

Results

While the discussion and analysis of the results of any model is a major undertaking, some generalities may suggest a context in which such analyses may be viewed. It is important that results of complex models should be compared with those obtained from simpler models in order to

240

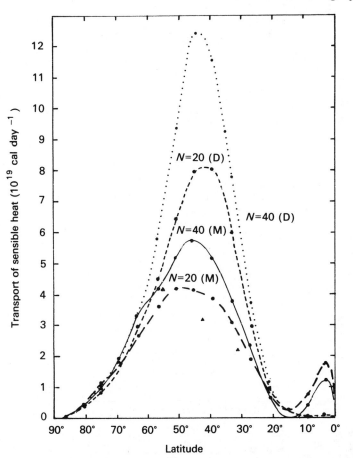

Fig. 5.4.1. Latitudinal distributions of poleward transport of sensible heat by the large-scale eddies in various models. The rates of transport in the actual atmosphere, which were obtained by Starr & White (1954), are also plotted as solid triangles. After Manabe *et al.* (1970). The model with $N = 40$ has twice the number of grid points as that for $N = 20$ (where $N = 40$ corresponds roughly to a grid length of 250 km). Notice that for the dry model atmosphere (labelled D) there is a substantial change in the energy transfer as the resolution is changed, even though both resolutions appear to be satisfactory from first principles. With the moist models (labelled M) there appears to be less change with resolution in spite of the fact that the scales now modelled include the much smaller dimensions of fronts and other cloud systems.

assess the additional information coming from the complication. Thus the high-resolution models produce realistic zonal-mean winds and temperatures, at least in the troposphere and using current radiative imbalance. Variation in the calculated heat fluxes as shown in Fig. 5.4.1 are nevertheless quite large. Since these features depend almost wholly on the horizontal variation of temperature they are reproduced just as well by the more implicit models and by the zonal-mean models discussed above.

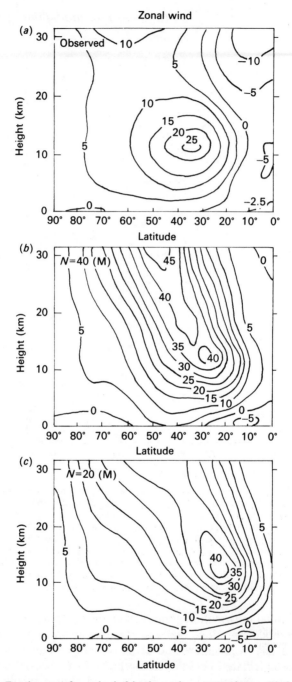

Fig. 5.4.2. Zonal mean of zonal wind in the various atmospheres. (a) Actual atmosphere (data source, Oort & Rasmusson, 1971; Batten, 1964); (b) the $N = 40$ (M) model; (c) the $N = 20$ (M) model. Units: m s^{-1}. After Manabe et al. (1970). In spite of looking quite realistic, the mean zonal wind is substantially larger than observed, implying that the model demands a horizontal temperature gradient 60% greater than the atmosphere. The observed decrease of wind above the tropopause is poorly reproduced.

The reversed (pole–equator) gradient of temperature in the lower stratosphere, associated with the decrease of zonal wind with height, is evidently a more subtle property. It is probably due to a reversed (equatorward) transfer of energy by motion of tropospheric origin penetrating into the lower stratosphere, but this demands rather good vertical resolution because the waves of baroclinic origin diminish in amplitude rather rapidly while the properties of forced waves depend critically on the absorption in the upper stratosphere (Shutts & Green, 1975). Most high-resolution models such as that shown in Fig. 5.4.2, do not give a strong enough thermal gradient here; consequently their jet stream tends to persist to too great a height. Manabe & Hunt (1968) show that doubling the vertical resolution does much to remove these discrepancies, which is consistent with the comments above on high-resolution models in the stratosphere.

When the constraint implied by the use of current radiation data is removed there is a general tendency for models to run off into ice ages. This could well be associated with the abuse of the upper boundary condition; which is always too firm and leads to total reflection of wave energy. A similar effect is that models run from real initial data, for a period of a month or so, tend to generate a 'blocked' zonal circulation much more readily than is observed.

Similarly, running a model with fixed input (like annual-mean radiation) does not allow it to hunt through various alternative states (like different resonance patterns) so that the system may settle down into one state whose selection amongst others depend weakly on the initial conditions. Again models run for a few months show disconcertingly large deviations from each other depending on subtle modification of initial conditions (Lorenz, 1975).

One aspect of the high-resolution models which is rather disturbing concerns their ability to represent the baroclinic waves that they have been developed to describe. Changes in horizontal resolution ($N = 20$ to 40), which appear superficially adequate to describe these systems, result in significant changes in the wave properties (Figs. 5.4.3 and 5.4.4). In general the models run with too little wave energy (and the waves work too efficiently). Similarly, when run for shorter forecast periods waves already present tend to die out unrealistically, and only reappear when the radiative imbalance has produced a rather extravagant temperature gradient (Gilchrist *et al.*, 1973).

While most of these discrepancies have been removed from the latest models this is not to say that the correction has been made on good physical grounds. Even the simpler high-resolution models are so complex, and the physical processes they represent so poorly understood, that there is great scope for inspired investigation of their properties and inspired empiricism in their construction.

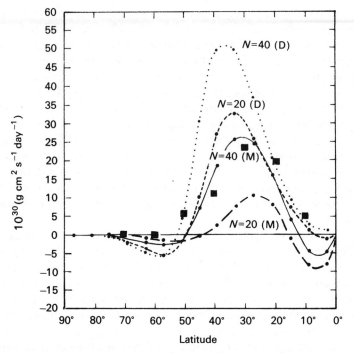

Fig. 5.4.3. Latitudinal distributions of poleward transport of angular momentum by the large-scale eddies in various models. The rates of transport in the actual atmosphere, which were obtained by Buch (1954), are also plotted (■). After Manabe *et al.* (1970). Surprisingly, in view of its secondary and rather subtle nature, the momentum transfer is reproduced rather well by the high-resolution moist model.

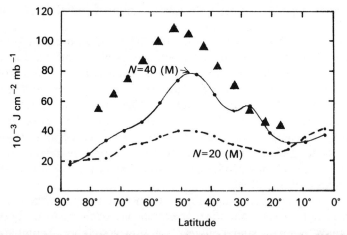

Fig. 5.4.4. Mean latitudinal distribution of eddy kinetic energy at the 500 mb level. (Actual atmosphere (▲), Saltzman, 1962). After Manabe *et al.* (1970). Even with high resolution the intensity of the eddies is substantially less than observed. This is coupled with the more intense gradients of temperature implied by the shears of Fig. 5.4.2 to reproduce the transfer shown in Fig. 5.4.1.

Climate models*

The basic parameters and processes which appear to be necessary for adequate climate modelling have been treated above. Here a very brief description of the basic methodology of climate modelling, and the kinds of models which have been produced is presented. More extensive reviews (to which the present author is indebted) include Joint Organizing Committee for GARP (1975), SMIC Report (1971) and Schneider & Dickinson (1974).

The basic interactions of the global environment which are relevant to climate and climate change have been discussed above and in Chapter 2. The time scale at which various parameters will interact with, or affect, the climate, to produce climatic change is shown in Fig. 5.4.5. Thus, if climate over a period of 10^4–10^5 years is to be studied, continental ice-sheet changes, isostatic adjustment, solar variability, volcanic dust, and changes in the earth's orbital characteristics must all be treated. On the shorter scale of 1–10 years, almost all of these parameters may be treated as constant and other parameters, which average out on the longer scale, become important.

Modelling over very long time periods is, at present, impractical for two main reasons. First, the very complicated General Circulation Models require so many computations that to simulate thousands of years of climate is not feasible with existing computers. These models, moreover, when used in climate studies, usually have unvarying boundary conditions, so that they run to a steady-state climate, applicable only for a certain season satisfying those boundary conditions. Secondly, the simpler climate models are, as yet, still subject to debate as to their physical reality and accuracy. There is no assurance that these models give realistic climates. Possibly the best current approach to long-term modelling is to use an atmosphere–ocean model, such as described by Manabe (1975) in which the atmosphere and ocean interact only at specified times and the computation proceeds at different speeds for each due to their different thermal relaxation times.

Another approach is to specify boundary conditions different from those now existing, and to use any climate model to generate the corresponding climate. This assumes that there is a unique climate for each set of boundary conditions, i.e. that climate is 'transitive'. There are, however, reasons for doubting whether the global system is transitive (Lorenz, 1975). If climate is intransitive, computer experiments such as J. Williams (1975) may generate only one of a possible number of climates. Sellers (1973) and others using very simple models have found that there were at least two steady-state solutions for the present-day solar constant. On

* Editorial contribution.

5.4 Models of climatic change

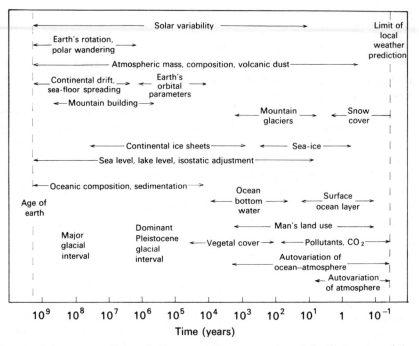

Fig. 5.4.5. Processes which probably act as climate controls and the likely range of time in which they are important. Many processes are involved in mutual feedback, both positive and negative and can involve the atmosphere, hydrosphere, cryosphere, lithosphere and biosphere. From *Understanding Climatic Change* (United States Committee for GARP, 1975).

the other hand, Paltridge (1975) found that his model 'did not appear to predict any form of climate intransitivity', and it is possible that intransitivity is more a property of the model or the computations than of the real climatic system.

Theoretical tools

Every climate model must take cognisance of the basic equations of meteorology. Whether they use the full equations, or simplify them to varying extents, depends on the modelling skill, climatic insight, purpose and prejudices of each modeller. These equations are the equation of motion in three dimensions, the equation of continuity, the equation of state, and the thermodynamic equation.

These six equations contain seven unknowns: the three velocity components (u, v, w), the three gas variables (pressure, temperature and density), and the heating term involved in the thermodynamic equation. This latter term may be a function of internal phase changes (condensation,

246

evaporation, melting, freezing), or of external processes such as radiative absorption and transmission, and sub-grid-scale parameterisation. The equations are further complicated by boundary effects near the ground, and these, too, need to be parameterised, or their computation explicitly stated. Full discussion of the basic equations is given by many authors: Haltiner (1971) is a good introduction.

Aims and means

The *aims* of climate modelling are clear. On the short-term basis of a season to five or ten years, the most important aim is the determination of the likelihood of a climate favourable or otherwise to food production (see Sections 7.1 and 7.4). Another reason for modelling is simply to extend our knowledge. Successful models generally imply that a reasonably realistic set of assumptions and/or parameterisations were taken in simplifying the basic set of equations. Unsuccessful models can in principle be analysed to see where they were deficient, and how their shortcomings may be overcome. In both cases, the total understanding of climate and climate theory is enhanced. Finally, a good climate model would allow simulated experiments with the atmosphere such as investigation of the effects of global pollution.

Two useful classifications of climate models are those of Adem (1975) and Schneider & Dickinson (1975). Adem believes that 'the thermodynamic energy equation was the basic one, to be used in a first approximation theory of climate', a belief shared by many climate modellers including Sellers (1973) and Budyko (1974). Schneider and Dickinson's classification is 'based essentially on the number of geometric degrees of freedom contained in the model' and includes all those models under Adem's scheme, as well as models which include the dynamics. The following is based on their classification.

(i) *Global averaging*

These models assume no horizontal redistribution of energy by winds or other processes but include such effects as radiative transfer, vertical distribution of absorbers, and clouds. The importance of surface albedo may be studied. For example, Budyko (1974) related the outgoing radiation from the earth to the average cloud cover, mean global temperature, surface albedo and incoming solar radiation, then studied the effect of changing solar radiation on ice cover (albedo) and global temperature. Other workers have studied changes in atmospheric composition, for example, Manabe & Wetherald (1967) prescribed solar radiation, albedo, surface temperature, lapse rate and concentration of optically active gases to investigate changes in CO_2 content. They found that the radiation

5.4 Models of climatic change

balance was so significantly altered by these changes that some feedback mechanisms had to be taken into account.

(ii) Modified global models

One way of improving global models, and thereby increasing their reality, is to incorporate feedback mechanisms. For example, changes in CO_2 concentration will affect atmospheric temperature and thus the moisture content of the atmosphere, leading to changes in cloud cover and albedo which will alter the input of solar energy. These changes will in turn modify ground and atmospheric temperature, and the cycle is repeated. The new steady-state situation (climate) may be substantially different to that deduced simply from temperature changes due to the initial variation in CO_2 content.

(iii) Thermodynamic models, with longitudinal and latitudinal variation

A further step in modelling complexity is taken by such models in which the main forecast parameter is normally surface temperature: such models usually also use the thermodynamic equation as their basic prognostic tool, and assume a balance between solar and terrestrial radiation inputs and outputs. These models also rely on a fair degree of parameterisation and thereby try to express some part of the calculations in terms of a simpler formula, or in terms of the values of the variables already computed. As an example, the surface pressure gradient may be treated as a function of the surface temperature gradient (Sellers, 1973).

These models may be functions of latitude only, or allow some long-itudinal variation as well, but all models depend on successful para-meterisation (Gates & Imbrie, 1975), and hence sometimes have been termed 'semi-empirical'. Models which allow variations both zonally and meridionally in their simplest form specify the longitudinally varying parameters and calculate the latitudinal variations. The essential feedback processes occur from the latitudinal computations. Some additional com-plexity was introduced by Sellers (1973) who allowed two points on each latitude circle, one for land, and one for ocean processes. Adem (1975) on the other hand parameterises both meridionally and zonally and thereby accounts for thermal transports by atmospheric and oceanic eddies in a model which extends to 10 km in the atmosphere (as one layer) and down to 50 m or so in the ocean.

(iv) Dynamic models with zonal or meridional averaging

Rather than have temperature as the basic variable, winds may be used, but here the process is reversed and now the thermal sources must be prescribed. These thermal sources may be external (solar or terrestrial radiation inputs at the boundaries) or internal (due to eddy fluxes, latent

248

and/or sensible heat effects, etc.). The solar radiation, for example, may be directly parameterised to the mean zonal temperature. Latent heat processes may be expressed in terms of the mean zonal wind, temperatures, and one meridional circulation for zonally averaged models. When the models allow latitudinal variations, the implication is that ocean–land contrasts are to be studied, and, in particular, the thermal differences (heat storage capacity, for example). Such models can simulate realistically the current latitudinal and annual variation of temperature (and the zonally-mean zonal and meridional components of the wind), given current radiation fields. It should also be possible to model some of the interaction between the atmospheric flow and the wind-driven currents of the ocean, preferably including cryospheric processes.

(v) *General circulation models*

Such models make use of the full set of equations and either compute or parameterise such effects as solar input, terrestrial radiation output, cumulus convection, phase changes in water, latent and sensible heat release, turbulent transfer and so on. In general, the parameterisation is left for those processes smaller than the grid-spacing of the models (typically 150–250 km), but there is no need for this to be so – parameterisation of some effects could also be carried out on a scale larger than the grid size. Models of this kind have commonly been confined to the atmosphere, with a fixed oceanic boundary, but combined models of the atmosphere and ocean have been developed (e.g. Manabe & Bryan, 1969). Missing, as yet, however, are models which treat atmosphere, ocean and the cryosphere.

The main drawback to these very general models is limited computational power. In spite of the very rapid advances made in the speed and data-handling capabilities of computers the complexity of general circulation and ice-mass models poses excessive demands on them.

Examples of two of the above approaches to climate modelling are presented below. The first is of the thermodynamic type, elaborated by Paltridge, whilst the latter, by Hunt (see Section 5.5), is an application of a general circulation model to the study of possible volcanic effects on climate.

Minimum entropy exchange, global dynamics and climate
G. W. PALTRIDGE

The basic model concept and development

The motions of the atmosphere and ocean obey basic physical laws such as the conservation of mass, energy and momentum. However, these laws have not proved sufficient for closure of the problem of atmosphere–ocean dynamics. It is not yet possible to predict *a priori* either the global mean climate or the global distribution of climate. Further, it is not possible to calculate with certainty the change in climate, which might occur as a result of a change in any of the external system parameters such as the energy output of the sun.

In an earlier paper, Paltridge (1975) related the discovery that global dynamics and climate may be 'controlled' by a simple minimum principle where the minimised quantity is related to the global rate of production of entropy. Application of the principle allows *a priori* prediction of the zonal mean temperature, cloud cover and meridional energy flow over the entire globe without recourse to specific description of atmospheric and oceanic dynamics. It may therefore be a powerful boundary condition for many of these dynamical problems. Further, the principle allows calculation of climate change on the basis that the dynamics of the system is something of a passive variable which alters appropriately so as to satisfy the criterion of minimum entropy production. Here we briefly describe the basic attack of the present model, the minimum entropy principle which it employs, and present some results. Full details of the model and its mathematics and physics are given in various papers by Paltridge (1974, 1975, 1976).

The work began with the hypothesis that there are sufficient degrees of freedom in the complexity of global dynamics for control by some minimum principle to be possible. There are many such 'laws' in physics. One characteristic they seem to have in common is that there is no way of telling beforehand which quantity is to be made a minimum. Here, in order to search for such a law, it was necessary first to develop a model of an atmosphere–ocean 'box' which was representative of the mean conditions of a latitude zone. It had to be internally mechanistic in the sense that only the net meridional energy flows into the box need be specified for the model itself to calculate the basic quantities of surface temperature T and cloud cover θ. Having devised such a model, an overall multibox model of the world could be assembled where the inter-box (i.e. meridional) flows of energy could be varied arbitrarily. The problem then reduced to a search for any single parameter of the overall system which was minimised for a unique set of the meridional flows – a set, furthermore, which was close to that observed on the real world and which led to the observed distribution of cloud cover and temperature. It was assumed that if an overall minimum principle is operative it should be sufficiently powerful to be apparent despite the inadequacies of the model employed in its search.

Initial evidence for the existence of such a principle was obtained with a three-box model of the mean annual hemisphere. It was found that a unique set of meridional energy flows close to those observed yielded a minimum in the quantity $\Sigma_i(F_{Si}/F_{Li})$ – i.e. in the sum over the three boxes of the ratio of absorbed solar energy F_S to the infrared or long-wave energy F_L radiated to space. The matter was pursued by examining the properties of a ten-box model of the entire globe. Results of this examination were reported by Paltridge (1975).

The ten-box model was based on energy balance requirements and consists of the formulation of three independent balance equations containing only three unknowns (cloud cover, θ, surface temperature, T, and latent and sensible heat transports, $LE+H$) apart from the meridional energy flow. Three basic equations were derived: (a) the energy balance of the ocean within a latitude zone, (b) the energy balance of the atmosphere within a latitude zone, and (c) parameterisation of latent and sensible heat flux in terms of surface temperature and cloud cover. The equations, parameters and constants involved are discussed by Paltridge (1974, 1975).

The equations can be solved to yield θ and T for any specified values of the meridional energy flows $LE+H$ (atmospheric flux) and O_N (net oceanic flux). As well, various parameters including the net solar energy input $F_S\!\downarrow$ and the long-wave energy output $F_L\!\uparrow$ at the top of the atmosphere could be calculated for each zone. A search was therefore made for some overall parameter of the entire global system which was minimised for *a unique set* of the meridional energy flows – a set which was close to those observed and which led to the observed mean annual distribution of zonal average cloud and of zonal average surface temperature.

It was found that minimising a quantity which had the dimensions of entropy exchange yielded extraordinary agreement between the observed and 'predicted' distributions of θ, T and meridional energy flow – see Fig. 5.4.6.

The model contained no direct specification of the system dynamics, since the cloud and surface temperature of each zone were calculated as passive variables purely from energy balance considerations. Apart from certain constants such as the albedo and emissivity of clouds, the input data consisted only of 'external' parameters – the solar constant and the mean meridional distribution of surface albedo. The implication of the results, if indeed they are accepted as more than coincidence, is that global dynamics is something of a passive variable which alters appropriately so as to satisfy the minimum principle. On this basis, application of the principle should allow *a priori* prediction of climate and climate change without recourse to detailed description of the internal processes.

In Paltridge (1975) the minimum principle was no more than a parameterisation which was apparently successful and potentially valuable but which had no obvious scientific basis. Such a situation is not unusual in

5.4 Models of climatic change

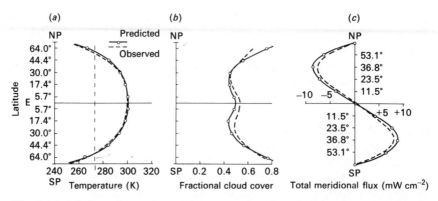

Fig. 5.4.6a, b & c. Comparisons between 'predicted' and observed north–south distributions of surface temperature, fractional cloud cover, and meridional (north–south) energy flux, respectively. The fluxes are normalised to unit surface area of each zone.

physics. However, in this case, and for at least two reasons, scientific backing is more than usually desirable. On the one hand the model itself contained some shaky physics and was therefore a doubtful medium for test. On the other, the only obvious comparison is with the 'one-off' example of the world itself. It is difficult to find other worlds where the physics is known sufficiently well to establish whether the principle is operative.

This problem is discussed in Paltridge (1976) where it is argued qualitatively that the earth–atmosphere system *should* adopt a format satisfying a principle of minimum entropy exchange. The latter also described an extension of the original two-dimensional, ten-zone model to a three-dimensional model where the earth is divided into 400 'boxes' on a 20 by 20 grid. The simple physics built into the individual boxes is much the same as in the previous paper (1975) and the deficiencies of that physics became both obvious and a considerable restriction.

However, the exercise has proved valuable. First, it appears possible to predict the individual atmospheric and oceanic cross-box energy flows without certain artificial constraints which were required in the earlier model. Secondly, a simple parameterisation of just one of the 'constants' built into the original physics allows reasonable simulation of the geographic distribution of cloud and surface temperature. Thirdly, the model provides an initial test-bed for an extension of the original concept – namely, to regard rainfall distribution as another variable which passively adjusts itself so as to minimise internal entropy production (as opposed to external entropy exchange). Introduction of this concept and an hydrologic cycle into the model produces a reasonable simulation of the global rainfall distribution.

252

Results

The predicted distributions of cloud cover and rainfall are given in Figs. 5.4.7 and 5.4.8. For comparison, the observed distributions of cloud cover and rainfall are shown in Figs. 5.4.9 and 5.4.10. The deficiencies of the model predictions are obvious enough – particularly in the case of cloud cover where the comparison is presented in the least flattering light of deviations from the zonal average. It should be remembered that the observations of cloud amount are as likely to be in error as are the model predictions. On the other hand, the simulation of rainfall is particularly pleasing bearing in mind the total lack of dynamic constraints and the large number of speculative assumptions in the model. It produces, for instance, low rainfall over the recognised desert areas, and belts of high rainfall in the tropics. It is important to note that low rainfall is generally associated with cloud deficit – a correlation which is not specifically built into the model.

Conclusion

The simple physics incorporated in the model has been stretched far beyond its capabilities in a preliminary attempt at simulating the geographical distribution of climate. Suffice it to say that the results are sufficiently realistic to lend strong support to the basic concept of treating dynamic processes as passive variables which simply adopt an appropriate format to satisfy thermodynamic requirements.

Some lines of future investigation are fairly obvious from this preliminary model. First, the concept of calculating cloud cover purely as a passive variable which satisfies energy balance would seem worth retaining. Secondly, sufficient physics should be built into the model to provide at least a simple parameterisation of the so-called constants involved in the thermodynamics. Thirdly, various restrictions imposed in order to simplify the model should be removed or modified. The most important task, however, is to engage in a full theoretical treatment of the various processes of internal entropy production.

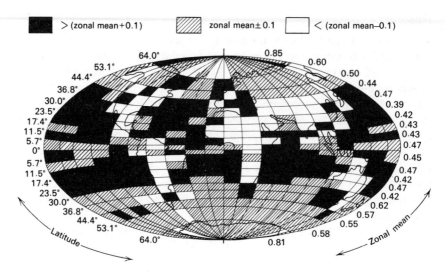

Fig. 5.4.7. Predicted fractional cloud cover in terms of deviation from the zonal means given on the right of the diagram. Annual mean case.

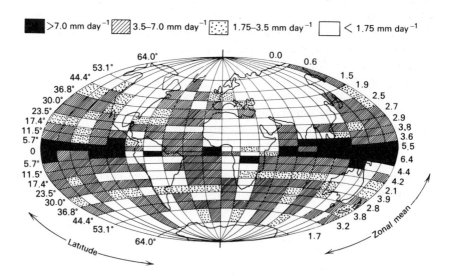

Fig. 5.4.8. Predicted rainfall distribution in four ranges. Zonal means (mm day^{-1}) are given on the right. Annual mean case.

254

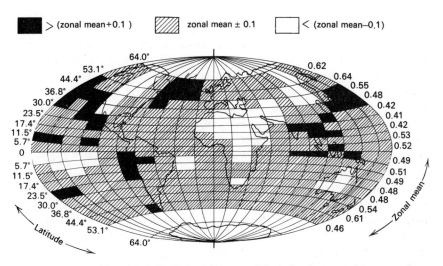

Fig. 5.4.9. Observed fractional cloud cover in terms of deviation from zonal averages given at right. Annual mean case. Polar region data not available. Data assembled from Berry *et al.* (1945) for overland, and from Miller & Feddes (1971) for over sea.

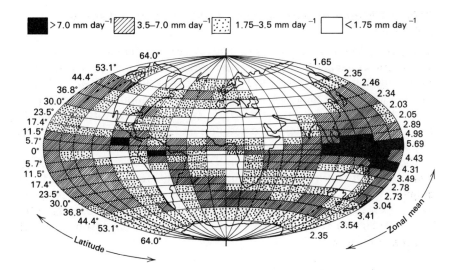

Fig. 5.4.10. Observed rainfall distribution in four ranges. Zonal averages (mm day^{-1}) are given on the right. Annual mean case. From Schutz & Gates (1974).

5.5   Volcanic events, climate, and climate modelling*

Much of this chapter has been devoted to various types of climate modelling within the cryosphere and atmosphere: and the example of a powerful new energy balance model has been presented. In this section, an example of the application of a complex general circulation model is given. The model is used to study the possible effects of a volcanic eruption which injects a large mass of debris into the stratosphere. One of the results of this computation is the calculation of reduction in the mean surface temperature of the hemisphere, a finding which is in agreement with deductions made from observational data. Here is in example of modelling and practical measurements confirming and reinforcing each other: observation provides a yardstick with which to measure the computer model's reality, and the latter, in turn, adds further credence and validity to the empirical results. But more importantly, the physics of the computer model allows many atmospheric processes (particularly the diffusion of the debris) to be studied in such detail as would not be possible by conventional means. The measurements and implications of eruptions are first given by Bray, and then a brief description of the numerical model, and its results, is presented by Hunt.

Volcanic eruptions and climate during the past 500 years
J. R. BRAY

Volcanic activity seems to have had a major influence on short-term temperature changes in the recent past (Arakawa *et al.*, 1955; Lamb, 1970, 1971; Mitchell, 1970; Bray, 1971; Reitan, 1974*a*; Pollack *et al.*, 1976). In this paper detailed analyses will be made of the influence of volcanism on temperature and crop yields since AD 1500 and an attempt will be made to combine long-term temperature and harvest data to give a relative temperature pattern for the Northern Hemisphere during that time.

The climatic data summarised in Table 5.5.1 include the two longest available temperature series (Central England, Manley, 1974; and Philadelphia, USA, Landsberg, personal communication); two other temperature series, one mainly European (Köppen in Arakawa, 1955), the other North American (New Haven); and grain harvest data from England, Europe and Japan.

Grain yield

The tendency for famines and poor grain harvests which occur as a result of cold summers to follow volcanic eruptions was noted for several Japanese famines by Arakawa (1955) and for a summary of the more

* Editorial contribution.

256

northerly European and Japanese data (Bray, 1971). This is confirmed by the data of Table 5.5.1, where it can be seen that there is also a tendency for the bad harvest years to run in series of two to four (which might be expected if these poor harvests were the result of several years of the cold springs and summers which tend to follow major volcanic eruptions). Thus, of the 44 bad and dearth grain harvests in England, 15 occurred up to 3 years after volcanic eruptions in which the Dust Veil Index (DVI) exceeded 1,000. Again, all but 3 of the 15 famine years followed eruptions $\geqslant 1,000$ DVI and all but 1 of the 12 dearth and 15 English famine years followed within 4 years of volcanic eruptions $\geqslant 500$ DVI. In northwestern Europe, 9 of the 22 poor grain harvests recorded by Le Roy Ladurie (1971) and Lamb (1959, 1970) from AD 1500 to 1970 occurred up to 3 years after volcanic eruptions $\geqslant 1,000$ DVI. And yet again, in Japan 31 of the 53 years with poor rice harvests from AD 1640 to 1970 occurred within 3 years after eruptions $\geqslant 1,000$ DVI, whilst 17 of the 24 Japanese famine years from AD 1640 to 1970 which occurred as a result of cold summers, were within 3 years after eruptions $\geqslant 1,000$ DVI. All of these relationships were statistically significant with chance probabilities from < 0.002 to < 0.03.

Temperature

Mean yearly temperatures before and after years with eruptions $\geqslant 1,000$ DVI have been studied for central England, Philadelphia, USA, and for five long-term New Zealand stations from 1868–1974. In all three areas there was a mean maximum yearly decline of 0.4–0.5 °C in the period up to three years following volcanic eruption. For large ash eruptions, considered separately, this decline was much greater. The mean central England temperature from 1669 to 1672 was 9.0 °C which was followed in the years after the 1673 eruption by temperatures of 8.7, 8.2, 7.9, 8.8 and 8.7 °C. The maximum yearly decline in the four years following the 1783 eruption was 1.6 °C for the central England data and 1.9 °C for Philadelphia. The eruptions from 1811 to 1815 were characterised by a maximum temperature decline of 1.8 °C for the central England data and 1.0 °C for Philadelphia.

If the poor crop dates summarised in Table 5.5.1 represent mainly years with below normal temperature, then these dates can be used to augment and extend the long-term temperature data. Thus if the harvest data in that table are considered to represent low temperature years, then from AD 1500 to 1970 there were 179 years which contained a total of 299 low-temperature observations. Of these 299 observations, 154 occurred within 3 years following the 32 volcanic eruptions $\geqslant 1,000$ DVI ($\chi^2 = 91.1$, $P < 0.001$) and 179 occurred within 4 years ($\chi^2 = 84.0$, $P < 0.001$). All but 3 of the 32 volcanic eruptions $\geqslant 1,000$ DVI were followed within 4 years by at least 1 low-temperature observation and there was a maximum of

257

5.5 Models of climatic change

Table 5.5.1. *Volcanic eruptions, poor harvests and low temperatures,* AD 1500–1970

Year	DVI	l.t.d.	Year	DVI	l.t.d.
1500	1,000	—	1648	—	ab
1501	—	a	1649	—	ab
1502	—	a	1650	500	—
1510	(300)	—	1658	—	a
1520	x	a	1660	2,100	—
1521	—	a	1661	—	a
1527	—	ab	1664	(500)	—
1528	—	ab	1665	—	g
1529	—	b	1673	1,000	a
1535	300	a	1674	x	ag
1538	x	—	1675	—	bcg
1539	50	—	1678	—	ag
1545	—	a	1680	1,400	—
1550	70	a	1681	—	c
1551	—	a	1683	—	g
1553–54	(1,000)	—	1684	—	g
1555	—	a	1688	—	g
1556	—	a	1689	—	g
1560	—	ac	1691	—	g
1562	—	a	1692	—	ag
1565	x	a	1693	800	abg
1569	10	—	1694	900	bg
1573	—	a	1695	—	acg
1577	x	—	1696	x	bcg
1581	x	—	1697	—	abg
1586	1,000	a	1698	x	ag
1587	x	—	1700	(100)	—
1590	50	—	1707	(1,300)	—
1593	1,000	—	1708	—	a
1594	—	a	1709	—	ab
1595	—	a	1712	300	—
1596	—	ab	1713	—	a
1597	300	ab	1716	x	g
1598	140	b	1717	(570)	—
1601	(1,000)	—	1721	(750)	—
1606–07	550	—	1724–29	420	—
1608	x	a	1725	—	b
1614	1,000	—	1727	70	a
1622	—	a	1728	x	a
1625	(800)	—	1730–33	(400)	—
1627	x	—	1732	x	c
1630	—	a	1735	x	—
1631	600	—	1739	+	—
1636	300	—	1740	—	abg
1638	500	—	1741	—	b
1640	500	c	1742	x	bg
1641	1,500	c	1744	400	—
1642	—	c	1749	70	h
1646	300	b	1751	20	g
1647	—	ab	1752	1,000	—

Year	DVI	l.t.d.	Year	DVI	l.t.d.
1754	300	—	1826	370	—
1755	1,600	c̱fh	1827	70	—
1756	x	a̱	1828	70	—
1758	—	fh	1829	370	fgh
1759	550	h	1830	90	—
1760	250	—	1831–33	(1,400)	—
1762	—	h	1832	—	hj
1763	(600)	cfh	1833	—	c̱h
1765	—	g	1834	—	c̱
1766	(2,950)	f	1835	(4,000)	chj
1767	x	fh	1836	x	c̱hj
1768	900	fh	1837	x	c̱fhj
1769	—	h	1838	x	c̱fgh
1770	x	gh	1839	—	ch
1766–71	(3,400)	—	1840	—	fgh
1771	—	fh	1841	—	h
1772	250	h	1843	x	chj
1774	—	h	1844	—	h
1775–77	(1,000)	—	1845	800	cfg
1776	—	h	1846	(1,000)	—
1777	—	ch	1849	—	h
1779	650	—	1850	x	cfh
1782	—	fg	1852	(550)	h
1783	2,900	c̱	1853	—	g
1784	—	cfghj	1855	(700)	fgh
1785	—	c̱fhj	1856	—	hj
1786	x	c̱fgh	1857	—	hj
1787	x	c̱	1858	20	h
1789	x	—	1859	—	h
1792	(20)	—	1860	x	g
1793	70	—	1861	(800)	—
1794	70	—	1862	—	h
1795	1,000	—	1864	—	f
1796	300	—	1866	x	c̱
1798	(300)	—	1867	(20)	c̱h
1799	600	fg	1868–70	(1,000)	—
			1868	—	c̱hj
1803–05	(1,100)	—	1869	x	c̱hj
1805	—	f	1870	—	f
1807–10	(1,500)	—	1871	—	fh
1808–09	—	f	1872	150	fh
1811	400	c	1873	30	hj
1812	600	fghj	1874	—	h
1813	(200)	h	1875	1,000	cfh
1814	(300)	cfgh	1876	—	h
1815	3,000	chj	1877	50	c
1816	—	ḇfghj	1878–81	(1,250)	—
1817	300	ḇchj	1879	—	cg
1818	—	hj	1881	—	c
1820–21	(500)	—	1883	1,070	fhj
1820	—	j	1884	—	cf
1821	300	hj	1885	300	hj
1822–24	(2,200)	—	1886	1,100	h
1823	x	hg	1887	—	g
1825	70	c	1888	(750)	ghj

259

5.5 Models of climatic change

Table 5.5.1 (*cont.*)

Year	DVI	l.t.d.	Year	DVI	l.t.d.
1890	170	—	1924	—	h
1891	—	g	1926	—	h
1892	100	cgh	1931	5	c
1893	—	hj	1932	70	—
1895	(1,300)	ch	1934	—	c
1896	—	c	1936	30	—
1898	140	—	1940	—	h
1899	10	—	1941	—	c
1901	—	h	1945	x	c
1902	(1,000)	c	1947	(70)	—
1904	40	hj	1951	20	—
1905	10	ch	1953	7	c
1906	110	c	1958	—	h
1907	500	h	1960	100	h
1912	500	cf	1962	—	h
1913	10	c	1963	800	gh
1914	40	—	1966	200	—
1917	—	ghj	1967	—	h
1919	—	g	1968	100	—
1921	200	—	1970	325	—

l.t.d. = low temperature data.

a = bad or dearth grain harvests, England (Hoskins, 1964, 1968), 1500–1759.

b = poor grain harvest, western Europe (Le Roy Ladurie, 1971; Lamb, 1959, 1970), 1500–1970.

c = poor grain harvests, Japan (Arakawa, 1955; Terada, 1972), 1640–1970, 1500–1970. Underlined crop data are those years in which famine occurred as a result of a cold summer.

f = temperature departure $\geqslant$ −0.4 °C from normal (Köppen in Arakawa, 1955), 1750–1920.

g = temperature $\leqslant$ 8.5 °C, central England (Manley, 1974) 1659–1973.

h = temperature $\leqslant$ 53.9 °F, Philadelphia USA (Landsberg, 1975) 1738–1967.

j = $T <$ 48.0 °F, New Haven 1781–1940.

+ = two eruptions occurred in 1739 (Tubachinskaya and Tarumai, Japan). It is evident from temperature, crop and famine data that at least one of these eruptions produced a large ash volume (1740 was the coldest year in England).

x = eruptions with unknown DVI (Dust Veil Index).

All DVI data are from Lamb (1970) with the addition of an estimate of 300 for AD 1510 from data in Thorarinsson (1967); underlined DVIs are *estimated* from temperature data. DVI values without parentheses are of greater reliability.

14 observations in the 3 years following the eruption of 1835. The close correspondence between volcanic eruptions and subsequent low temperatures is clearly seen in the data presented, especially following the larger eruptions and those occurring in more or less successive years.

Glaciation

A temporal relationship between volcanic eruptions and increased periods of glacial advance has been noted for the recent 'Little Ice Age' (Bray,

260

Table 5.5.2. *Phases of volcanic eruptions, low temperatures and glacial advances in the Northern Hemisphere, AD 1500–1970*

Volcanic eruptions		Low-temperature data		Glacial advance phase in Alps	
Date	DVI	Date	No. obs.	Date	No. refs.
1500	1,000	1501–02	2	*c.* 1500	1
1553–54	1,000	1555–56	2	—	—
1586	1,000	1586	1	—	—
1593–98	1,440	1594–98	7	—	—
1601	1,000	—	—	1595–1620	5
1614	1,000	—	—	—	—
1625	800	—	—	1628	1
1636–41	3,400	1640–42	3	—	—
1646	300	1646–49	7	1640–50	1
1660–64	2,600	1661–65	2	—	—
1673	1,000	1673–78	8	—	—
1680	1,400	1681–84	3	1676–81	3
1693–98	3,250	1693–98	16	1710–28	2
1707	1,300	1708–09	3	—	—
1739	?	1740–42	6	1741–43	2
1752–55	2,900	1755–59	7	—	—
1766–71	3,400	1766–72	11	—	—
1775–77	1,000	1776–82	5	1770–80	4
1783–86	3,470	1783–87	15	—	—
1795–99	2,200	1799	2	—	—
1803–10	2,600	1805–09	3	—	—
1811–17	4,800	1811–18	24	1814–22	4
1822–24	2,200	1823–25	3	—	—
1831–35	5,400	1832–41	25	—	—
1845–46	1,800	1845–50	7	1835–55	5
1852–55	1,250	1852–60	12	—	—
1860–62	1,500	1860–67	6	—	—
1868–70	1,000	1868–74	14	—	—
1875	1,000	1875–79	7	—	—
1883–86	2,470	1883–88	12	1875–90	1
1902–07	1,660	1902–07	7	—	—
1912	500	1912–13	3	1910–26	1
1963–66	1,000	1963–67	3	—	—

1971, 1974a), the period from AD 860 to present (Lamb, 1970), the past 40,000 years (Bray, 1974*b*), and the entire Quaternary (Kennett & Thunell, 1975). Any testing of the timing of volcanism and glaciation must be based on exact information for the periods of glacial advance. There is only one area for the past 500 years where such information is fairly complete – the European Alps. To eliminate bias in the choice of advance dates, Table 5.5.2 summarises the results of five studies of Alpine glacial advance phases (Kinzl, 1932; Matthes, 1943; Grove, 1966; Le Roy Ladurie, 1971;

Patzelt, 1973). These results show a total of 12 advance phases from AD 1500 to 1970 of which the advances from 1595 to 1620, 1770 to 1780, 1814 to 1822 and 1835 to 1855 were the most prominent. The timing of these advance phases is shown in Table 5.5.2 in relation to a listing of volcanic eruptions ≥ 1,000 DVI together with the number of low-temperature observations from Table 5.5.1 immediately following each eruption. In several cases, additional eruptions < 1,000 DVI have been included when they were proximate to major eruptions.

The data in Table 5.5.2 tentatively suggest that the Alpine glacial advance phases follow series of volcanic eruptions and that they also follow the low temperatures which seem to accompany increased volcanicity. Because of this apparent tendency of glacial phases to follow a series of eruptions, it is difficult to assess the exact short-term relationships between volcanism and glacial advance. The majority of glacial advances during the 1814–22 phase occurred around 1818–20 which was 3–5 years after the 1815 eruption and 7–9 years after the 1811 eruption. The majority of advances during the 1835–55 glacial phase occurred around 1843–50, 8–15 years after the massive 1835 eruption. Many of the ice advances during the 1595–1620 phase culminated around 1600–5, which was 14–19 years after the 1586 eruption and 7–13 years after the 1593 eruption. These lag periods are similar to the observed intervals between the onset of snow accumulation in a firn field and subsequent glacial advance at the terminus. These intervals have been reported as from 3 to 6 years for shorter glaciers on steep beds (Faegri, 1948; Flint, 1971; Manley, 1961) and from 7 to 18 years for longer glaciers (Bray & Struik, 1963; Faegri, 1948; Flint, 1971; Manley, 1961).

Conclusions

There are significant tendencies for low temperatures and for poor grain harvests in the higher latitude regions of the Northern Hemisphere to follow volcanic eruptions. The low grain yield dates combined with the low-temperature data gave a general temperature pattern for the AD 1500–1970 period which was significantly associated with volcanic eruptions ≥ 1,000 DVI. The greater the eruption and the more these eruptions were part of a closely spaced series, the longer lasting and more intense were these low-temperature phases. Many of the major glacial advance periods since AD 1500 may have been triggered by the shorter ablation periods caused by the cold resulting from large eruptions and especially from a series of large and closely spaced eruptions. This conclusion does not imply that volcanism caused the 'Little Ice Age', but rather that volcanism may be related to the timing of the Holocene ice advances and may have contributed to the lowered global temperatures which accompanied these phases.

A simulation of the possible consequence of a volcanic eruption on the general circulation of the atmosphere

B. G. HUNT

Volcanic eruptions are one of the most spectacular of all naturally occurring phenomena on Earth. Generally their impact is very localised, where it tends to be disastrous. More rarely an eruption is of sufficient magnitude to produce global effects resulting from the transport of debris to the high atmosphere.

Previous attempts to simulate the atmospheric consequences of an eruption have been extremely limited. For example, Batten (1974) used a two-layer general circulation model to investigate the influence of a *fixed* mid-latitude 'stratospheric' dust cloud on the atmosphere; MacCracken & Potter (1975) incorporated a *fixed* stratospheric aerosol loading, corresponding to that caused by the Mt Agung eruption, into a rather elaborate zonally averaged model; and other calculations have been made with radiative–convective equilibrium-type models to explore the possible effects of aerosols in the atmosphere (e.g. Reck, 1974; Weare *et al.*, 1974). In the present model the volcanic debris was realistically diffused outward from an initial tropical source, thus producing a variable response in both position and time. Also the three-dimensional general circulation model used here was capable of a much more adequate representation of the basic atmospheric structure than those involved in the previous studies.

The model experiment

The general circulation model was hemispheric with 18 vertical levels distributed between the surface and 37.5 km, each level having approximately 1,200 grid points. The model was integrated for fixed annual mean conditions without orography or land–sea contrast, but including the hydrologic cycle. Surface temperatures were determined from an instantaneous balance between the various heating and cooling components. The radiation scheme was reasonably comprehensive, incorporating solar and long-wave radiation determined by fixed climatological distributions of water vapour, carbon dioxide, ozone, clouds and surface albedo. The convective adjustment scheme in the model was based on a relative humidity criterion of 100% for the onset of convective instability and precipitation. Further details of the model formulation are given in Hunt (1976a) and Manabe & Hunt (1968).

It was decided to try to simulate the consequences of an eruption similar to that of Krakatoa which occurred in 1883 at 6° S, 105° E. A large eruption such as this was desirable because of the existence of signal-to-noise problems in model experiments (e.g. Chervin & Schneider, 1976).

5.5 Models of climatic change

In addition a reasonable amount of information was available concerning the debris uplifted to the stratosphere by Krakatoa (Deirmendjian, 1973). A low-latitude site was preferred because these produce more significant atmospheric effects than those at high latitudes (Lamb, 1970). Finally, considerable experience had been gained in an earlier experiment (Hunt & Manabe, 1968), involving the large-scale diffusion of passive radioactive debris from a tropical source in a very similar model.

In order to reduce truncation errors the more realistic point source was replaced by an initial distribution of volcanic debris in a uniform zonal ring centred on the equator at a height of about 23 km. Because of the equatorial 'wall' in the model, subsequent diffusion was constrained to one hemisphere.

Two modifications were made to the control model in order to represent the volcano. The first required the inclusion of an additional mixing ratio to simulate the geographical distribution of the debris. The second involved coupling the optical properties of the debris with the radiation, in order to permit an interaction between the debris and the thermodynamical properties of the model.

The equation controlling the debris distribution (Hunt & Manabe, 1968, p. 505) was time-integrated along with the other prognostic equations using a time-step of 10 minutes. The debris was transported by winds generated in the model, and thus the subsequent distributions were obtained without any recourse to the use of large eddy diffusion coefficients such as are required in simpler models. Any debris which reached the lowest model level (0.85 km) was assumed to be immediately washed out of the atmosphere; no other sink was incorporated into the model. This procedure probably resulted in a slight overestimate of the tropospheric residence time for the volcanic debris.

Specification of the initial number densities for the debris, and the coupling of the optical properties of the debris with the radiative code in the model, required definition of an optical model. The model chosen was due to Deirmendjian (1969) and its present application is described in Hunt (1976b). The radiative calculation was performed once per day using the local debris concentration to estimate total backscattering. Resulting radiative tendencies were maintained constant between radiative calculations.

The experiment was started from day 254 of the control run of Hunt (1976a) which, however, extended past this time (assumed to be when a state of quasi-equilibrium was reached). The initial distribution of volcanic debris was inserted instantaneously into the fully developed, three-dimensional flow field of the model, and permitted to interact with it immediately. The volcano model was integrated for a total of 150 days to give an overlap of 130 days with the control experiments. An additional

264

experiment was run for about 14 days starting from precisely the same conditions as the volcano model, but omitting the coupling of the debris with radiation.

Large-scale diffusion of the volcanic debris

A very limited discussion will be presented here: more detail is given by Hunt (1976*b*).

Figure 5.5.1 (*a*)–(*e*) shows synoptic distributions at alternate levels from 1 to 9 after 90 days. Levels 1 and 3, characteristic of the middle stratosphere, differ essentially from the lower levels. Maximum debris concentrations were located in the tropics, and a monotonic poleward gradient was maintained at all times. These levels exhibited the expected low wave numbers and lack of synoptic activity of the actual stratosphere. Level 5 at this time showed characteristics of both middle and lower stratosphere, viz. maximum mixing ratios at low and middle latitudes and more synoptic activity. At levels 7 and 9 the maximum mixing ratios were located in mid-latitudes and the minimum in the tropics, the reverse of levels 1 and 3. These synoptic distributions were clearly associated with higher wave numbers.

A comparison of the computer surface-pressure distribution (not shown here) with the debris distributions for levels 7 and 9 indicated that, in general, high mixing ratios were associated with surface lows and lower values with surface highs. Such correlations are well known for atmospheric ozone (Dobson *et al.*, 1927; Reed, 1950). Hunt & Manabe (1968) in their Fig. 27 presented a breakdown of the transport mechanisms responsible for these correlations and their analysis is equally applicable here.

A cross-section of the mean latitude–height distribution of the debris-mixing ratios at day 70 (Fig. 5.5.1 (*f*)) shows data similar to those in the synoptic distributions at day 90.

Further examination of the poleward flux of volcanic debris averaged over the last 10 days of the experiment, showed a dominance of transport by large-scale eddies, with the mean motions being relatively unimportant outside the tropics.

Comparison of the fluxes for the interactive and non-interactive volcanic experiments showed them to be substantially different, in particular for the mean meridional motions, due to the influence of the volcanic debris on solar radiation.

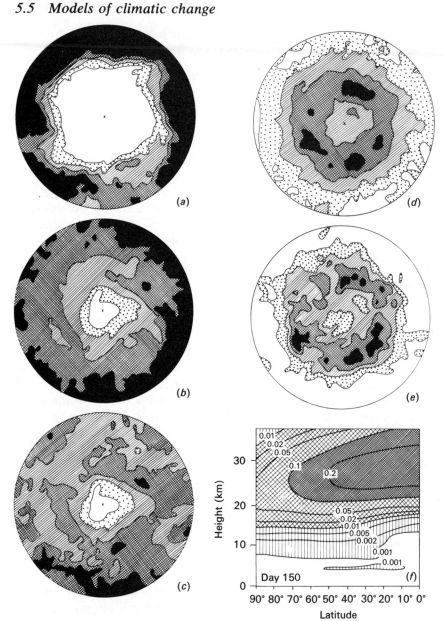

Fig. 5.5.1. Instantaneous synoptic distributions of the volcanic debris mixing ratio at alternate levels from 1 to 9 in the model after 90 days of integration are given in (*a*)–(*e*). These levels correspond to heights of 37.5, 25.6, 20.5, 16.8 and 13.7 km respectively. In these figures the circumscribing circle identifies the equator and the cross the pole. The variation in the mixing ratio represented by the shading varies with height, ranging between a factor of 3 and 10. Largest mixing ratios are indicated by darkest shading. In (*f*) the zonal mean latitude height distribution of the volcanic debris mixing ratio is shown after 70 days of integration. (Units: 10^{-6} g g^{-1}).

266

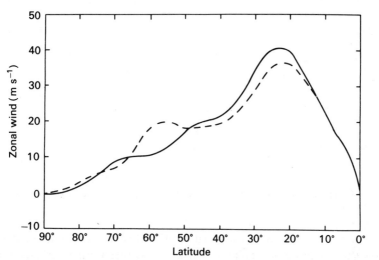

Fig. 5.5.2. The zonally averaged zonal winds at 12.3 km, the height of the subtropical jet stream, for the control and volcano experiments (—— control, – – – – volcano, averaged from 50–60 days).

Large-scale atmospheric consequences

Changes in the zonal wind attributed to the volcanic debris are indicated by latitudinal variation of the zonally averaged wind at the height of the subtropical jet stream for the period 50–60 days (Fig. 5.5.2). Overall, there was a reduction in the zonal wind intensity during the volcano experiment, particularly for the jet stream itself. The full data indicated that such a reduction was typically produced in the 50- to 100-day period of the experiment with little difference being observed prior to this period. Subsequently the intensity of the jet in the volcano experiment exceeded that in the control run. The nature and magnitude of the differences in the wind intensities between the two experiments were such that they reasonably can be attributed to the influence of the volcanic debris rather than random perturbations. The vertical and mean meridional velocities also exhibited variations between the control and volcano experiments. These variations were basically changes in the intensity rather than the structure of the mean meridional cells of the general circulation.

In general it was difficult to identify dynamical variations between the two models which could unequivocally be associated with the volcanic debris. However, comparison of the two outputs (Hunt, 1976*b*) does show a clear suppression of the normally marked pulsations in the hemispheric mean upwards flux of geopotential energy by the large-scale eddies, which is an important observed feature of the real atmosphere (Webster & Keller, 1974).

267

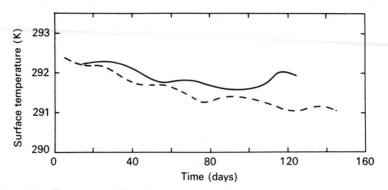

Fig. 5.5.3. Comparison of hemispheric mean surface temperatures based on averages over 10-day periods (—— control, – – – – volcano).

Since surface pressure is the most readily observed of all atmospheric variables it would be particularly useful if a perturbation associated with volcanic eruptions could be identified from the current experiment. Unfortunately, the normally large day-to-day synoptic scale variations in surface pressure, particularly polewards of 30° latitude, mask any systematic change due to the volcanic debris, even using 10-day means.

Changes in temperature, particularly at the surface are considered most important in the experiment, because of their significance in an agro-economic sense.

A comparison (not shown here) of instantaneous synoptic surface temperatures (calculated on the assumption of a balance between the various heating terms with no storage in the earth) for the volcano experiment at days 0 and 90 revealed that significant cooling occurred at all geographical regions but particularly in the tropics.

In Fig. 5.5.3, a comparison is made between the volcano and control experiments for the hemispheric mean case. Over most of the integration the temperature difference was only 0.2–0.3 K but towards the end of the experiment it approached 1 K. This anomalous behaviour resulted principally from a warming in the subtropics of the control model. Although the average net hemispheric cooling was slight, this result does highlight the potential climatic importance of large volcanic eruptions, as has been stressed by Landsberg & Albert (1972) for the particular case of the Tambora eruption of 1815. Lamb (1970, 1971) has discussed the scant data available which suggests that hemispheric coolings of the above magnitude are not unrealistic. When compared with the cooling of 0.6 K which has occurred in the Northern Hemisphere between 1958 and 1963 (Starr & Oort, 1973), the possible impact of a large eruption is seen to be quite significant, though lasting atmospheric effects of single eruptions are improbable (Hunt, 1976*b*).

6. Modification of climate

6.1 The biosphere, atmospheric composition and climate
I. E. GALBALLY & J. R. FRENEY

The biosphere can influence climate through the gaseous and aerosol composition of the atmosphere, and the state of the land surfaces with respect to albedo, thermal capacity, water-holding capacity and aerodynamic roughness.

People are a minor constituent of the biosphere but can have a potentially large effect on climate by many of their activities. These include:

(i) changing the atmospheric composition with pollutants;

(ii) modifying the biosphere by deforestation, chemical pollution, etc.; and

(iii) generating heat by fossil fuel combustion and nuclear fission processes.

Incoming solar radiation drives the atmospheric circulation and produces the climate that we know. About 30% of the incoming solar radiation is reflected back to space, about 20% is absorbed by clouds, gases and particles in the atmosphere, and the remainder is absorbed at the earth's surface. Our discussion will be confined to gases and particles in the atmosphere that absorb and reflect incoming or outgoing radiation and that participate in cloud formation. Only those effects, which are important on time scales not much longer than human lifetimes are discussed.

Of the atmospheric constituents, ozone, carbon dioxide, aerosols and cloud condensation nuclei are important in determining the earth's climate. Each of these constituents has an important source or regulating mechanism in the biosphere and the possible variability of these constituents due to the biospheric influence is generally unknown.

Here we attempt to quantify the various mechanisms of biosphere–climate interaction. These estimates should be regarded as indicators of whether or not any of the various mechanisms merit further study because of its possible influence on climate.

Ozone

Ozone absorbs incoming solar radiation of wavelength shorter than about 300 nm and thus shields the earth's surface from short-wave ultraviolet

269

6.1 Modification of climate

Table 6.1.1. *Relative importance of ozone destruction processes in the present atmosphere*

Ozone decomposition at the earth's surface	1%
Ozone–atomic oxygen recombination	~20%
Catalytic destruction by water vapour	10–20%
Catalytic destruction by odd nitrogen oxides (NO, NO$_2$, etc.) and possibly chlorine oxides	60–70%

radiation. This absorption converts about 2% of the energy in the incoming radiation into heat for the stratosphere (Kondratyev, 1969). A change in the stratospheric ozone amount would change the heating rate of the stratosphere and through this could have effects on stratospheric and tropospheric circulation. Also a change in the ultraviolet radiation reaching the earth's surface (due to a change in the ozone amount) could have effects on those biological systems that are exposed to sunlight (Grobecker, 1975).

Atmospheric ozone is formed by solar ultraviolet radiation (< 242 nm) dissociating molecular oxygen. This ozone is being continuously destroyed by various mechanisms. The relative importance of the various destruction mechanisms is determined by atmospheric transport processes, for example, the rate of ozone destruction at the earth's surface is regulated by the transport of ozone from the stratosphere to the troposphere. These ozone destruction mechanisms and their relative importance in the present atmosphere are given in Table 6.1.1. The amount of ozone present in the atmosphere is the result of an equilibrium between the production, transport and destruction processes.

Changes in the rate of any of these destruction processes will lead to changes in the equilibrium ozone amount. The chemistry of these processes has been extensively discussed elsewhere (Chapman, 1930; Hunt, 1969; Galbally, 1971; Johnston, 1974; Crutzen, 1974; Rowland & Molina, 1975).

Reaction with the odd nitrogen oxides (i.e. those nitrogen oxides with only one nitrogen atom, NO, NO$_2$, etc.) provides the major sink for atmospheric ozone. These odd nitrogen oxides are formed in the stratosphere from nitrous oxide, N$_2$O (McConnell & McElroy, 1973), which comes from biological denitrification processes in soils and oceans. A large number of soil and water bacteria are able to respire using nitrate and nitrite instead of oxygen (Schlegal, 1974). Nitrogen is the final product of denitrification but under certain conditions N$_2$O and NO, which are intermediates in the denitrification reaction

$$NO_3^- \rightarrow NO_2^- \rightarrow NO \rightarrow N_2O \rightarrow N_2$$

are liberated and reach the atmosphere (Renner & Becker, 1970). A factor

270

which is most conducive to nitrous oxide evolution in soil is a high soil moisture content (Arnold, 1954).

There have been measurements of the annual production of N_2O in three soil types and these have been extrapolated to give a global soil source of N_2O of about 2×10^{13} g N_2O yr^{-1} (Schutz *et al.*, 1970).

Ocean surface waters, with respect to the atmosphere, are supersaturated with N_2O and it is estimated that the ocean–atmosphere flux of N_2O is 14×10^{13} g N_2O yr^{-1} (Hahn, 1974). However, these results are weighted heavily towards three high concentrations in the tropical Atlantic and thus the global oceanic–atmospheric flux of N_2O may be as low as 5×10^{13} g yr^{-1}.

The photochemical destruction rate of N_2O in the atmosphere is about 2×10^{13} to 3×10^{13} g N_2O yr^{-1} (Bates & Hays, 1967). A comparison of the production and destruction rates suggests that there may be some unidentified sink for atmospheric N_2O and that the atmospheric lifetime of N_2O is between 20 and 40 years.

The N_2O concentration in the surface layer of the ocean (0–100 m) is about 0.5 μg l^{-1} (Hahn, 1974) and thus the surface layers of the world's oceans contain a total of about 2×10^{13} g N_2O. It is evident that the ocean–atmosphere flux of N_2O is large compared with the N_2O content of the ocean surface waters. In this layer the N_2O must have a turnover time of less than a year suggesting that there must be some equilibrium between biological N_2O production and loss of N_2O to the atmosphere.

From this discussion it is apparent that the major destruction mechanism for atmospheric ozone, catalytic destruction by nitrogen oxides, is driven by a biological process, denitrification. Year-to-year changes in the biological activity of the oceans (see Ryther, 1966) or changes in the ocean surface temperatures could change the oceanic–atmospheric flux of N_2O leading to long-term variations in the total atmospheric ozone content.

Mankind can cause an increase in atmospheric nitrous oxide by adding nitrogen compounds to the environment. In 1966–7, 2×10^{13} g N yr^{-1} was added to soil as fertiliser (Byerly, 1970) and much of this nitrogen would end up as nitrate. A further 2×10^{13} g N yr^{-1} was accidentally converted to nitrogen oxides in fossil fuel combustion processes. These nitrogen oxides after release to the atmosphere, and conversion to nitrate within the air, would have ultimately reached the soil and oceans through dry deposition and rainfall (Galbally, 1975). Broadbent & Clark (1965) suggest that 10–15% of this total yearly mineral nitrogen input is lost by denitrification. This can occur within small pockets of microbial activity or within the larger sized soil aggregates in presumably well-aerated soils. If we assume that 10% of the man-made nitrogen is converted in the soils and oceans by denitrifying bacteria to N_2O and N_2, and that there is equal partitioning between N_2O and N_2 we calculate that in 1966–7 mankind could have

271

6.1 Modification of climate

effected the release of 3×10^{12} g N_2O yr^{-1} to the atmosphere. This source is a few percent of the estimated natural emissions of N_2O. If production of fertilisers and fossil fuel combustion continue increasing at current rates then within two or three decades mankind could be contributing significantly to the N_2O cycle in the atmosphere leading to a reduction of the ozone content of the atmosphere.

Aircraft that fly in the stratosphere directly inject nitrogen oxides from their exhausts. With increased stratospheric flights this source of nitrogen oxides can lead to a reduction in the equilibrium ozone amount (Johnston, 1974; Crutzen, 1974; McElroy et al., 1974; Grobecker et al., 1974). It is estimated (Grobecker et al., 1974) that the operation of 100 Concorde/Tu 144 type aircraft would lead to a 0.4% reduction in total ozone in the Northern Hemisphere, and operation of 100 Advanced SSTs would lead to a 1.7% reduction.

Stratospheric ozone can also be catalytically reduced by chlorine (Molina & Rowland, 1974). Currently there are no measurements of the basic reactants, chlorine and chlorine oxide (ClO) in the stratosphere so the significance of this mechanism cannot be determined. Methyl chloride, the most likely source of natural chlorine in the stratosphere is present in the troposphere in approximately 10^{-9} v/v mixing ratio (Lovelock, 1975). Methyl chloride is relatively insoluble in water, hence is not washed out by rain, and can probably penetrate the troposphere to the stratosphere. One of its sources appears to be biological activity in the ocean (Lovelock, 1975) although there is scant knowledge of this.

The man-made chlorine compounds, Freon 11 ($CFCl_3$) and Freon 12 (CF_2Cl_2), have been detected in the stratosphere (Lovelock, 1974; Krey & Lagomarsino, 1974; Hester et al., 1975; Schmeltekopf et al., 1975) where they can be photodissociated giving chlorine, thus causing a reduction in the ozone equilibrium level. The problem of fluorocarbons in the environment and their possible depletion of atmospheric ozone has been extensively reviewed by Rowland & Molina (1975). Two points seem important here. The atmospheric lifetime of the fluorocarbon gases are 20 years or more, so even when further release is completely stopped, any detrimental effects on the ozone layer will continue for 20 years or more. Furthermore it takes some time for the fluorocarbons to diffuse from the ground-level source to the 25–30 km level, where their photolysis products catalytically destroy ozone. If fluorocarbon production rates were held constant at present levels, the ozone depletion could be a few percent at the turn of the century, rising to a steady state of the order of 10% next century.

Analysis of global distributions of total ozone for the period 1957–70 reveals an upward trend of about 7.5% per decade in the Northern Hemisphere and about 2.5% per decade in the Southern Hemisphere

(London & Kelley, 1974). The record for a single station at Arosa, Switzerland (Birrer, 1974), analysed in 10-year periods from 1926 to 1972 reveals trends varying between extremes of +6% per decade to −9% per decade with no significant trend over the whole period. Analysis of these trends is complicated by Kulkarni's (1973) finding that of an apparent trend of +6% per decade of total ozone at Brisbane, Australia, 3% could be due to a change in the haze scattering in the atmosphere.

There is a problem of separating human effects from such presumably natural long-term fluctuations in total ozone content. Until we can quantitatively explain the natural year-to-year and longer time period fluctuations of atmospheric ozone in terms of the production, transport and destruction processes, we are in a poor position to determine any human effects on the ozone layer.

Carbon dioxide

The earth's accessible reservoirs of CO_2 are in the atmosphere, the oceans and in the living and decaying biomass. There is continuous exchange between these reservoirs, and mankind is continuously releasing the CO_2 stored in fossil fuels.

The important problem here is to determine the reasons for, and the effects of, an observed rising concentration of atmospheric CO_2. To understand this increase we must know the exchange between these reservoirs and the proportion of carbon dioxide released from fossil fuels which remains in the atmosphere.

Plants remove carbon dioxide from the atmosphere by photosynthesis and release carbon dioxide by respiration. The short-lived land biota (lifetime $1 \sim 10$ years) have a carbon content of perhaps one-tenth of the atmosphere's and the long-lived biomass (lifetimes $10 \sim 100$ years) has perhaps twice the carbon content of the atmosphere (Keeling, 1973a). The biosphere would respond to an increase in atmospheric CO_2 by growing faster, if no other nutrients are limiting, leading to an increase in the storage of carbon in the living and decaying biomass.

During this century the global land biomass could have been decreased due to deforestation or chemical pollution, e.g. acid rain, or it could have increased due to increased use of plant fertiliser and the increased atmospheric CO_2 concentration. Pearman & Garratt (1972) suggest that during this period deforestation of tropical areas could have increased the atmospheric CO_2 concentration by 0.3 ppm per year, this being about 40% of the observed increase. Farmer & Baxter (1974) present evidence (from carbon isotope studies on trees rings) that there could have been a CO_2 increase of perhaps 14 ppm due to enlargement of the world's farmland at the expense of forest and grassland, between 1900 and 1920. In the clearing process wood and litter either were burnt or allowed to decay, thus releasing CO_2 to the atmosphere.

Hall *et al.* (1975) have determined the last 15 years' record of biotic metabolism in the Northern Hemisphere by examining the annual cycles in the CO_2 record. They found no trend in the semi-annual net photosynthesis or semi-annual net respiration. Unfortunately they do not indicate what actual change in the metabolic rates or in the biomass could have been detected by their analysis. Keeling (1973*a*) has analysed various global environmental models of CO_2 and finds the maximum likely increase in the biomass between late last century and 1954 to be 4%. An actual increase of up to 2% during that period would be sufficient to resolve all the observational facts regarding CO_2 with the predictions from the models. Such an increase in the biomass is undetectable with current measurements.

The ocean surface waters to 100 m depth have a carbon content approximately equal to the atmosphere's (Pearman & Garratt, 1972). In ocean surface water inorganic carbon is present as dissolved carbon dioxide 0.5%, bicarbonate ions 82%, and carbonate ions 18%, giving a total dissolved inorganic carbon content of 2.01×10^{13} moles 1^{-1} (Broecker, 1974). The CO_2 partial pressure in equilibrium with the sea water is determined by the alkalinity,* the total dissolved inorganic carbon, and the temperature of the sea water. A 10% increase in the CO_2 partial pressure in equilibrium with sea water leads to a 1% increase in the total inorganic carbon content of the water. Excellent discussions of the complexities of these processes have been presented by other authors (Broecker, 1974; Keeling, 1973*a*).

If the total inorganic carbon content of sea water is constant, the partial pressure of CO_2 in equilibrium with the water rises 4% per 1 °C rise in water temperature (Broecker, 1974). The average sea-surface temperature of the North Atlantic ocean decreased by up to 0.3 °C between the early and late nineteenth century (Lamb, 1972*a*) and by 0.6 °C over the period 1951–72 (Wahl & Bryson, 1975). Such changes would appear to be significant for the global balance of uptake and release of CO_2, and have been suggested to explain aspects of the global atmospheric CO_2 data.

Careful measurements at the remote observatory on Mauna Loa, Hawaii, showed that the annual average atmospheric concentration of CO_2

---

* The alkalinity is equal to the difference in charge between all cations and anions (excepting HCO_3^- and CO_3^{2-}) in sea water and is a comparatively fixed quantity. Sea water is electrically neutral, therefore the sum of charges on HCO_3^- and CO_3^{2-} per unit volume must equal the alkalinity. This means that atmospheric CO_2 is in equilibrium with HCO_3^- and CO_3^{2-} in sea water via the reaction

$$H_2O + CO_2 + CO_3^{2-} \rightleftharpoons 2HCO_3^-$$

This constraint limits the absorption of CO_2 by ocean water.

increased from 317 ppm by volume in 1959 to 327 ppm in 1971 (data Keeling *et al.*, 1972; with 1974 WMO scale and carrier gas corrections). Measurements from aircraft over the North Atlantic and from ground stations in Alaska and the South Pole show similar concentrations and similar rates of increase during this period. Analysis of measurements of atmospheric CO_2 for the period 1857–1906 give a concentration of 293 ppm compared with 319 ppm for 1907–56 (Bray, 1959). It is now generally accepted that the atmospheric CO_2 concentration was probably about 290 ppm in the late nineteenth century (Keeling, 1973*a*) and that the CO_2 concentration in the atmosphere is currently rising. The increase in mass of atmospheric CO_2 between 1959 and 1969 was equivalent to approximately 50% of the CO_2 released by human activity (Ekdahl & Keeling, 1973).

It would appear that during this century, through fossil fuel combustion, limestone burning and deforestation, mankind has put into the atmosphere perhaps twice as much CO_2 as can be accounted for by the observed increase in atmospheric CO_2. This additional CO_2 must have been absorbed by either an increase in the land biomass or loss of CO_2 to the oceans, or both.

The future growth of atmospheric CO_2 is determined by industrial CO_2 emissions and the response of the land biomass and the oceans. Environmental models of CO_2 developed so far have been extensively reviewed by Keeling (1973*a*) and Bacastow & Keeling (1973) and results from one such model are presented in Table 6.1.2. Of particular interest is the partitioning of CO_2 between the ocean, atmosphere and biosphere. This partitioning is based on assumptions concerning biospheric response to an increase in atmospheric CO_2. As Bacastow and Keeling note, the predictions of growth of the biosphere should be treated with considerable caution. Based on an annual growth rate of fossil fuel usage of 4.5%, and the maximum likely uptake of CO_2 by land biota, the atmospheric CO_2 content will have increased by about one-third (to 400 ppm) at the turn of the century and will have doubled (600 ppm) by early next century (about AD 2040).

The climatic effect of such an increase is difficult to predict. Manabe & Wetherald (1975) used a simplified general circulation model of the atmosphere with fixed cloudiness and calculated the temperature change due to doubling the atmospheric CO_2 from 300 to 600 ppm. They estimated a general warming of about 2 °C in surface air and a more active hydrologic cycle. At high latitudes this warming should be larger (up to 10 °C at 80° N) due to recession of the snow boundary. An important factor that is not yet included in the model is a mechanism for change in cloud cover. A 2.4% increase in amount of low cloud over the globe could lower the surface temperature by 2 °C negating the carbon dioxide

6.1 Modification of climate

Table 6.1.2. *Predictions of atmospheric CO_2 for the years* AD *2000 and 2040 from coupled ocean–atmosphere models allowing for biota increase* (*Bacastow & Keeling, 1973*)

The man-made input and atmospheric CO_2 are expressed as fractions of the pre-industrial atmospheric CO_2 content (6.2×10^{17} g C)

Variable	CO_2 content			
	1954	1970	2000	2040
Man-made CO_2 input	0.11	0.18	0.54	2.48
Atmospheric CO_2	1.05	1.08	1.26	2.39
	Cumulative partitioning of man-made CO_2 input (%)			
Atmosphere	44	46	50	56
Biota	41	40	36	33
Oceans	15	14	14	11
Ocean surface water pH	8.26	8.25	8.19	7.93

warming (SMIC, 1971). Hence we can only say that increasing the CO_2 content of the atmosphere will change the mechanism for heat balance in the atmosphere.

There is a need to improve our knowledge of the uptake of CO_2 by land biota and the oceans, and the effect of atmospheric CO_2 on global climate within the next few decades.

A side effect of this CO_2 increase is that the oceans will gradually become more acidic (see Table 6.1.2) and early next century the sea will become undersaturated with respect to aragonite (Fairhall, 1973). Aragonite is a crystalline form of calcium carbonate used by some marine organisms to build their shells or skeletal structures. Both the increased acidity and the undersaturation could cause serious stress in the ocean biological community.

Atmospheric aerosol

Particles participate in the radiation balance of the atmosphere by scattering and absorbing incoming solar radiation and by absorption and emission of terrestrial infrared radiation (Junge, 1975). In clear sky conditions the background burden of 'atmospheric aerosol' (i.e. the suspension of particles in the air) reflects back to space between 1 and 2% of the incoming solar radiation (Braslau & Dave, 1973). Also the aerosol particles absorb up to 5% of the incoming solar radiation for the heating of the troposphere (Kondratyev, 1973). The ratio of this absorption to backscatter along with the albedo of the underlying surface determines whether increased atmospheric aerosol will warm or cool the earth–atmosphere system (Chýlek & Coakley, 1974).

276

The emissivity of the aerosol particles is high in the 8–12 μm atmospheric 'window'. From emissions in this window, aerosol could account for 2–5% of the total cooling in the tropics and up to 25% in the polar regions (Junge, 1975).

While the aerosol might cause heating of the atmosphere, it can only increase surface temperatures if the aerosol layer is in convective contact with the earth's surface (Russell & Grams, 1975). Hence stratospheric aerosol, regardless of its absorption must lead to a cooling effect at the earth's surface.

The effectiveness of atmospheric aerosol for scattering and absorption of sunlight is determined by the size-number distribution of the aerosol and the refractive index of the aerosol material. More than 90% of these optical effects are caused by the particles of radius 0.1–4 μm (Quenzel, 1970; Jaenicke, 1973). Particles smaller than 0.1 μm are ineffective at scattering light, and those greater than 4 μm are too few in number to contribute greatly to the scattering and absorption.

Various measurements of size-number distributions of atmospheric aerosol particles in the 0.1–4 μm radius range at remote continental and oceanic locations, show 10–100 particles cm^{-3} (Blifford, 1970; Junge, 1972a; Meszaros & Vissy, 1974). Similar measurements in the Mojave Desert, California, show concentrations from 60 to 260 particles cm^{-3} in clean air and up to 3,000 particles cm^{-3} on occasions when polluted air is advected from the coastal cities (Sverdrup *et al.*, 1975).

This background tropospheric aerosol comes from the sea, soil dust and gas-to-particle conversion within the atmosphere. The order of importance for composition of background tropospheric aerosol is sea salt < mineral dust < sulphates, etc. (Delany *et al.*, 1975; Cadle, 1973). The composition of tropospheric aerosol is roughly proportional to the source strengths once these are corrected for loss of sea salt and mineral dust in the atmospheric surface layers.

Sea salt

Sea-salt particles are produced by bursting air bubbles on waves in the ocean. The concentration of sea-salt particles is a significant fraction of total particle number only in the region over the ocean below the low cloud level (Woodcock, 1953; Dinger *et al.*, 1970; Meszaros & Vissy, 1974). Annual production of atmospheric sea-salt particles at the ocean surface is about 10^{15} g NaCl yr^{-1} of which perhaps 10^{14} g yr^{-1} or less, reach the free atmosphere about the cloud layer or penetrate over land (Eriksson, 1959). The production and removal of sea-salt particles are determined by present climatic processes, e.g. wind speed and rainfall, and as Junge (1975) has stated 'the sea spray aerosols are unlikely to be the reason for any climatic variations'.

277

6.1 Modification of climate

Mineral dust

Mineral dust is produced by wind erosion of soils. 'Soil moisture, wind velocity, roughness and vegetative residue are extremely important parameters in determining the availability of soil particles for removal on a non-vegetated piece of land' (Gillette *et al.*, 1972). Measurements during a dust storm indicated there was 3.5×10^{12} g dust in the 1–10 μm radius range in the dust cloud and the surface production was 3×10^{-6} g cm^{-2} s^{-1} (Gillette, 1974). Judson (1968) estimated that 10^{10}–10^{12} g of soil particles per day are added to the atmosphere, but later results (Gillette, 1974) suggest that the upper value is more likely to be correct.

The pervasive nature of mineral dust in the atmosphere is shown by various studies. Junge & Jaenicke (1971) made some measurements of particles over the Atlantic west of Africa when the local visibility decreased from 100 to 20 km. They found about 200 particles cm^{-3} of Saharan dust present with a size range of 0.3–20 μm. Delany *et al.* (1967) determined that such Saharan dust is transported from Africa to the Barbados across the Atlantic by the Trade Winds. Griffin *et al.* (1968) found similar transport of soil clays from the deserts of Western Australia out over the Indian Ocean by southeasterly winds.

There is considerable possibility of some relationship between mineral dust production and climate change. Small displacements of climatic zones or overgrazing and other agricultural activities can convert grasslands into rapidly eroding areas. This has happened since Roman times to some border areas of the Saharan and Arabian deserts (Huzayyin, 1956). Bryson & Baerreis (1967) have discussed similar effects in the Rajasthan Desert, northwest India. More recently, albedo contrasts between denuded soils from overgrazing with nearby regions covered by natural vegetation in the Sinai region have been reported (Otterman, 1974).

It has recently been suggested that the high albedo of the desert surface leads to a net radiative heat loss relative to its surroundings and to descending air over the region, thus reducing rainfall and ensuring the continuance of the desert climate (Charney, 1975; Charney *et al.*, 1975). Mineral dust in the air over the desert will tend to counteract this process as the dust will lead to a reduction in the albedo of the desert–atmosphere system (Russell & Grams, 1975). However, this whole question is speculative and it is not obvious, at present, what the relationship is between mineral dust and climate.

Gas-to-particle conversion

The major source of background aerosol in the atmosphere is gas-to-particle conversion. Sulphate, ammonium, nitrate and organic compounds are the dominant components of these aerosols.

In spite of a vast amount of literature on the subject, very little is known about the cycle of sulphur in the atmosphere. The gases sulphur dioxide, hydrogen sulphide, dimethyl sulphide, other organic sulphides, and particles of sulphuric acid, ammonium bisulphate and ammonium sulphate have been identified in the atmosphere (Breeding *et al.*, 1973; Charlson *et al.*, 1974; Meszaros & Vissy, 1974; Rasmussen, 1974; Bigg, 1975).

Junge (1972*b*) has pointed out that man-made emissions to the atmosphere amount to perhaps 4×10^{13}–7×10^{13} g S yr^{-1} whereas precipitation and dry deposition remove 1×10^{14}–3×10^{14} g S yr^{-1} from the atmosphere and the source of the remainder is currently unknown. It has been suggested that biological production of H_2S is the major source of atmospheric sulphur (Hitchcock, 1975). Volcanic activity is a minor source for tropospheric sulphur, but it appears to be the major source for the sulphate layer in the stratosphere (Castleman *et al.*, 1973; Lazrus & Ganrud, 1974; Cadle, 1975).

It is assumed that the various sulphur compounds emitted to the atmosphere are oxidised, to sulphur dioxide, and then converted to sulphuric acid. Sulphates, either as sulphuric acid or ammonium salts can then condense to form atmospheric aerosol (e.g. see Clark & Whitby, 1975).

The main sources of ammonia in the atmosphere are biological decomposition of organic material on the earth's surface, decomposition of urea or ammonium fertilisers (Simpson & Freney, 1974), grazing systems (Denmead *et al.*, 1974) and combustion processes. Healy *et al.* (1970), after studying various sources concluded that animal urine was the major source of ammonia in the United Kingdom. Direct measurements of the flux of ammonia into the atmosphere from grazed pastures (Denmead *et al.*, 1974) also suggest that animal wastes in grazed areas are the major sources of atmospheric ammonia.

The ammonium concentration of ocean surface water is biologically controlled (Riley & Chester, 1971), and the ammonium content is probably between ten and one hundred times the atmospheric content of ammonia. The pH and ammonium concentration of ocean water are such that the oceans regulate the ammonia concentration in the air over the oceans. This, no doubt, is an important aspect of the cycle of ammonia through the atmosphere.

Ammonia in the atmosphere reacts with sulphuric acid forming ammonium bisulphate and ammonium sulphate (Charlson *et al.*, 1974; Meszaros & Vissy, 1974) and about half the aerosol in the optically effective range appears to be one of these two ammonium salts. A considerable fraction of the remainder appears to be sulphuric acid.

Nitrate aerosol is formed from nitrous oxide or nitric oxide in the atmosphere. The processes whereby these two gases are formed from nitrate by denitrifying bacteria are discussed above. Galbally (1975) estimates the natural NO_x production in the Northern Hemisphere to have an upper limit of 3×10^{13} g N yr^{-1}. This is in good agreement with the deposition of nitrate in rain water, a major sink of atmospheric nitrate. Production by mankind of nitrogen oxides by combustion processes might at this stage be half the natural emissions. In the Northern Hemisphere nitrate aerosol and deposition have probably increased significantly during the last few decades, due to the increasing human contribution.

The other source of background aerosol is organic vapours emitted from plants (see Went, 1964, 1966; Rasmussen & Went, 1965). The major compounds emitted by vegetation are terpenes, including α-pinene, β-pinene, limonene, isoprene and camphor (Rasmussen, 1972; Tyson *et al.*, 1974).

There are varying estimates of the organic component of background aerosol from 5% by mass (McMullen *et al.*, 1970), to 25–30% (Winkler, 1974). Hoffman & Duce (1974) found the organic carbon content of marine aerosol at Bermuda was about 0.3 μg m^{-3} or up to 20% of the mass of sea-salt aerosol in light wind conditions. Rasmussen (1972) has made a comprehensive study of the world-wide emissions of terpenes and estimates the annual rate to be 20×10^{12}–500×10^{12} g yr^{-1}. These figures when corrected for the molecular weight of terpenes compared with air are 100×10^{12}–$2,000 \times 10^{12}$ g yr^{-1}. As Rasmussen (1972) writes, 'In truth a world-wide terpene emission estimate is an extremely tenuous figure and presently only of value as a resource guideline.'

Consideration of the sources and sinks of background tropospheric aerosol, summarised in Table 6.1.3, suggest that the major sources of tropospheric aerosol are biologically controlled. It is evident that none of these processes are well understood and that currently we have little if any idea as to what changes we could expect in atmospheric aerosol due to natural fluctuation in the biosphere. Also the radiation properties of background aerosol are virtually unknown. These appear to be important areas for future research.

Cloud condensation nuclei

Clouds in the atmosphere reflect back to space about 20% of the incoming solar radiation so any process that modifies cloud properties (or cloud amount) is of major importance in climate.

Cloud condensation nuclei, CCN, are those particles in the atmosphere

Table 6.1.3. *Sources and sinks of background tropospheric aerosol.*
Units 10^{12} g yr^{-1}, *calculated as chemical species indicated in parentheses*

Component	Source	Sink
Sea salt (NaCl)	Ocean bubbles 1,000 in atm. surface layer 100 in troposphere	Rainout and dry deposition, ~ 100
Mineral dust (silicates)	Soil erosion, 3–300	Rainout and dry deposition, 3–300
Sulphates (S)	Man-made combustion, 40–70 Biological release probably as H_2S, balance, i.e. 100–200	Rainout and dry deposition, 130–260
Ammonium (N)	Man-made combustion, 4 Animal wastes on grazing land, 100–200?	Rainout and dry deposition 30–60
Nitrate (N)	Man-made combustion, 15 Biological denitrification, 20–30	Rainout and dry deposition, 20–40
Organic (C)	Biological release, 100–2,000	?

that will freely grow into cloud droplets by diffusion of water vapour once the atmosphere has become minimally supersaturated. Measurements in clouds show that the water vapour supersaturation is generally less than 1% (Warner, 1968). Typically there are of the order of several hundred CCN cm^{-3} in continental air and less than one hundred CCN cm^{-3} in maritime air (Twomey & Wojciechowski, 1969).

The composition of CCN is difficult to determine because they occur mainly in the size range 0.01–0.1 μm with a mass about 10^{-16} g (Twomey, 1971). Indirect evidence of their composition is available from evaporation studies, where the particles are heated to various temperatures and the number that disappear during the temperature change are noted. It appears that most of the nuclei are composed of volatile compounds with boiling characteristics similar to NH_4Cl, $(NH_4)_2SO_4$, and H_2SO_4, although a few percent of the particles had the characteristics of NaCl (Dinger *et al.*, 1970; Twomey, 1971). It appears that the particles are formed in the near-surface air, and that the rate of production is greater in air over the continents compared with that over the oceans. Twomey (1971) found no evidence of direct production of the nuclei at the soil surface.

The global production rate of CCN from natural processes is estimated to be 10^{21}–10^{22} nuclei s^{-1} (Pruppacher, 1973; Twomey, 1974), while anthropogenic sources appear to contribute 10^{19}–10^{20} nuclei s^{-1} (Twomey, 1974; Auer, 1975). The total mass involved in CCN is of the order of 10^{12} to 10^{13} g yr^{-1}.

The information available indicates that CCN are formed from those gases in the air, SO_2, NH_3, etc., that have biological sources. It is obvious from the discussion above on aerosol that there is more than sufficient material cycled through the troposphere per year to produce this mass of CCN. It seems worth inquiring why there are not more CCN in view of the surplus amount of sulphate and ammonium available.

An increase in cloud condensation nuclei due to natural processes or pollution can lead to an increase in the solar radiation reflected by clouds and consequently have climatic effects. The water vapour content of clouds is determined by large-scale motions. For a given water vapour content the number of CCN determines the droplet size within the cloud, and the number and size of the droplets determines the reflectance of the clouds. When the number of CCN is increased the fraction of reflected solar radiation is increased. Twomey (1974) suggests that the reflection of solar energy by clouds already may have been increased by cloud nuclei of anthropogenic origin. He suggests that a 3% increase in CCN, perhaps that amount already caused by mankind, would lead to a further 0.5% of the incoming solar radiation being reflected back to space. Such an effect would be very important in determining the earth's climate.

Conclusions

It is apparent from the above discussion that ozone, carbon dioxide, aerosols and cloud condensation nuclei all have roles in climate, although the quantitative aspects are not yet determined.

It is also apparent that:

(i) the concentration of atmospheric ozone is intimately connected with biological production of N_2O;

(ii) a small currently unmeasurable change in the biomass could have a large influence on atmospheric CO_2;

(iii) atmospheric aerosol is mainly formed from biological products; and

(iv) cloud condensation nuclei come from the same compounds.

It will be possible to predict the mean concentrations and variability of these atmospheric constituents and their effect on climate only when there is a better understanding of the role of the biosphere in determining the composition of the atmosphere.

6.2   Atmospheric carbon dioxide: recent advances in monitoring and research
G. I. PEARMAN

The carbon dioxide theory of climatic change has possibly been the most commonly discussed theory of its kind. It proposes that changes in atmospheric CO_2 concentrations disturb the global temperature by changing the so-called 'greenhouse' effect of that gas.

The theory is quite old, but received a revival by Plass in 1956. At that time there was evidence that global mean temperatures had been rising since the beginning of the century and as fossil fuel usage had increased during the same period, the inferences were obvious.

More recent temperature data (e.g. Lamb, 1973; Reitan, 1974a) suggest that since the 1940s there has been global, or at least Northern Hemispheric cooling. It has been necessary, therefore, to concede that the cause of climatic variations is not quite that simple. One needs to involve at least one other factor to explain the observed temperature trends.

It has been common practice to suggest that particulates may be the other factor (e.g. Bryson & Wendland, 1970) although authors subscribing to such an interpretation generally conclude that in the future the CO_2 effect will dominate (e.g. Mitchell, 1972). Simplistic arguments suggest that increasing atmospheric particulate loading will cause global cooling, but such conclusions can and have been challenged (e.g. Gribbin, 1975).

There are a number of relatively distinct areas of endeavour relating to the study of atmospheric CO_2. These are:

(i) monitoring of the trend in atmospheric concentration;

(ii) describing the magnitude of the various CO_2 reservoirs and the exchange rates;

(iii) predicting future fossil fuel production rates and atmospheric CO_2 concentrations;

(iv) predicting biological, climatic and oceanic responses to the future changes (see e.g. Fairhall, 1973).

The literature which has appeared recently dealing with these topics is voluminous and could not be adequately reviewed here. This paper therefore reports on a restricted number of recent advances which tend to reflect more on the studies we have been involved with in Australia.

Atmospheric monitoring of CO_2

At the time of Plass' (1956) paper, there was no way of knowing if atmospheric CO_2 concentrations had increased. Earlier measurements were reviewed (see Pearman & Garratt, 1972, for discussion) but due to the methods used and the location of measurements, the conclusions were not definitive. In 1957, Dr C. D. Keeling of the Scripps Institution of Oceanography established baseline (background) CO_2 measuring stations in Antarctica and at Mauna Loa, Hawaii. Since that time several additional CO_2 monitoring stations have been established (see Miller, 1975; Pearman, 1975). At most of these stations CO_2 is measured continuously in air near the ground and the data are selected according to meteorological conditions to ensure that they are representative on a large scale. The Swedish and Australian programmes depend on aircraft sampling.

6.2   Modification of climate

This network of stations has evolved as a result of the interests of the various research institutions, and it is likely to expand over the next decade with the development of the proposed network of World Meteorological Organization baseline monitoring stations (World Meteorological Organization, 1974a).

Interpretation of the baseline data requires high-precision interstation calibration techniques to ensure the necessary degree of comparability between stations. In the past few years it has been discovered that baseline measurements as previously reported in the literature must be corrected for the following reasons.

(i) The carrier gas effect: non-dispersive infrared gas analysers used for CO_2 measurements measure incorrectly when the carrier gas of the calibration mixture differs from that in the gas sample being measured. Errors of up to about ± 4 ppmv (parts per million by volume) can result from this effect which is discussed in more detail by Bischof (1975), Pearman & Garratt (1975) and Pearman (1977).

(ii) Absolute calibration: up until 1974, all published data were expressed in the 1959 manometric scale. More detailed manometric determinations by the Scripps Institution of Oceanography of the concentrations in the standard gas tanks, indicate that the 1959 scale should now be corrected to a new calibration scale which will be referred to as the 1974 WMO CO_2 calibration scale. This correction makes a difference of < 0.2 ppmv at atmospheric concentrations.

(iii) Tank drift: the Scripps Institution has also indicated that a small change occurred in the standard tanks over the period 1958 until the recent manometric measurements and this means that the global trends in CO_2 concentration over that period, i.e. ~ 15 ppmv, is in error by ~ 1 ppmv.

(iv) Non-linearities: all non-dispersive infrared CO_2 analysers are non-linear in their response to CO_2 and, in addition, the 1959 manometric scale is now known to be non-linear with respect to actual concentrations. Pearman (1977) has shown that these non-linearities can result in errors of several tenths of a ppmv.

Considering each of these factors, it should be obvious that the published CO_2 data are in error in both an absolute and relative sense. In Fig. 6.2.1 I have attempted to correct the global data for each of these errors, where the relevant information is available.

Pearman (1977) reports an intercomparison of existing stations performed in late 1974. That study suggests that even after applying the above corrections, discrepancies of 1–2 ppmv still exist, and that at present this represents the relative accuracy of the CO_2 network. The observed differences in concentration between the stations shown in Fig. 6.2.1 may or may not be meaningful.

Nevertheless it is still generally accepted that at the turn of the century

284

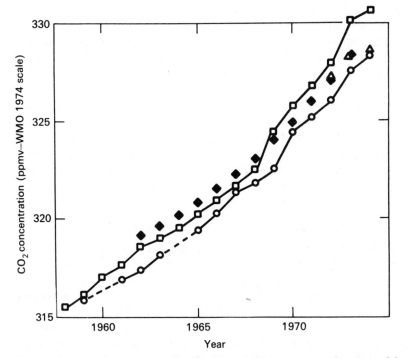

Fig. 6.2.1. Atmospheric carbon dioxide concentration as measured at several baseline monitoring stations. North Atlantic data (◆) from Mr W. Bischof (personal communication); Mauna Loa (□) and Antarctic (○) data from Dr C. D. Keeling (personal communication); Bass Strait data (△) from the CSIRO project in this laboratory. Carrier gas corrections (see text) have been applied of +3.9 ppmv for each set of data except the Australian set where the appropriate correction was −4.4 ppmv. Complete comparability is not possible because of the unknown pressure effects on the carrier gas correction for the high-altitude station of Mauna Loa. Concentrations have been adjusted to comply with the scale adopted by the World Meteorological Organization.

the concentration was about 290 ppmv and since that time a steady increase has occurred to the present level of about 330 ppmv. Recently, however, Farmer & Baxter (1974) and Freyer & Wiesburg (1973, 1976) have concluded that increased land cultivation during the first three decades of this century raised the atmospheric CO_2 content at a rate much faster than possible as a result of fossil fuel combustion. Their conclusions are based on the observation of similarities in trends of carbon-13/carbon-12 ratios in rings of a number of Northern Hemisphere trees.

An alternative explanation for the isotope variations might be that large-scale changes in planetary temperatures have influenced the isotope fractionation which occurs during the photosynthesis process. This is the interpretation placed on isotope variations observed by Libby & Pandolfi (1974) and it would appear to be the assumption underlying the use of stable

285

6.2 Modification of climate

isotope ratios for the correction of radiocarbon measurements for dating purposes. Certainly measurements made on rings of Southern Hemisphere trees (Pearman *et al.*, 1976) show a high correlation between long-term (decadal) temperatures and the stable carbon isotope ratios. However, it is not possible at this stage to establish whether or not the temperature correlation is due to a direct effect on photosynthetic fractionation or a result of indirect effects on carbon reservoir exchanges. The direct temperature explanation is not supported by evidence of Troughton *et al.* (1974) and Freyer & Wiesberg (1976) who have shown experimentally that in the laboratory a number of non-arboreal species show a temperature dependence of fractionation which is small and of an opposite sign to that suggested by Libby & Pandolfi (1974).

The use of large-scale gradients to infer fluxes of CO_2

Bolin & Keeling (1963) summarised data on the latitudinal variation of tropospheric CO_2 from the North to the South Poles and found that the mean annual CO_2 concentration was 1 ppmv higher in the Northern Hemisphere than in the Southern Hemisphere. From the foregoing considerations it should be clear that to some extent such conclusions must now be in doubt. However, accepting for the moment the existence of such a mean gradient and assuming that the increase in concentration in both hemispheres is due mainly to Northern Hemisphere combustion of fossil fuels, we can consider the interhemispheric exchange of CO_2.

Allowing that $\sim 50\%$ of the CO_2 produced is taken up by the oceans, the north–south flux of CO_2 required to increase the mass of CO_2 in the Southern Hemisphere by the equivalent of 1 ppmv per annum is 0.8×10^{16} g CO_2. This implies a large-scale interhemispheric horizontal eddy transfer coefficient $K \simeq 2 \times 10^{10}$ cm^2 s^{-1} which is consistent with Bolin & Keeling's (1963) work.

However, the concentration of carbon dioxide in both hemispheres varies seasonally, with the Northern and Southern Hemisphere cycles being out of phase by ~ 6 months and being of different amplitudes. Figure 6.2.2 compares the average Bass Strait annual cycle with those observed at Mauna Loa and the South Pole. Thus the monthly mean interhemispheric gradient varies considerably with time of year, and interhemispheric exchange must depend on the degree of coincidence between this cycle in gradient and the intensity of interhemispheric mass exchange. Furthermore, the concentrations vary with altitude, as does the direction of mean mass flow. Using the seasonal interhemispheric mass flow data of Newell (1974b) and the seasonal variation in the vertical CO_2 distribution in the Northern Hemisphere as given by Bolin & Bischof (1970) it is possible to show that even without an interhemispheric annual mean

286

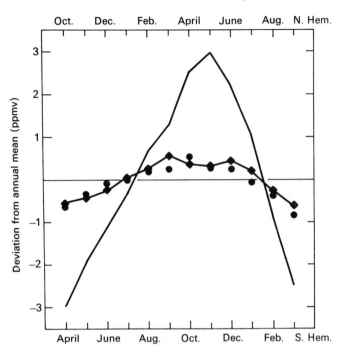

Fig. 6.2.2. Mean annual cycle of atmospheric carbon dioxide concentration at three baseline stations. Data for the years 1972, 1973 and 1974. Note six-month displacement of time scales between Mauna Loa and Southern Hemisphere data. —— Mauna Loa, ◆ South Pole, ● Bass Strait.

concentration gradient, interhemispheric CO_2 fluxes of the magnitude calculated above can occur. However, the flux would be into the Northern Hemisphere. Applying an annual mean concentration difference between hemispheres of ~ 1–2 ppmv is sufficient to accomplish a southward flux of $\sim 0.8 \times 10^{16}$ g CO_2 per year.

The significance of these calculations is that if large-scale gradients are to be interpreted in this way, there are several requirements in the studies, not widely accepted in the past.

(i) Annual mean interhemispheric gradients are going to be small, < 1–2 ppmv, in fact smaller than the present interstation precision.

(ii) The annual cycle in concentration at different altitudes in the atmosphere is required.

(iii) The nature and timing of the interhemispheric mass exchange process needs to be appreciated more fully.

Details of the annual cycle of CO_2 concentration are not only important in this context but they serve as a further indicator of the rates of atmospheric transport and the seasonality of the inter-reservoir exchanges

287

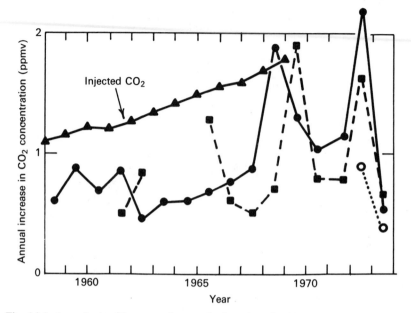

Fig. 6.2.3. Annual rate of increase of atmospheric carbon dioxide concentration as observed at Mauna Loa (●), Antarctica (■) and Bass Strait (○). Also shown is the rate of increase expected if all fossil fuel CO_2 produced each year remained in the atmosphere (▲).

of CO_2. It is generally accepted that the Northern Hemisphere annual cycle in CO_2 uptake results from the seasonal nature and extent of the biospheric CO_2 uptake in that hemisphere. Attempts to model the global annual cycle (Machta, 1972) using biospheric data have suggested discrepancies in predicted and observed results. These can be explained as being due to the inadequate biospheric data and the involvement of both biospheric and ocean exchanges in the annual cycle.

Using aircraft data collected over the Tasman Sea and Bass Strait, Garratt & Pearman (1973) were able to show the likelihood of mid-latitude ocean effects of ∼ 0.5 ppmv as a result of seasonal temperature changes in the mixed layers of the ocean. These studies are continuing at present with the aim of adequately describing the inter-reservoir exchanges and the factors that influence these exchanges.

Apart from the uncertainties indicated above in the interpretation of annual cycles in concentration and hemispheric differences, what can be said about the rates of increase in concentration at the various stations? In Fig. 6.2.3 the annual production of CO_2 from fossil fuel combustion (Keeling, 1973b) is used to calculate an annual atmospheric concentration increase, assuming all the CO_2 were to remain in the atmosphere. These estimates are then compared with the observed rates of increase at Mauna Loa, the South Pole and Bass Strait. It is possible to conclude that on

288

average approximately one half of the injected CO_2 has remained in the atmosphere. However, the apparent variability of this fraction is disturbing. In the case of the Mauna Loa data where the annual mean value is determined from an annual cycle of $\sim \pm 3$ ppmv, it is possible that small changes in this cycle from year to year could cause significant variation in the annual mean values. Despite this, during the earlier part of the Mauna Loa record, particularly the period 1962–8, there was a considerable degree of consistency, such that the annual increase appeared to increase concurrently with the increase in fossil fuel production. The South Pole, which shows an annual cycle of $\sim \pm 0.5$ ppmv does, however, also demonstrate the same degree of variability in the rate of observed annual increase of concentration. The coincidence in the 1972–3 increase, the near coincidence in the 1968–70 period and the relatively small annual increase which occurred from 1973 to 1974 at Mauna Loa, Bass Strait and the South Pole, further support the suggestion that the observed year-to-year variations in the rate of increase may be at least in part real, and reflect global variations.

Such year-to-year imbalances between what is injected into the atmosphere and what remains to be measured represent huge masses of CO_2 of order 10^{16} g CO_2. However, the annual turnover of CO_2 between the atmosphere and oceans is believed to be about two orders of magnitude larger than the annual injection of fossil fuel (see e.g. Plass, 1972). Thus annual imbalances of $\sim 1\%$ in this exchange could result in the observed variability in the apparent uptake of fossil fuel CO_2.

Prediction of future fossil fuel CO_2 production and atmospheric concentration

If all of the fossil fuel produced in recent years for energy purposes is assumed to have been completely oxidised, it is possible to estimate the annual production of CO_2. Relatively small additional amounts are produced due to the kilning of limestone (Keeling, 1973*b*) and natural gas flaring (Rotty, 1974).

A large number of predictions have been made as to the rates of production in the future. The most recent of these is that of Darmstadter & Schurr (1974). Their predictions are forward to the year 1985, and for the purpose of this estimate I have assumed a continuation of the 1985 rate of growth of energy production to the year 2000. These authors predict a slightly higher rate of growth in total energy production ($\sim 5\%$) than most previous authors, despite a full recognition of the 1974 'energy crisis' and the expected changes in attitudes. Based on these predictions the fossil fuel production is likely to quadruple by the year 2000, see Fig. 6.2.4.

6.2 Modification of climate

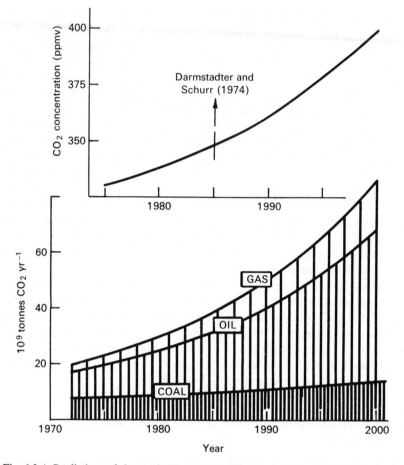

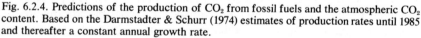

Fig. 6.2.4. Predictions of the production of CO_2 from fossil fuels and the atmospheric CO_2 content. Based on the Darmstadter & Schurr (1974) estimates of production rates until 1985 and thereafter a constant annual growth rate.

We have already seen that the increase in the observed atmospheric CO_2 concentration is equivalent to about 50% of that which would be expected if all the CO_2 were to remain in the atmosphere. The balance is assumed to have been absorbed by the oceans and biosphere. In Fig. 6.2.4, it is assumed that this buffering effect removes 54% (based on 1972 estimates) of the injected CO_2, and that the proportion removed will remain constant until the year 2000. The analysis indicates that concentrations by the year 2000 will be 20% greater than at present.

Other attempts to predict future CO_2 concentrations consider the possible interaction of the CO_2 reservoirs, but generally produce results quite similar to those of Fig. 6.2.4 (e.g. Machta, 1972; Bacastow & Keeling,

290

1973; Hoffert, 1974; Smil & Milton, 1974). The model of Cramer & Myers (1972) predicts an atmospheric content of only 350 ppmv by the year 2000, because of its assumptions regarding the response of the biosphere to the increasing concentrations.

It is difficult to assess whether or not the biosphere has responded in the past to increasing atmospheric CO_2. Because of the magnitude of the biosphere ($\sim 6 \times 10^{18}$ g CO_2 – Bacastow & Keeling, 1973) an annual increase in its mass of about 0.2% would be sufficient to account for the $\sim 50\%$ of fossil fuel CO_2 which is not remaining in the atmosphere. There is no direct way that we can detect such small changes in magnitude. Revelle & Suess (1957) and Broecker *et al.* (1969) argue that it is unlikely that the biosphere has or will be increased in size as a result of increasing atmospheric CO_2. Bacastow & Keeling (1973), like most authors who have attempted to model global CO_2, have included in their model a biospheric response to CO_2 concentration. However, these authors accept that 'many biologists believe the land biota to be static or shrinking'. Their 'preferred' models predict a near doubling of the biota in the next 100 years and this they believe to be unrealistic. Hall *et al.* (1975) have examined the annual cycles of atmospheric CO_2 in the Northern Hemisphere and have concluded that there has been no observable increase in the magnitude of the biospheric exchange since the commencement of measurements at Mauna Loa.

The temperature effect of increasing CO_2

Ultimately it is the effect that changing CO_2 concentrations may have on the surface temperature of the earth that motivates our research of atmospheric CO_2. Many attempts have been made to model the atmospheric response to CO_2 increase and it is not intended to review all of these models (see Schneider, 1975). In Table 6.2.1 the results and some general comments relevant to some of these attempts have been summarised.

In the earlier models, vertical columns of atmosphere were considered in which realistic radiative energy budgets were established to produce realistic surface temperatures. 'Convective adjustments' were added to allow energy to be transported vertically (as if by convection) and thus obtain realistic lapse rates. The need, however, has been for models that satisfactorily allow for the three-dimensional transport of energy and the description of a number of obvious feedback mechanisms. From Table 6.2.1, it can be seen that the most recent models have included large-scale atmospheric circulation and feedback between the atmospheric temperature, water vapour and snow cover. In these regards, the model of Manabe & Wetherald (1975) is the most inclusive. However, the model still fails to allow for adjustments in cloudiness and the heat storage and

291

6.2 Modification of climate

Table 6.2.1. *Summary of some of the attempts to model the effects of changing the atmospheric carbon dioxide to half or double its present concentrations*

Temperatures are in °C and are generally estimates of the surface atmospheric temperature change

Author	Temperature change		Comments
	Double CO_2	Half CO_2	
Plass (1956)	+3.6 +2.5	−3.8 −2.7	Clear skies, 'reasonable average cloud distribution' } All other components constant – no feedbacks
Kaplan (1960)	Smaller than Plass (1956)	−1.8	Improved treatment of radiative flux, average cloudiness, no feedbacks
Möller (1963)	+1.5 +10		Fixed absolute humidity } Fixed relative humidity } Allowance for overlap of CO_2 and H_2O absorption bands – cloudiness as for Kaplan (1960)
Manabe & Strickler (1964)	—	—	Inclusion of 'convective adjustment' to approximate the upward transfer of heat by atmospheric motions. Improved experimental data and computation for absorptivity
Manabe & Wetherald (1967)	+2.92 +2.36 +1.36 +1.33	−2.80 −2.28 −1.30 −1.25	Clear skies } Fixed relative Average cloudiness } humidity Clear skies } Fixed absolute Average cloudiness } humidity Continuation of Manabe & Strickler (1964) study
Gebhart (1967)	+1.2	—	Allowance for near infrared absorption – tends to compensate for infrared effects
Rasool & Schneider (1971)	+0.6 +0.7	−0.6 −0.7	Fixed absolute humidity } Fixed relative humidity } Average tropospheric temperatures thus underestimate surface temperatures
Sellers (1973)	+0.1	−1.0	Replaced by Sellers (1974) with improved emissivity formulation and parameterisation to facilitate model equilibration
Sellers (1974)	+1.3	−1.6	One ocean and one continent distributed latitudinally as for the real earth. Snow and ice cover as feedback. Parameterised meridional heat and potential energy flux
Rakipova & Vishnyakova (1973)	+1.3	—	Horizontal macroturbulence, condensation and evaporation, surface heat exchange. Two-dimensional model
Manabe & Wetherald (1975)	+2.9	—	Simplified three-dimensional general circulation model. Idealised topography, no heat transport by oceans, large-scale eddies and heat transport calculated explicitly. Shows 7% more active hydrologic cycle with doubling of CO_2, but fixed cloudiness, i.e. no cloud feedback. Large effect due in part to snow-cover feedback.

transport by the oceans. It predicts a 7% increase in the hydrologic cycle with a doubling of the CO_2 concentration, but does not allow for a concomitant increase in cloudiness. Plass (1972) has pointed out that a 1% increase in cloudiness in the troposphere could compensate for any temperature increase that might result from CO_2 increases up to the year 2000. Reitan (1974a) has shown that the estimated change in temperature due to increased CO_2 can be halved if the heat going into the oceans is considered.

Manabe & Wetherald (1975) recognise the limitations of their model. They conclude: 'In view of the many simplifications adopted for the construction of the model, the results from this study should not be regarded as a definitive study of this problem.'

Considering all of these factors it would appear that at present our best estimate of the temperature rise which might be expected with a doubling of the atmospheric CO_2 concentration is about 2–3 °C. Also remembering the assumptions in the predictions of future CO_2 concentrations, our best estimate is that the increase of global surface temperature due to CO_2 increase by the year 2000 should not exceed about 0.5 °C.

7. The effect of climatic change and variability on mankind

7.1 Cultural and economic aspects

The grand speculations*

The role of climate, and of change and variability of climate, on human evolution, development and prosperity is an intriguing and important subject. It is, however, fraught with the dangers of subjective speculation.

Aristotle (c. 384–322 BC) asserted that those who live 'in cold climate and in Europe are full of spirit, but wanting in intelligence and skill; and therefore they retain comparative freedom, but have no political organisation, and are incapable of ruling over others. Whereas the natives of Asia are intelligent and inventive, but they are wanting in spirit, and therefore they are always in a state of subjection and slavery. But the Hellenic race, which is situated between them, is likewise intermediate in character, being high-spirited and also intelligent.'

Other philosophers and chroniclers of the ancient Mediterranean civilisations of Greece and Rome held similar views, but some two thousand years later, when the United States and the British empires were in the ascendant, rather different 'theories' were widely held to be true (Kevan, 1971).

In 1915, Huntington produced a map of 'human energy' applicable to 'the races of Western Europe' on the basis of climatic conditions (Huntington, 1915). 'The places shaded black have a climate favourable to a very high degree of energy in people of European races. The next darker degree of shading indicates places where high energy would be looked for, although not the highest... The most noticeable feature is the group of two large black areas in the United States and part of Southern Canada on the one hand, and in Western Europe on the other. The remaining high areas, three in number, are surprisingly limited. One in Japan and Korea... the New Zealand area extends over into the south-east corner of Australia and is probably essentially correct... Finally, although the South American area certainly should be placed on the map, its exact extent is doubtful.'

* Editorial contribution.

294

This analysis was quoted with approval by the Australian physiographer Griffith Taylor (1916) who had earlier published a similar map of regions of Australia 'suitable for *close white settlement*' (Taylor, 1911). Taylor commented 'It will be seen that Huntington classifies more than half of Australia as unlikely to develop a powerful white civilisation. This is simply due to Australia's position in the Trade Wind belt; if she were 10 degrees further south we could hope to rival the United States!'

The Commonwealth Meteorologist, H. A. Hunt, was more pragmatic in introducing Taylor's 1916 report: 'Mankind has adapted himself to the less congenial climates...and our architects can no doubt render much service with respect to the housing problem in the Tropics...a dwelling in our hot humid areas seemingly requires only sun and weather walls...the comparatively inexpensive wind electrical generating mills now on the market would conduce to better health and comfort, for they would not only supply a cooler light...but in addition would furnish power for electric fans, electric irons, etc.'

Greater scepticism was expressed by Dordick (1953) who said of Huntington's conclusions that they had 'not been attained by means of experimental evidence...Instead his evaluation of humid tropical climates is the result of an inadequate analysis of historical, biological and economic characteristics presented by communities of the humid tropics, and is based upon reasoning, which has failed to distinguish between cause and effect and which has oversimplified the relationship between man and his environment.' On the basis of various climatological and physiological data, Dordick concluded that 'the climate of lowland New Guinea appears to be compatible with the efficient performance of severe physical and complex mental activities by acclimatized white men...[although] occasionally the limits of atmospheric tolerance may be exceeded...[and the latter] must be countered by various cultural measures which civilized man has developed.'

Others have contributed to the debate as to how climatic change may have influenced history. These include Carpenter (1966), Claiborne (1970), Le Roy Ladurie (1971), Lamb (1968), Manley (1958), Singh (1971), and Dansgaard *et al.* (1975). Such questions are taken up later in this chapter by Bryson and Anderson, and more obliquely by Maunder and Charlesworth. For the most part, however, all these writers have concerned themselves with Northern Hemisphere peoples and history.

Clearly one major influence on southern peoples has been the changes in sea level, particularly at the end of the Pleistocene. These cut off the mainly overland migration and trade routes from Asia to Australia through Indonesia and the Torres Straits and isolated the Tasmanians beyond Bass Strait. More distantly, the cutting of the Bering Straits land bridge between

7.1 The effect on mankind

Asia and the Americas ended the successive waves of overland migration which were eventually to push southwards as far as Patagonia.

The roles of the Torres Straits as a bridge and barrier to genetic, linguistic and cultural flow is examined by Kirk, Wurm and Golson respectively in Walker (1972), while Mabbutt and C. White in Mulvaney & Golson (1971) discuss Aboriginal adjustments to changing climates in the arid inland of Australia and in northwest Arnhem Land respectively. The large changes in coastal alignment consequent on sea-level changes in the areas of the Arafura Sea and Gulf of Carpentaria must have profoundly affected Aboriginal life through local climatic influences, changes in the availability of sea foods, and forced migration. Golson (1972) and Jones (1975) also touch on the climatic aspects of Australian and New Zealand pre-history, while climatic factors, notably the large year-to-year variability of rainfall, have been shown to have a profound effect on spatial organisation among the !Kung Bushmen of the Kalahari Desert (Lee, 1972).

More detailed analyses

Whatever the state of such evidence and speculation about the effects of climate on the grand scale of human affairs, at the more practical level instanced by H. A. Hunt (above) there is abundant evidence of the effects of weather and climate on human comfort, agriculture, industry and settlement.

For example, Maunder (1970) estimated the difference in the income of dairy farmers in the South Auckland province of New Zealand due to an abnormally 'wet' or 'dry' January (i.e. a departure from the average rainfall by one standard deviation) at about $NZ 2m. Mason (1966) estimated that the total economic value of weather *services* (not of weather or climate itself!) in the United Kingdom was at least £50m to £100m per year, for a total cost of £4m. Berggren (1975) reviews a whole range of economic benefits of climatological services.

The very important effect of climatic change and variability on crop production in marginal areas is well documented for Iceland where Fridriksson (1973) showed from actual data that a drop of 1 °C in mean annual temperature led to a reduction in hay (dry matter) yield by one tonne per hectare (average yield 3–5 tonnes ha^{-1}). Australian crop production is generally more critically dependent on rainfall, leading to a major preoccupation in agricultural meteorology with the subject of drought (e.g. Foley, 1957; Gibbs, 1975) and its amelioration by irrigation, cloud-seeding, or long-range prediction. The World Meteorological Organisation (1975 b) has also devoted much attention to the effects of drought on agriculture.

Analysis of long-term yield trends for wheat, maize and soybean in the

296

United States over an 80-year period (National Oceanic and Atmospheric Administration, 1973) suggested that technological change accounted for 70–80% of yield variance over time, and weather variability for 10–20%. In contrast, an analysis of Australian crop yields by Russell (1973) showed that 60–80% of the variance was due to weather variability, while technology trend accounted for only 10–30%. The reason for this difference is that a much higher proportion of Australia's crops are grown in climatically marginal areas. Those limited areas of Australia with a better climate or irrigation have yield levels and rates of increase which compare well with those of the United States. The implications of such results for Australia in the context of possible long-term trends in climate are discussed briefly by the Australian Academy of Science (1976).

Historical data and crop–climate relationships, where these are available, can be used to build models of crop productivity under varying climatic conditions, as instanced below by the work of Arnold and Galbraith. However, 'second-order' effects may well mean that estimates based on these direct relationships represent maximum effects. Different farming strategies and crop varieties, and changed price structures, may well tend to minimise losses to the farmer, although not always to the national or world economy, or to the consumers.

The effects of climate on human industrial activity and settlements, e.g. on road construction (Maunder *et al.*, 1971), the design of urban stormwater systems (Wooldridge, 1971), and building design (World Meteorological Organization, 1970), are extensive and widely reported in the literature. Walsh has studied the effects of climatic variability and change on the thermal performance of some typical Australian dwellings, and his results are presented in the second case study below.

Case study one: climatic change and agriculture in Western Australia
G. W. ARNOLD & K. A. GALBRAITH
Some general considerations

Agriculture has developed in the Mediterranean-type climatic areas of Western Australia and is based on annual crops and pastures that grow in winter and spring. The growing season varies from 4 to 8 months, depending on rainfall. Varieties of crops have been developed for different areas on the basis of climate and soil. Successful pasture legume varieties will sustain quite large year-to-year variations in climate and be maintained in the long term but they may fail to persist if consistent changes in climate occur.

In the higher rainfall areas of the southwest of Western Australia a reduction in winter rainfall would have either little or a beneficial effect on agricultural production because plant growth is restricted due to water-logging of the soils. However, a 10–20% reduction of rainfall in other areas would reduce farm productivity and would mean that most of the area currently receiving less than 300 mm rainfall would go out of agricultural production. This is an area of 1.3 m ha^{-1} with 1,070 farms producing 21% of Western Australia's wheat crop besides other grains and livestock products. Based on prices received in 1973–4 the total loss of production would be $114m annually. To this must be added further losses in production from reduced rainfall in all areas below 600 mm. Thus a decrease in rainfall would have a large economic impact, at least in the short term. In the long term a change in crop type or pasture species may reduce the effect.

Analysis of likely effects using simulation models

A simulation model of sheep production on subterranean clover pasture (Arnold & Campbell, 1972; Arnold *et al.*, 1974) was used to examine the effects of changes in rainfall and its distribution and of changes in mean ambient temperature in localities in Western Australia. These were Bakers Hill, with a mean annual rainfall of 660 mm, Esperance, which has a mean rainfall of 673 mm, and Merredin, where the rainfall averages 328 mm.

For each locality ten years' actual climatic data were used. The sheep production model was run for ten years for each of four stocking rates at each locality to obtain curves relating stocking rate to various parameters of the ecosystems. Standard deviations of these parameters were then calculated. These normal climate curves were used to find stocking rates where production per sheep began to decline. This rate is usually close to the economic optimum rate.

The model was then re-run for each locality at the optimum rate but daily average temperature ±1 °C from the actual temperature, and rainfall as ±10% of actual rainfall on each rainy day. Standard deviations were again calculated.

Economic analyses were then made as reported by Arnold & Bennett (1975). A sheep farm of 750 ha was assumed on which 20% of the sheep were replaced each year at a cost of $15.2 per head with the sale value of sheep ($0.21 kg^{-1}) being replaced depending on their weight. Wool was valued at $2.03 kg^{-1}. Extra feed needed to prevent sheep dying in periods of pasture shortage was valued at 2.1 cents per head per day. The standard deviations of the values for sheep and wool sold and extra feed used were taken as the farmer's risk in running his enterprise. This is important since risk as well as profit will be considered by farmers in determining at what stocking rate to run their farm.

Table 7.1.1 gives the biological results in detail for Bakers Hill. Table

Table 7.1.1. *Effects of climate changes at Bakers Hill, Western Australia, on production from a subterranean clover pasture*

	Normal climate	Av. temp. +1 °C	Av. temp. −1 °C	Rainfall +10%	Rainfall −10%	Normal climate
Stocking rate (sheep ha⁻¹)	14	14	14	14	14	12
Clover seed set (m)	174	248	187	233	153	271
SD	52	47	89	38	79	41
Pasture yield (1,000 kg ha⁻¹)	4.3	4.7	4.1	4.7	4.1	5.0
SD	630	290	1,250	310	630	323
Max sheep wt. (kg)	56.3	58.1	53.4	58.0	54.0	58.2
SD	2.3	1.9	7.9	2.5	5.7	2.4
Wool (kg)	4.6	4.8	4.4	4.8	4.4	4.9
SD	296	227	760	260	677	288
Days hand feeding	33	9	54	10	68	12
SD	41	18	68	20	70	19

Table 7.1.2. *The effects of climate changes on profit* and risk from sheep production at Bakers Hill, Esperance and Merredin ($'000)*

		Normal climate	Av. temp. +1 °C	Av. temp. −1 °C	Rainfall +10%	Rainfall −10%	Normal climate
Bakers Hill	Stocking rate (sheep ha⁻¹)	14	14	14	14	14	12
	Income	68.0	79.0	57.0	78.0	55.0	66.0
	Risk	16.4	10.0	35.0	11.0	32.0	10.0
Esperance	Stocking rate	10	10	10	10	10	9
	Income	49.9	52.8	45.3	54.9	47.8	48.0
	Risk	17.3	16.4	18.6	14.1	17.0	11.0
Merredin	Stocking rate	7	7	7	7	7	5
	Income	31.5	30.5	17.6	27.6	13.8	20.0
	Risk	10.5	9.0	22.4	16.2	19.4	8.0

* Profit here is income from sales less variable costs.

7.1.2 gives the profits and risks for Bakers Hill together with Merredin and Esperance. At lower stocking rates (12 sheep ha⁻¹, see right-hand columns), the ecosystem is much less sensitive to variations in climate. This is because even though plant production may be affected by differences in climate the sheep are rarely short of pasture and maintain their production.

At Bakers Hill the predicted changes in pasture yield are no more than 10% but the amount of clover seed produced, vital for annual regeneration of the pastures, is increased 50% by a 1 °C increase in temperature. The number of days of hand feed was the variable most sensitive to climate variations. For all parameters the percentage change in the standard deviations was much greater than that in the mean values. Thus increases in both temperature and rainfall produced a more stable system equivalent to that at a 25% lower stocking rate of 12 sheep ha⁻¹. A 1 °C change in temperature and a 10% change in rainfall have equal effects. A decrease in either reduces profit 18% and doubles risk.

At Esperance, which has a longer growing season and milder climate than Bakers Hill, the effects of climate were small. But in the low rainfall of Merredin they were large. There a 10% drop in rainfall halved profit and doubled the risk. To retain the same risk with reduced rainfall means that stocking rate would have to be dropped to 5 sheep ha⁻¹ and profit would be reduced by 35%. However, a 10% increase in rainfall had virtually no effect. This is at first sight surprising but is explainable. Increasing rainfall during the short growing season is not of use since there are few periods of moisture stress. If the increase were at either the beginning or end of the growing season they would, theoretically, be of much greater value because they would lengthen the growing season.

Another simulation model (Arnold & Galbraith, 1974) was used to predict the effects of climate changes on grain yields of lupins at Geraldton (mean rainfall 460 mm). This crop is of growing importance in the area. With a crop, some of the effects of climate can be controlled by varying the time of seeding the crop. Three planting dates were examined using the normal climate and the variations used for pastures. The effects were similar at each planting date with increases in temperature and rainfall increasing yield from 13–22% and decreases decreasing yield by 9–24%.

Changes in the frequency of rainfall were examined by allocating the rain to fall once every 14, 7, 5 or 3 days. This had little effect on crop yields on either heavy or light soils but quite drastic effects on pasture and animal production. The reason for this is that the nature of rainfall events is more significant than amount of rainfall, particularly at the beginning of each growing season. The germination of seeds requires the soil surface (and the seed) to be wet for a certain period of time. The seed is thus sensitive to false starts to the growing season signalled by short periods of wetness. A large fall of rain in a day or several smaller falls over two to three days will cause germination and growth but the same amount of rain spread at 3- or 4-day intervals will not maintain soil moisture at the surface high enough for germination.

The simulation studies appear to give realistic results and illustrate that pasture ecosystems are adapted to present climate, particularly patterns of rainfall. It is concluded that any changes in these patterns could have serious consequences for agricultural production under current pasture ecosystems in a Mediterranean-type environment.

Case study two: climatic variability and the thermal performance of buildings
P. J. WALSH

Every building modifies the outdoor environment to produce an indoor thermal environment measurable in terms of indoor temperatures or of heating or cooling loads ('load' implies here an energy input over a specified period). The problem of effective thermal design of a building may be stated then in two parts. First, to provide at all times of occupation and where possible without artificial heating or cooling a comfortable indoor thermal environment. Second, where conditioning of the indoor environment is needed, to provide it with minimum energy usage and from a heating or cooling plant of the minimum necessary capacity. Obviously the indoor thermal environment is affected by building plan as well as the climate of the location. Important climatic parameters include dry bulb temperature, direct and diffuse solar irradiances, wind velocity and cloud cover.

The observational record on which climatic statistics are based is generally about thirty years in length, and the climate of the location is felt to be specified by such a record. The 'typical' variability of the climate is then that which occurred in the observational record. However, increasing evidence is being accumulated to show that significant climate changes of natural cause may have time scales greater than the length of such a record and yet less than the projected lifetime of many domestic and commercial buildings. Furthermore, man-made influences, particularly those associated with large urban centres, can alter the climate of a location.

Thermal performances of some typical Australian dwellings have been assessed for two locations over periods of several years. It must be emphasised that these periods are much shorter than those generally deemed necessary to assess typical variability, due to lack of radiation data of the required precision. Nevertheless the results obtained do indicate some clear trends.

Typical variability

In the consideration of building thermal performance at a given location it is the extent of the variability in the local climatic record, i.e. the extremes in climate, that are of most relevance to the building designer. Monthly means are often not useful design statistics due to frequent and large departures from these values.

As an example, Fig. 7.1.1 shows distributions over the 6-year period 1966–71 of hourly outdoor dry bulb temperatures for Melbourne, Aus-

301

7.1 The effect on mankind

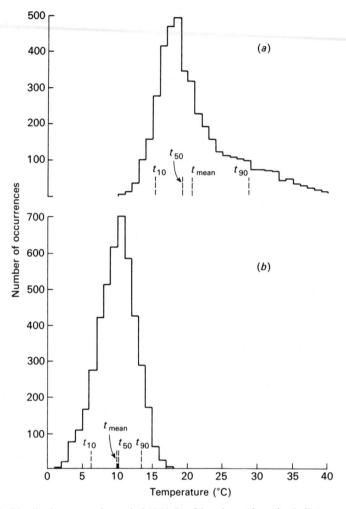

Fig. 7.1.1. Distributions over the period 1966–71 of hourly outdoor dry bulb temperatures in Melbourne for the months of (a) February, (b) July. The vertical axis indicates number of occurrences of hourly temperatures within unit temperature range.

tralia, for February and July. It is noteworthy that the February distribution is far more skewed than the relatively normal July distribution. Useful design temperatures would be t_{90} (for summer) and t_{10} (for winter) as determined from the February and July distributions respectively. Although these are temperatures exceeded on 10 % and 90 % of occasions in the months concerned, they are also temperatures exceeded on roughly 2 % and 97 % respectively of occasions throughout the year.

Computer technology has now made it relatively easy to incorporate

302

coincident values of all the relevant parameters into the evaluation of building thermal performance. Thus thermal design may be based on calculated indoor temperatures or sensible heating or cooling loads, and differences produced by various building types may be readily established. Furthermore, it becomes possible to make thermal calculations over periods of several years and hence to assess the typical variability of building thermal performance.

The calculation technique outlined by Muncey *et al.* (1970) and modified by Scanes (1974) has been utilised here. As input data it requires details of building layout and construction, solar absorptances and emittances of outdoor surfaces, shading devices, any internal loading, and coincident hourly values of the five climatic parameters previously mentioned. Climatic data covering the 6-year period 1966–71 for Melbourne (latitude 37° 50' S), and the 4-year period, February 1966–January 1968 and November 1970–October 1972 for Port Moresby (latitude 9° 48' S), have been compiled and the above technique used in the evaluation of the thermal performance of some typical Australian dwellings. These periods were limited by the available solar radiation data.

Three dwelling types distinguished only by outer wall material of asbestos cement, brick veneer, or cavity brick, and considered representative of Australian houses have been used in the calculations (Walsh, 1976).

It is necessary for calculated indoor temperature data to be presented in a simplified yet meaningful manner, to assist in the choice of building design. Ballantyne (1975 *a*, *b*) has described a method by which calculated indoor temperatures are rank-ordered for each hour of the day and month of the year, and then either ten or ninety percentile temperatures presented as a grid, with columns representing values for each month, and rows representing values at evenly spaced intervals during the day. Ten or ninety percentile temperatures correspond approximately to temperatures exceeded on 27 days per month and 3 days per month respectively and give a measure of extreme indoor temperatures experienced in winter and summer respectively. Isotherms corresponding to transition temperatures from one thermal sensation to another (e.g. 'neutral' to 'warm') can be plotted on the temperature grids to indicate as zones those times of year and day when various thermal sensations will be experienced at the given probability level.

Figure 7.1.2 shows ten percentile temperature grids for Melbourne over 6 years, (*a*) outdoors, (*b*) indoors, asbestos cement, (*c*) indoors, brick veneer, (*d*) indoors, cavity brick construction, with blinds open to maximise solar heat gain. Similar grids have been calculated for Port Moresby and for various assumptions as to shading by blinds.

At those times when either unacceptably cold or unacceptably warm conditions prevail for at least 3 days per month, the need for artificial

7.1 The effect on mankind

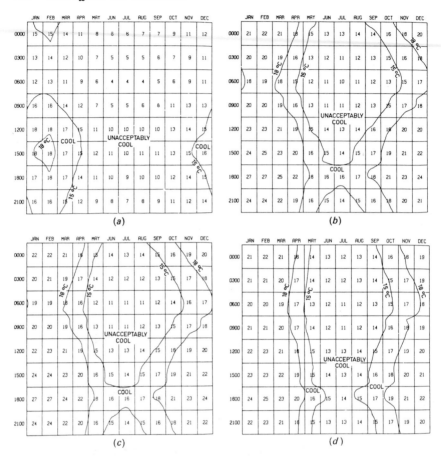

Fig. 7.1.2. Ten percentile temperatures (°C) in Melbourne over six years: (*a*) outdoor, (*b*) indoor asbestos cement, (*c*) indoor brick veneer, (*d*) indoor cavity brick; all dwellings with blinds open. Temperatures colder than those shown at a given month and hour of day will occur on only three occasions per month.

heating or cooling respectively is suggested. Use of the heavyweight cavity brick construction significantly reduces the length of the necessary heating and cooling seasons in Melbourne but has little effect on the Port Moresby cooling season which is year-long for all three construction types.

Energy requirements for heating and cooling

The results of calculations of hourly heating and cooling loads over several years may be used for two purposes as described by Ballantyne (1975*b*). First, by preparing load grids similar to the temperature grids, the hourly rate of heating or cooling to be exceeded at a certain frequency

304

Table 7.1.3. *Annual heating or cooling loads for each of the three standard construction types. Units: giga-joule*

| | Melbourne | | | | | |
| | Asbestos cement | | Brick veneer | | Cavity brick | |
Year	Cooling	Heating	Cooling	Heating	Cooling	Heating
1	8.3	35.4	7.5	33.4	5.6	41.7
2	9.5	31.9	8.6	29.9	6.4	36.3
3	12.0	33.6	11.1	31.9	9.5	40.7
4	9.1	34.6	8.4	32.7	7.0	41.1
5	7.8	34.6	7.0	32.8	4.7	42.0
6	10.3	33.1	9.5	31.5	7.7	40.8

| | Port Moresby Cooling | | |
Year	Asbestos cement	Brick veneer	Cavity brick
1	73.4	72.3	82.7
2	71.4	70.2	80.1
3	79.9	78.8	92.4
4	74.9	73.5	84.9

of occurrence can be obtained and thus heating or cooling plant size determined. Secondly, an assessment may be made of energy requirements for heating or cooling and of the possible extremes in such requirements.

The importance of such methods is highlighted by the fact that about 10% of Australia's energy usage is for heating or cooling of buildings, most of this being for space heating of dwellings. The so-called 'energy crisis' has thus made a considerable impact on the heating and air-conditioning industry and a large part of its research effort is presently being devoted to the development of computer-based energy calculation techniques.

Table 7.1.3 shows annual loads for each of the three construction types. Whereas the cavity brick construction has the highest energy requirement for heating in Melbourne or cooling in Port Moresby, it has the lowest for cooling in Melbourne. This is due to the high diurnal swing in outdoor temperature that occurs during the Melbourne summer, thus allowing heat-storage capacity to exert an influence. In approximately steady-state conditions, however, only resistance to heat flow determines energy requirements.

It is also apparent from Table 7.1.3 that annual energy requirements for cooling in Melbourne are far more variable than Melbourne heating requirements or Port Moresby cooling requirements. Thus for the cavity

7.1 The effect on mankind

brick construction the amount by which the greatest annual load exceeds the smallest is 16% for heating in Melbourne and 15% for cooling in Port Moresby; however, it is 102% for cooling in Melbourne. This excessive variability in cooling demand is not yet of great significance since cooling demands in Melbourne are small in absolute terms. Nevertheless it could cause problems for power authorities if the use of air conditioning for summer cooling in such temperate climates were to become widespread, since peak seasonal demand is difficult to estimate accurately.

Further, excessive energy demands due to abnormal heating or cooling seasons could feasibly create difficulties in countries importing even a small proportion of such energy requirements. Manley (1957) shows that in the United Kingdom the range of variation in demand for fuel in heating is about the same as it was then found necessary to import, viz. 4% of national needs, so that climate variations could directly affect the balance of trade. Such is not the case for Australia where ample coal is available to supply electricity needs and where, at least for the present, natural gas is plentiful and only small amounts of oil need be imported.

Urban effects

Building designers and climatologists need to be aware of urban effects on thermal performance of buildings for two reasons. Firstly, in establishing design criteria for new cities or towns undergoing urban development; secondly, so that the applicability of data gathered in the centre of a large city to a suburban locality (or vice versa) may be determined. Clearly the effects are unimportant if they are exceeded by yearly or longer-term variations. Landsberg (1970 b) quotes average changes in climatic elements of an urban region in comparison with surrounding rural environment. Some of these, for example, increases in winter average minimum temperature of 1–2 °C, are of the order of typical or longer-term variations, and hence need to be considered.

In attempting to evaluate the effects of the urban environment on building thermal performance certain approximate methods can be readily applied. Boyd (personal communication) suggests that a change in any climatic element could be directly applied to the corresponding design weather element. Thus, if the city centre is 1 °C warmer than the surrounding rural or suburban area and the variability is known (or assumed) to be about the same, then the external design temperatures in the city centre should be 1 °C higher.

A useful and simple method for evaluating heating loads is the degree-day concept. This is a unit, based upon temperature and time, used in estimating fuel consumption and specifying nominal heating load of a building in winter. If the chosen base temperature is B, then for any one day when

306

the mean temperature *M* is less than *B*, the number of degree-days is *B–M* and is assumed directly proportional to the nominal heating load for that day. O'Brien (1970) has presented heating degree-days for some Australian cities using a design temperature of 70 °F (21.1 °C) and a base temperature of 65 °F (18.3 °C).

Landsberg (1970 *b*) suggests that heat generated by urbanisation may reduce the number of heating degree-days by 10%. This represents a significant saving in energy costs. (However, as many Australian city office buildings require year-round cooling the effects of urbanisation could be counterproductive.)

Natural long-term variability

Since most buildings today are constructed with the intention of at least a fifty-year life span (a period longer than that over which typical climatic variability is often assessed) it follows that natural long-term changes in climate are of some relevance to building designers, the extent of that relevance again depending on the degree of change. Prediction of possible change is fraught with difficulty and therefore never confidently undertaken. It would probably be reasonable to apply suspected changes directly to external design values, while thermal performance during past abnormal years would possibly be representative of future changes. Year 1 of the Melbourne data assessed in Table 7.1.3 might be taken as representative of a future situation in which mean temperatures have dropped by about 0.5 °C.

The degree-day method, too, provides a means of assessing changes in heating requirements. Manley (1957) has used a variant of it, namely a method of 'degree-months' to examine variation of the heating-season (September–May) energy requirement since 1700 in the United Kingdom. He found that the seasonal requirement over the coldest decade exceeded that for the warmest decade by about 25%. Further, the requirement for the coldest individual heating season exceeded the average requirement by 36% and exceeded that for the warmest heating season by 85%. No analysis of long-term Australian data has yet been undertaken to indicate whether such extremes could occur in this continent.

In order to gauge the effects of a decrease in mean outdoor temperature, the data used by O'Brien (1970) to calculate degree-days for Melbourne were re-examined with base temperature increased by 0.5 °C and by 1 °C. These gave increases in heating requirement of 9.6% and 19.6% respectively. Such increases are only crude estimates of possible future variations in heating requirements but indicate that they could significantly affect future energy-consumption patterns.

The philosophy of building climatologists is often that the climate of

7.1   The effect on mankind

a locality, once specified by observation over, say, thirty years, remains 'constant' apart from typical variability. Yet evidence from historical records shows that for several decades there may be significant departures from this 'constant' climate.

Such departures do have implications for building designers, for while occupants may simply endure occasional periods or seasons of extreme indoor conditions, some modification of life style and habits might follow longer abnormal periods. Thus if comfort requirements remain constant, fuel requirements could in the future alter significantly.

A grand synthesis?*

Case studies such as the above illustrate the possibilities and the limitations of the necessary first-order study of the impact of climatic change and variability on mankind. An overall picture can only be built up from many such case studies, each of which has its own difficulties with availability of data, assumptions, and questions of interpretation. Isolating critical climatic parameters such as rainfall or temperature at a particular point in a growing season is not always straightforward, and often meteorological data is not available in the most appropriate form.

The World Meteorological Organization and many other bodies are making large efforts in these areas (see e.g. World Meteorological Organization, 1974 b). Beyond these first-order problems, however, lies the problem of drawing the threads together to allow for the real situation where the usual assumption of 'other things being equal' is not true. Economic forces, substitution effects, strategies and political decisions are all relevant at this second-order level. It has been touched on by a growing number of workers, for instance Taylor (1974), Newman & Pickett (1974), International Federation of Institutes for Advanced Study (1975), Collis (1975), Kellogg & Schneider (1974), Rockefeller Foundation (1974) and Harris (1976).

Two important climatic factors deserve special emphasis in any such grand synthesis. One concerns an adequate understanding of the large-scale spatial patterns of climatic variability, as discussed in Section 4.3 above. This has obvious application in that it tells us that drought or floods are likely to hit certain large areas simultaneously. If we can spread our investments, insurance risks, or storage capacities over different areas likely to experience opposing or non-correlated tendencies we will be more likely to survive. Priority might well be given to transport, trade or aid links between such areas so as to facilitate transfer of stock, grain or water to or from drought- or flood-afflicted areas. On a world scale, special

* Editorial contribution.

thought should be given to these patterns in relation to national boundaries, regional co-operation and relief operations.

The second crucial factor concerns rates of climatic change. If climatic changes are small, and of opposite sign in different areas or seasons, it would appear intuitively that *in the long-run* the various local and seasonal effects would be as likely to be beneficial as adverse, and thus have a small overall effect. However, it is far from obvious that the beneficial and adverse effects will balance out *during the period of adjustment* of the economies of the various areas concerned. The need for changes in capital works and equipment, irrespective of the direction of climatic change, will make the rate of change a critical factor. In economic terms, the tolerable time scale for significant climatic change will thus be clearly related to the useful lifetime of mankind's agricultural, industrial and civic capital investments. If some of these investments are to be rendered obsolete by climatic change, these must be replaced or 'written off' on the same time scale as the climatic changes.

Such considerations highlight the importance of climatic studies aimed at a greater understanding of possible trigger mechanisms such as Antarctic ice surges, and at increasing our knowledge from the palaeoclimatic record of rates of change. The severe consequences of any sudden climatic transition, such as those contemplated by Flohn and others, makes their *possibility* a matter of concern even if the *probability* of such an occurrence in the next decade or century is rather small.

Even much smaller changes, such as have occurred in historic times, will become more critical as long as growth in population and in the utilisation of available land and water resources continues. Fresh water supplies for irrigation, industry and cities depends on the fine balance between precipitation and evaporation so that quite small changes in precipitation, temperatures, or average wind speeds can produce much greater proportional changes in available water. That is one lesson to be learnt from the huge fluctuations in closed lake systems during the relatively stable Holocene Period (see Section 3.5), and it means that mankind is much more critically dependent on a stable climate than is generally believed.

We must be prepared to evaluate climatic changes by considering their regional ecological and economic impact, rather than the more remote and extreme possibilities of large global changes such as the melting of ice caps or the commencement of a new ice age. This will involve complex interdisciplinary problems, but it is only through such studies that we can hope to comprehend fully the importance of climatic change.

We must move on from grand speculations about climatic change as a factor in human history to a grand synthesis based on observed facts, brought together into an interactive model network which will allow rational predictions to be made. Subjective political and other factors may render results from such models unreliable *as predictions*, but raise the possibility of their use for good or ill *as aids in politico-economic decision-making*, as is discussed below by Maunder.

7.2   Climate and the historians
J. L. ANDERSON

The weather, as a causal factor, has not been entirely ignored in historical studies. In agricultural, economic, or even military history, the occurrence of rain or drought, heat or cold – regarded as being fortuitous and exogenous – is at times incorporated in the explanation of given events: harvests in the eighteenth century (Ashton, 1959), for example, or the outcome of the battle of Waterloo in 1815 (Fuller, 1970). More broadly, the role of climate together with other elements of physical geography is acknowledged in discussions of such matters as settlement patterns, crop distributions and trade routes. However, climatic change as an explanatory variable has been neglected, or at best approached with extreme caution by historians generally. It is the purpose of this paper to suggest why this has been so.

Perhaps the simplest reason for this neglect lies in an intellectual over-reaction against monocausal theories of history. It would be uncharitable to suggest that another reason for an uncertainty, scepticism, or even ignorance on the part of historians about the nature and extent of climatic change is that much of the literature on the topic is located in scientific rather than historical publications. The cause is surely deeper: the hazardous nature of an essay into cross-disciplinary studies. For example, in reviews of *Discontinuity in Greek Civilization*, while Carpenter (1966) was commended for making climatic change fundamental to his work, the scientist reviewers nevertheless exposed some basic inadequacies in the thesis (Lamb, 1967a; Wright, 1968). Finally, even if there were no dispute about the nature of a certain climatic change, interpretation of this in terms of its effect on man is fraught with difficulty.

Problems of data

The first problem confronting the historian is that of the reliability of his evidence, in terms broadly of representativeness, accuracy and objectivity. It is on the strength of this foundation that his understanding of what happened in the past must rest.

Direct evidence of weather must necessarily be documentary, the description of weather in literary or numerical form recorded at the time of its occurrence. Qualitative data such as state, monastic, manorial and personal papers and chronicles, and even weather diaries, may not be reliable for a number of reasons. Amateur observers in any science tend to note the exceptional rather than the commonplace, and may present a picture of climate based on unrepresentative extremes. This problem would be compounded, and the value of the evidence correspondingly lessened, in so far as the chronicler used weather data to further his own purposes: an incompetent reeve accounting for poor returns perhaps, or a monk intent upon providing colourful corroboratory evidence of Divine wrath. In more general terms there is necessarily a subjective element in descriptions of weather, 'cold' or 'dry' meaning different things to different observers, or to the same observer at different times.

Quantitative data of historical climate have, in addition to accuracy, their own criteria for acceptability. Figures should be comparable between areas and altitudes; continuous, in order to reveal oscillations or turning points in trends; comprehensive in coverage of the elements of meteorology; and specific in terms of the element recorded. Of course this represents an ideal, and an insistence such as Le Roy Ladurie's (1959) on the rigorous application of these criteria is a counsel of perfection, probably impossible to achieve given the circumstances of data generation, preservation, and construction. However, the aim is consistent with his general thesis that a scientific history of climate, free from anthropocentric prejudice, must be produced before a history of the relationship between climate and man – 'ecological history' – is attempted (Le Roy Ladurie, 1971).

Even if all of the foregoing criteria were met in any practical sense by construction of annual series, brief periods of extreme temperatures, or untimely precipitation, may be ecologically critical (Post, 1974), but escape the statistical record, and therefore avoid detection.

Indirect evidence of weather, and by aggregation climate and climatic change, may be documentary, or made available by observation of the landscape or by more sophisticated scientific methods. The crucial problem of reliability that follows from the necessity to infer climatic cause from observed effect is that of separating the climatic from other possible causes. An observable or recorded onset of aridity may have resulted from human activity, rather than from a general climatic change (East, 1962; Raikes, 1967). An example of the more general problem of cause and effect is the flooding of the Netherlands in the fourteenth and fifteenth centuries, which is suggested to have been caused by depressed economic conditions and consequent neglect of the dykes, rather than by a change in climate (Slicher van Bath, 1963). Again, it has been suggested that the extension of limits of cultivation of certain crops, for example, olives in France (Le Roy Ladurie, 1971) and grain in England – *c*. 1800 – resulted from economic rather than climatic factors (Manley, 1962).

Although for some locations, data which are continuous through time are available for such events as the thawing and freezing of water bodies, or dates of blossoming, they are at best partial indicators of weather in general, unless it is assumed that these incidents are necessarily representative of the pattern of weather for the season or year concerned. Similarly, unrepresentativeness in an areal sense may arise in an index based on accounts of harvests (Jones, 1964; Manley, 1962).

Additional problems are presented by indicators which are analytically more remote from the actual weather. A recorded increase in the number of prayers offered for rain in Barcelona in the sixteenth century may indicate an increase in aridity, or piety, or both (Le Roy Ladurie, 1971). Comparing that series with one for grain prices in that city does not completely resolve the problem (Claiborne, 1970).

Finally, for an assessment of the meaning and implications of much of the information discovered by such sciences as palynology, dendrochronology and glaciology, the historian must rely upon the investigations, evaluations, and interpretations of his scientific colleagues. This necessary reliance on secondary sources, not personally verified, tends to run counter to the historian's training, and in addition may present him with the dilemma of which scientific 'authority' to accept.

Problems of interpretation

Even if the existence and nature of a climatic change can be established, the question of the significance of the data must be answered, i.e. what effect would the change have had on the actions of men? Le Roy Ladurie (1971) highlights the difficulty in his conclusion that 'in the long term the human consequences of climate seem to be slight, perhaps negligible, and certainly difficult to detect'.

Although even in the third quarter of the nineteenth century the size of the harvest was suggested to be the largest single determinant of trade fluctuations (Bagehot, 1915), the question of whether poor harvests have acted to stimulate or retard economic growth in the long run is not resolved. A good harvest means an increased demand for labour, and increasing real wages as food prices fall. This may then create a demand for manufactured goods; or conversely may permit increased leisure, and reduce demand generated by agricultural investment. An *alternative* model suggests that although a poor harvest will cause a fall in real wages, and by inference some fall in demand from that source for manufactured goods, the most important source of demand for the products of nascent manufacturing industries – in eighteenth-century Britain for example – was farmers' purchasing power (Deane & Cole, 1967). This could be increased as a result of poor harvests, because given an inelastic demand for grain their reduced output would, in total revenue terms, be much more than offset by the rise in price. Of course if harvests are poor, and other things remain constant, national output is lowered: the key to the unresolved argument as to which of the alternative models will apply in any given case depends on the extent of the fall in the national (or total) income, and how the nature of the change in income distribution will affect demands for manufactured goods and provide an impetus to economic progress generally.

Similarly, ambiguity is suggested by Le Roy Ladurie (1971) to exist in relation to explanations of migrations in terms of climatic change, presumably on the grounds that one effect cannot be explained by two different causes; it being contradictory to see migrations as having been occasioned by both amelioration and deterioration. A contrary view which resolves such inconsistencies may be illustrated by a Cartesian diagram with emigration, e, and climatic anomaly, c, as the axes; these being related through a constant, m, by the equation $e = mc^2$. Migrations resulting from an increased cold, damp, or drought are intuitively acceptable. On the other hand, climatic amelioration could plausibly lead to a breakdown in those cultural institutions which had restricted population numbers to levels appropriate to the less favourable environment, this eventually resulting in population pressure and out-migration.

A further reason why historians have been less than unanimous in their acceptance of the importance of climatic change in history perhaps lies in the small variations involved. Changes such as that in England of about 1.2–1.4 deg C in annual mean temperature and 10% in rainfall (Lamb, 1965 b) are not, at first sight, impressive save to the climatologist. This view appears to ignore the point that small variations in annual averages can indicate changes which are quite significant – a 15–20% change in the average length of growing season in England, for example (Lamb, 1966 a). A recent change of comparable magnitude resulted in the growing period being reduced by two weeks in England, and the frequency of snow on the ground in Midland England doubled (Lamb, 1969). In areas or situations that were latitudinally, altimetrically, or economically marginal, this effect would be magnified. An example of the last of these may be found in the late medieval period, when an expanded population was extracting little more than subsistence from the land despite a maximum effort at the given level of technology, and was accordingly vulnerable to the vagaries of the weather and to climatic deterioration generally.

Another point is that annual averages which show small fluctuations in their trend may mislead scholars into underestimating effects of change.

313

7.2 The effect on mankind

Whether the series relates to meteorological conditions or economic indicators such as grain prices, 'good' years may offset 'bad' years only in an arithmetic sense; a working population weakened or decimated by famine, its seed stocks depleted, is unlikely to revive in response to a few good seasons. As early as Genesis it was recognised metaphorically that seven lean cows could devour seven fat cows, with the former showing no benefit.

Perhaps the most important consideration is that an 'index' of climate based on annual averages does not reveal the variability of weather patterns associated with different degrees of continentality of climate. Utterström (1955) ascribes the violence of crop fluctuations in the fourteenth and fifteenth centuries to continentality of the climatic regime, which combines wider ranges of temperature and precipitation with a reduced reliability, and a shift to summer rain. In this connection it may be noted that in Britain a warm and wet summer 'is the greatest weather hazard for the sheep industry' (Jones, 1964) and poor wheat harvests may be expected to result from severe or at least changeable conditions in winter and spring, and wet summers and autumns (Jones, 1964; Titow, 1960). These matters provide an analytical connection between variability in weather, which the historian cannot avoid recognising, and long-run climatic change which acts, with less recognition, as an analytically remote but nevertheless effective causal mechanism.

Methodology: science, history, and economic history

If climatologists and historians work independently, but without a clear demarcation of their respective areas of research, there is a risk of conclusions being drawn on the basis of circular reasoning; the climatologist deriving climatic data from chronicles, the historian then using the climatologist's conclusions to explain the documentary material. However, in most historical studies inadequacies of data necessitate extrapolation and interpolation, or, more generally, argument by inference, and a systematic historical geography of climate can provide a useful check on the reliability of documentary evidence, or perhaps suggest a deterioration or amelioration significant in the long run but not perceived directly and recorded by contemporaries.

It is doubtful whether the separation that exists between climatologists and historians has been caused simply by the difficulties and hazards of cross-disciplinary studies already outlined. A more fundamental reason may be suggested. While both study cause and effect, the climatologist, as a physical scientist, is both systematic and uniformitarian, studying and explaining processes the laws of which are assumed to be generally applicable in both time and space. By contrast the historian is more

314

closely concerned with particular rather than general relationships, studying the interaction of factors in order to solve unique problems specific
in time and to place, and so may not be equipped philosophically to
address himself to the study of long-run movements in history and the
relationships of those with climatic change.

The economic historian is usually less inhibited in this respect, and is
perhaps the scholar most qualified to bridge the gap between climatology
and history. The subject matter of economic history typically involves
thought in terms of statistical aggregates – cultures and classes rather than
individuals – and the focus of attention is characteristically on the longer-
run patterns of change, rather than on a specific event. This necessarily
involves treating certain factors as variables that to the historian, because
of a legitimately 'smaller scale' approach to the past, would be analytical
constants. Although the economic historian may adopt as a working
hypothesis the view that economic effects generally have economic
causes, he is normally obliged to consider what are termed 'non-economic'
factors; demography, disease, war, religion, and the physical environment
in which the relevant community is located. Evidence in the records of
a number of contemporaneous movements in economic variables in
adjacent but otherwise dissimilar societies, such as the sixteenth-century
rise in population in Europe, or the 'general crisis of the seventeenth
century', is persuasive evidence of a general cause: the role of climate
must at least be investigated.

Finally, the economic historian, examining long-run movements in the
economic foundations of societies, is more likely to be involved, explicitly
or implicitly in a search for ultimate rather than proximate causes, pushing
as necessary his research into causality up to and possibly beyond the
conventional boundaries of his discipline. (Thus to an economist interested
in grain prices, harvests are a proximate cause, weather an ultimate one:
to the meteorologist, weather is an effect, its proximate cause being
atmospheric dynamics, its ultimate cause solar radiation, and so on.) This
element of research dynamics, coupled with an increasing availability of
evidence of climatic change, is leading to an increasing recognition of the
role of climate in economic history.

Conclusion

A scepticism about the importance of climatic variations may be the result
of the training of historians, and the frequently narrow focus of their work.
More fundamentally it may be seen as an expression of an intellectual 'two
cultures' dichotomy, a problem becoming more acute as the seemingly
exponential expansion of information makes the crossing of boundaries
even within a single discipline an arduous and scholastically perilous

enterprise. In relation to the physical environment which necessarily conditioned men's actions in the past, the existence and nature of topographic changes may be indicated by or inferred from old maps and chronicles, or detected by modern air-photography techniques. The evidence of climatic change is more ambiguous, subtle, and elusive and, even once determined, its effects in both biological and economic terms are open to differing interpretations.

However, despite these difficulties, in view of the mass of data being generated by the research of historians of climate and allied scientists, the scholar interested in the larger aspects of economic and social change must return to his sources, asking of them new questions, seeking new correlations, and analysing anew the relationships involved. 'Climatic history is a young subject full of unsolved problems, and what it will finally reveal is uncertain; but it is too important to be ignored by the economic historian' (Davis, 1973).

7.3   Cultural, economic and climatic records
R. A. BRYSON

A global view of cultural change from the climatic perspective

At least two extreme views of cultural change may be postulated. One might be that cultures contain the seeds of their own change or the change of those with which they come in contact, that new ways of life are generated primarily by diffusion of ideas, by force, or by invention within a culture. On the other hand one may postulate that an important factor in cultural change is environmental change, such as soil degradation, climatic change, changing water levels and the like. Rather than take an extreme view that either one or the other of these mechanisms is nearly the sole cause of cultural change, one might take the more moderate view that both are operative, sometimes in combination, sometimes not.

One might then try to devise methods for sorting out these two basic mechanisms of cultural change, at least in a general way. It would seem reasonable that such things as a Spenglerian 'natural aging' of a culture would result in a random time distribution of cultural change if we consider all cultures in the world as our statistical universe. It would not seem likely that there would be synchronism in cultural rise and decay in various parts of the world, at least prior to rapid global communication, if this were the operative mechanism. Similarly there would be no globally preferred time of arrival of a life-style-changing idea, such as the bow and arrow, at the site of the various cultures. Diffusion is a time-transgressive process.

By contrast, climatic change and similar geophysical events tend to be nearly synchronous globally (Wendland & Bryson, 1974). The sign of the change may be different in different places, one place getting dryer when another gets wetter. The magnitude of the change may also be vastly different, for example, the large high latitude temperature change from glacial to post-glacial compared with relatively constant tropical temperatures. Eustatic sea-level changes must also be globally synchronous. If climatic changes are rapid and interspersed with relatively stable times, then there should be globally preferred dates of climatic change. It is possible, statistically, to sort out these preferred dates from those generated by progressive or random changes (Bryson *et al.*, 1970). It is also possible to sort out globally preferred times of culture change, presumably related to times of environmental change, from internally generated (random) or diffusive changes (progressive).

Thus Wendland & Bryson (1974) have inferred times of climatic change from an analysis of over 800 radiocarbon dates associated with pollen maxima, sea-level heights, maximum extent of glaciers, and top and bottom surfaces of peat beds. Choosing these extrema ensured that only data associated with first- or second-order discontinuities in these records were selected. The histogram of these 815 dates was analysed into a number of normal distributions, and the latter were interpreted as indicating the times of global climatic change. To ensure a more homogeneous population, and to increase the signal-to-noise ratio of the analysis, only pollen data were used in a second analysis. This yielded the dates in the left of Table 7.3.1: the 'major' and 'minor' climate changes were designated on the basis of the number of radiocarbon-dated events marking any discontinuity. While this Northern Hemisphere analysis may be refined by further work, there are still too few Southern Hemisphere dates available to do a separate analysis to determine whether the dates for the Southern Hemisphere are the same as those derived from the whole body of data.

Times of maxima of cultural change rates can also be inferred from radiocarbon (^{14}C) dates. While the difficulty of identifying the beginning or ending of *one* culture from a series of ^{14}C dates is clear, when the series is associated with *many* cultures, times of common discontinuity within all the data may become apparent. Wendland & Bryson (1974) gave a method of analysis which could be used to identify the most favoured and synchronous times of discontinuity from ^{14}C dates. They suggested a contingency table as an analysis format and a maximised chi-square test to identify common discontinuities. The table's abscissa is labelled in radiocarbon years BP, and the ordinate contains a list of all cultures represented by the gathered data. Radiocarbon dates associated with a given culture are located at the intersection of the appropriate culture and the ^{14}C age on the contingency table. A total of about 4,800 dates associated

317

7.3 The effect on mankind

Table 7.3.1. *Dates of climate change, and of maxima of cultural change rates*

Globally preferred dates of climatic change as determined by biological and geological indicators (radiocarbon years BP)				Dates of globally synchronous maxima of culture change rates, statistically determined (radiocarbon years BP)			
Wendland & Bryson (1974)		Frenzel (1975)		Chi-squared method		Cumulative change method	
Major	Minor	Major	Minor	Most significant	Less significant	Clear	Less clear
—	—	500	—	—	670	600	—
850	—	—	—	830	—	850	—
—	—	—	—	—	1,100	—	1,100
—	—	—	—	1,260	—	1,250	—
1,680	—	1,600	—	—	—	—	—
—	—	—	—	—	1,820	1,800	—
2,760	—	2,500	2,800	2,510	—	2,250	—
—	—	—	—	3,110	—	3,100	—
—	3,570	3,600	—	—	3,600	—	3,600
—	4,240	4,150	—	4,230	—	4,300	—
5,060	—	4,900	—	—	—	—	—
—	—	—	—	—	5,510	—	5,500
—	6,050	5,950	—	5,900	5,790	5,900	—
—	6,910	6,950	6,500	—	7,200	—	7,200
—	7,740	7,650	—	—	7,790	7,800	—
8,490	—	8,500	—	—	—	—	—
9,300	—	9,300	—	9,530	—	—	c. 9,500± 200
10,030	—	—	—	—	—	—	—

with 401 cultures were assembled primarily from the journal *Radiocarbon* through volume 13. Only those cultures with five or more dates were retained for the analysis since too few dates associated with a given culture may not delimit the temporal extent of the culture. The final array contained about 3,700 dates associated with 155 cultures.

Another method is to tabulate all the initial and terminal dates of each culture and plot the cumulative number of cultural beginnings and endings from some arbitrary initial point such as 10,000 years ago. The result is a graph consisting of straight-line segments with surprisingly well-defined discontinuities. This is, of course, a method well-known to climatologists for it is the double-mass method for determining discontinuities in rainfall records. The results of this analysis agree surprisingly well with the chi-square analysis, and both are shown in the right of Table 7.3.1.

While the cultural dates for the Southern Hemisphere are sparse, the most significant dates of cultural change determined globally appear to

318

Table 7.3.2. *Southern hemisphere and equatorial culture termini compared with globally significant dates (T, terminal date; I, initial date)*

Culture	European contact	600	850	1,100	1,250	1,800	2,550	3,100	3,600	4,300	5,500	Not within ±100 years of global date
South America												
Belen	T		I									
San Miguel	T		I									
Nericaqua		T			I							
Tiahuanaco	?					I						T450
Chibcha		T				I						
Marajoara		T										I2,000
Gabbanzal		T										I3,850
Atacamian			T			I						
Lauricocha				T								I5,150
Formative				T				I				
Bahia Caraquex							I					T2,150
Valdivia												T2,950, I5,150
Chavin							T					I3,750
Australia												
Bondaian			T						I			
Africa												
Zimbabwe		T	I									
Kalomo			T			I						

apply quite well. In Table 7.3.2 all the available initial and terminal dates for non-Mediterranean Africa, South America and Australia are compared with the dates in Table 7.3.1. One would expect about one-third of the cultural end-points (as currently known), to fall within ±100 years of the specified dates by chance. Actually about three-quarters of the dates fall within these limits. This is really quite surprising, for the probability of having found the earliest manifestation of a culture, or the last, is quite small when a limited collection of dates is available. However, the archaeologist is more likely to devote the cost of a radiocarbon date to the bottom and top of his sequence than to intermediate dates.

Two general conclusions may be reached from the above evidence. First, the timing of synchronous culture history events in the Southern Hemisphere is roughly the same as in the Northern Hemisphere. Second, there appears to be rather good correspondence between dates of cultural and climatic change. This correspondence is surprisingly good in light of the probable error in the dating of both sets of changes. This in no way contradicts the existence of other origins of culture change. It is likely that those cultures which changed at the times of climatic change were those in marginal areas or under other pressures as well. Many cultures persisted through dates at which many others changed. Certainly a great deal more research is indicated.

7.3 The effect on mankind

Specific cases

Mycenae

Plato, in *Timaeus*, tells of a conversation between Solon and an Egyptian priest (Carpenter, 1966). In essence the priest tells Solon that occasionally a drought or flood comes along that so disrupts a society that it loses even the art of writing. This comment was meant to indicate to Solon that the Greeks did not know the great and wonderful things that their forebears had done because they had been subjected to such a catastrophe. Carpenter used this story to introduce his book *Discontinuity in Greek Civilization* (Carpenter, 1966) which dealt with the disappearance of the Mycenaean culture about 2,900 radiocarbon years ago (1300–1200 BC, calendar date).

Carpenter postulated that since in a lifetime of work on Mycenae he could find no evidence to support the orthodox explanation of Dorian conquest, there must have been some other cause. He proposed a protracted drought. However, from the culture history of the whole region he was forced to conclude that the drought affected the southern Pelopponesus, Crete, Boeotia and Anatolia but not Rhodes, Attica or Thessaly. It should have afflicted Euboea, Phokis and the Argolid, but not the western slopes of the Panachaic mountains and Kephallenia. This is indeed a complex pattern of drought and no-drought to have persisted off and on for perhaps a century or two.

The time of the Mycenaean demise was a time of significant culture change world-wide. It was also the time of onset of climatic instability that finally settled down in what we have called the subatlantic climatic episode (Baerreis & Bryson, 1965) in which the Graeco–Roman, Classic Mayan, Han and Dorsett cultures flowered.

The problems for the palaeoclimatologist are whether such a drought pattern is possible and whether such a pattern actually existed at the time. The principal components of the rainfall, or Palmer drought index, for the modern instrumental record indicate that a drought with the pattern postulated by Carpenter is a normal mode of behaviour of the climate (Bryson *et al.*, 1974). Winter 1954/5 was dominated by this pattern, and indeed the pattern of precipitation anomalies for the period November 1954 to March 1955 is very close to that suggested by Carpenter as characteristic of the time of Mycenaean decline. There is no meteorological reason why such a pattern could not be much more frequent in one century than another. Indeed, weather types, both daily and monthly, often show large variations in frequency of occurrence (Blasing, 1975).

The question as to whether the time around 2,900 years BP was the onset of such a period of frequent such patterns is more difficult. However, Bryson *et al.* (1974) showed that use of any one of several circulation-typing schemes will give a close relation between the general hemispheric circulation type and the particular drought pattern in the eastern Mediterranean. This makes it possible to use data from a variety of locations to check for consistency with the type associated with a particular pattern. In the case of Mycenae, a score of different pieces of evidence from Africa and Europe all were consistent with Carpenter's postulated drought pattern and showed conditions after 2,900 years BP similar to the winter of 1954/5. The drought pattern suggested by the winter 1954/5 is not indeed anything quite as simple as a bodily displacement of the whole subtropical arid (or Trade Wind) zone, but the fact that it is concentrated on a particular range of longitudes, and even shows correspondence of smaller-scale detail over Greece, lends conviction to the thesis that this is something like what happened around 1200 BC. Until contrary evidence is turned up it would appear that Carpenter's hypothesis is the most viable one available.

The result of the climatic variations starting about 2,900 years BP appears to have been a major rearrangement of cultures in the eastern Mediterranean. It was after the depopulated Peloponessus was once again fruitful that the Dorians moved into the region (Carpenter, 1966).

There is a specific warning for palaeoclimatologists in this case. Since climatic variations may produce negative anomalies in some regions and positive anomalies nearby, there is obviously some point between with no change. Clearly one cannot then argue that, because a particular site of field investigation shows no change at a particular time, no change was occurring in the whole region. Indeed, the field palaeoclimatologist's delight, a nice continuous undisturbed core, may be selected because it lies on just such a nodal line between regions of opposite change!

Mill Creek

Table 7.3.1 shows that in the twelfth century AD (800 years BP) there was a climatic variation associated with a time of culture changes. Palaeo-pedological investigations along the Arctic tree line in Canada show that the summer circulation patterns shifted southward in central North America (Sorenson *et al.*, 1971). This should result in a southward displacement of the summer westerlies from southern Canada into the northern United States (Bryson, 1966; Baerreis & Bryson, 1968). As a consequence the northern plains should have been subject to frequent dry winds descending the east face of the Cordillera and spreading eastward to displace the usual northward flow of moist tropical air. Using modern data stratified into weak westerly flow across the United States and somewhat stronger flow that would be associated with a moderate southward displacement, one finds that there should have been up to 50%

reduction of mid-summer rains in parts of central North America (Baerreis & Bryson, 1968, p. 7). Such a reduction should have been disastrous to peoples living in marginal areas within the region, and a significant pressure for change on those in less marginal locations.

To test this hypothesis, excavations were carried out on village sites of the Mill Creek culture in northwestern Iowa. The choice was relatively simple. The maize farmers (non-irrigated) who had lived farther west in a drier habitat prior to this time actually disappeared. The Mill Creek people, in a less precarious location, were close to the prairie–short-grass boundary and survived into the fourteenth century. It was to test the hypothesis of a drought impacting their economy that the excavations were carried out.

In summary, the results of the investigation were as follows:

(i) within a few decades in the twelfth century the vegetation changed from tall grass prairie on the uplands with forest in the larger valleys to steppe-like vegetation and essentially only phreatophytes along the streams in all but the major valleys;

(ii) the meat diet of the people shifted towards more bison (grazers) and less deer (browsers), the shift being from 97% deer to 67% bison in the more sensitive locations;

(iii) pottery styles and apparently population distribution changed at the same time (Baerreis & Bryson, 1967);

(iv) farther east there were major changes in the middle Mississippian culture.

The significance of this event to the present time is that the region that was affected by drought from the twelfth to the fourteenth century is now the spring wheat, maize and soybean region of the United States. What has happened, climatically, is clearly possible. The mid-western drought of 1974 was very similar in distribution to the 200-year drought around 830–600 years BP.

The Indus culture

From about 4,100 to 3,600 years BP there are numerous radiocarbon dates referred to the Indus culture (Harappan) in northwest India and Pakistan. Four dates extend beyond this time to 3,100 years BP. Then there is a hiatus until the major body of Painted Grey Ware dates starting about 2,500 years BP. These are all globally significant culture change dates (Table 7.3.1). During the last ice age, when the polar regions were very cold and the circumpolar vortex was very large year-round, dunes had formed in Rajasthan. There was no monsoon. When the ice-age climate ended 10,800 years ago, the monsoons began, the ground-water level rose and fresh-water lakes formed between some of the dunes.

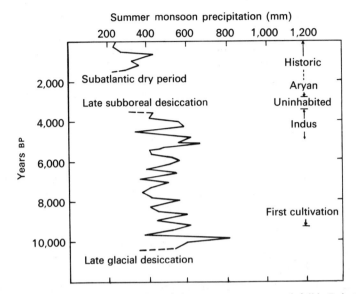

Fig. 7.3.1. Tentative reconstruction of the summer monsoon rainfall in Rajasthan over the past 10,800 years, based on fossil pollen accumulated in a lake. Singh *et al.* (1972), converted to climatic profiles by the transfer equation method of Webb & Bryson (1972). The lake contained fresh water until shortly before it dried up about 3,500 years ago.

Using pollen profiles from these lakes (Singh *et al.*, 1972, 1973) and the analytical method of Webb & Bryson (1972) the Holocene rainfall history of northwestern India was reconstructed (Fig. 7.3.1). There was no significant rainfall during the cold, late Pleistocene, and the region was a desert of drifting dunes. This figure shows very clearly that the area at the edge of the Indian monsoon is climatically hazardous at best, century-long average rainfall varying by a factor of two. A millennium or so after the advent of the post-glacial monsoons (about 10,000 years BP), the ground water had risen, lakes had formed in the low areas and there was agriculture in the area. The pollen record begins to show cultivated grains by 9,400 years ago, a significant date, during a time of increasing monsoon rainfall. The monsoon rainfall record is hard to interpret because it reflects both the general hemispheric temperature level and the rate of change of hemispheric temperature (Brinkmann, personal communication, 1975). When high-latitude temperatures are low, the hemispheric average temperature is relatively low and the circumpolar vortex is expanded. This tends to displace the subtropical anticyclones equatorward and suppress the monsoon. However, when the temperature is falling the continents cool faster than the sea, changing the land–sea contrast and also affecting the monsoon rainfall.

The Indus culture flourished during a time of reliable, good summer

323

monsoons plus ample winter rainfall. Then the rains declined, the lakes turned salt and dried up, and the culture disappeared from what had been its heartland. The dunes once again became mobile, burying towns of the Indus people. After a period of 600 years during which the region was unoccupied, a different people moved into the region. The consequences of such a failure would be greater today in this most densely occupied of all deserts.

Modern cases

To examine the impact of climatic variation at the present time we will consider two different kinds of grass and their response to climate: the hay of Iceland, and the wheat which provides about a quarter of the world's grain.

The climatic record of Iceland reconstructed by Bergthorsson (1969) shows generally higher temperatures in the few centuries after Viking settlement almost 1,100 years ago. There was a colder spell around the fourteenth century but the coldest period was between AD 1600 and the

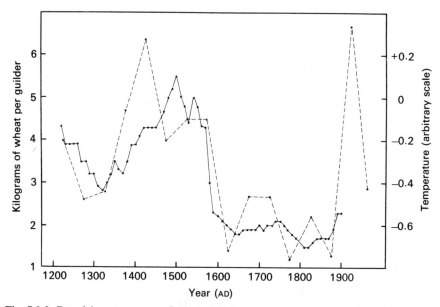

Fig. 7.3.2. Decadal mean amount of wheat that could be purchased for one Dutch guilder (average of England and Holland) compared with estimated North Atlantic temperatures, $\bar{T}$ (—•— kg/guilder, --•-- $\bar{T}$). The North Atlantic temperatures are plotted to an arbitrary scale and are the 50-year means of Bergthorsson's (1969) Icelandic temperature series but using an earlier tabulated version. This comparison was suggested by Leona Libby (personal communication, 1974). The height of the temperature peak 1400–1450 is questionable and the last 50-year period is incomplete.

Table 7.3.3. *Gross spring wheat income losses in the United States due to deviations of temperature, T, and precipitation, R, from normal (millions of dollars)*

The model assumes approximately 500×10^6 bushel crop, \$3.50/bu price, and that all other factors are constant

Anomaly period	$\Delta T = $ +1 °C	$\Delta T = $ −1 °C	$\Delta R = $ +1 mm	$\Delta R = $ −1 mm	$\Delta R = $ +20%	$\Delta R = $ −20%
Preseason	—	—	+7	−8	+21	−30
April	+40	−40	+3	−3	+22	−25
May	−22	+13	+2	−2	+4	−37
June	−70	+70	+2	−2	+37	−44
July	−78	+92	—	—	−2	−2
Whole season, i.e. all periods	−131	+136	+14	−15	+82	−139

early twentieth century. The cold periods were times of frequent famine, for both the fisheries and the hay which supports the animals on land are sensitive to climate in the marginal position of Iceland. In particular, the hay is more temperature-limited than water-limited, about 80% of the yield variance being related to temperature after adjustment for fertiliser application and condition of the pastures at the end of the previous season (Bryson, 1974 b). The sensitivity of yield to summer temperature, as derived from the yield equation is about 15% per °C. The decline of temperatures in the recent decades has apparently reduced yield nearly a quarter. Since hay is the basis of practically all land-based agriculture, the impact on the Icelandic economy has been severe. A return to the temperatures of the last century would produce 40% reductions in yield.

Most people do not depend on agriculture that is as sensitive to temperature as Icelandic pastures. However, in the higher mid-latitudes there is still strong response to climatic variation. Consider Fig. 7.3.2 which shows the course of North Atlantic temperatures since AD 1200, as indicated by Bergthorsson's reconstructed Icelandic temperatures (Bergthorsson, 1969). Plotted on the same graph is the amount of wheat that could be purchased for a Dutch guilder (allowing for inflation). The fivefold variation appears to follow the temperature quite closely with a decreasing lag.

To put the climate and wheat yield on a more precise basis we may calculate the impact of a small climatic anomaly by using the spring wheat model devised by the Climate/Food Research Group of the University of Wisconsin, Madison (1975). The model, which considers monthly temperatures and rainfall, preseason moisture, technological trends and soil type, reproduces past yields per planted acre with about 97% accuracy for

325

the United States spring wheat region. Using this model, Table 7.3.3 was calculated. This table shows that while a temperature anomaly of 1 °C for each month of the growing season would only change the yield by about 7.5%, half the Icelandic hay sensitivity, that is still $136m. However, a 20% reduction of rainfall each month would cost another $139m in gross value. Considering the profit margin of the wheat farmer, the impact on his net return would be far greater than 7–10%. High-technology agriculture is not immune to climate variation.

From 1945 until at least 1972/3, a return to climates like those of the AD 1600–1900 period has been taking place as evidenced by declining Northern Hemisphere temperatures (Mitchell, 1963; Reitan, 1974 a), a cooling of the North Atlantic (Rodewald, 1973), an equatorward retreat of the monsoon in West Africa (Bryson, 1973 a, b) and Japan (Asakura, 1972), and increased snow and ice cover over the Northern Hemisphere in 1971/2 (Kukla & Kukla, 1974).

As the climate changed during this time, the world of mankind has changed dramatically as well. The world-wide spread of antibiotics and insecticides greatly reduced suffering from infectious diseases and malaria, but has aided the doubling of the inhabitants of our finite world. For many years world reserves of food grains diminished after the post-World War II recovery peak. There was a short-lived increase with the introduction of the higher yielding rice and wheat of the 'Green Revolution' but then the decrease resumed (Fig. 7.3.3).

Whether or not the climate trend of the 1950s and 1960s is continuing, it is clear that even today, with high technology, we are not immune to the effects of climatic variation, at least economically, and at most in the possibility of widespread famine. Could anyone argue that a return to the monsoon failure of 3,600–2,900 years BP would fail to have dramatic consequences? What has happened, can happen. It is one role of the palaeoclimatologist to assess these possibilities, and of the climatologist to assess their probability in the future.

Conclusion

Most but not all palaeoclimatic investigations have reached the conclusion that Holocene climatic fluctuations are of relatively small magnitude. Robust cultures in good locations could probably weather any of those which have occurred. However, it appears that in marginal situations, culturally or climatically, the larger fluctuations of the Holocene have been sufficient to trigger culture change. This has probably been a result of change in the economic base, agriculture, hunting-gathering, or herding, either by removing it entirely (Indus culture) or reducing its carrying capacity (Mycenae, Mill Creek). The response whether *in situ* modification

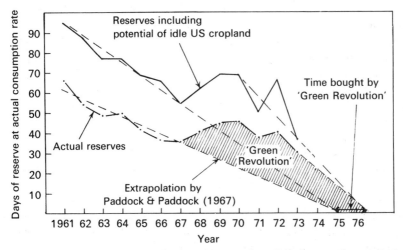

Fig. 7.3.3. World food grain reserves, year by year, since 1961 show a rather steady decline with only a minor offset produced by the 'Green Revolution'. The graph shows that the world consumed more than it produced from 1961 to 1967, then again after about 1970 (from Bryson, 1974*b*).

of lifestyle, migration, or literal disappearance of the people must be determined case by case. The current world food grain reserve situation confronts contemporary man with the urgent need to make use of the lessons of the past, and to plan adequately for the future.

7.4 Economic and political issues
W. J. MAUNDER

Weather and climate as variable resources

During the last few years, individual nations and the international community have become aware of and accepted the view that there are certain physical limits to the availability of natural resources and food. During 1973, the world began to receive sharp warnings of some of these limits, particularly concerning the supply of energy and of food, and it became apparent that our ability to make efficient and humane use of the world's resources through existing institutions was more limited than we had thought. As a consequence, in 1974 three important world conferences took place: The World Population Conference, The World Energy Conference, and The World Food Conference, and in these a start was made to study realistically the fundamental changes that will be required to ensure the optimum use of the world's resources to meet the demands of the future.

Included among these resources are those of the atmosphere. One of

327

the major problems in regard to this resource, however, is the promotion of a definition that clearly recognises the atmosphere as a resource which is variable. This variability can be evaluated in ways which are useful to planning and management, and certainly in regard to food production this knowledge must be used for intelligently planning national and world food security programmes if these are to be successful. 'Climatic change' is thus paramount in any consideration of weather and climate as a resource.

Several aspects of the term 'climatic change' have to be studied before any worthwhile analysis of the problem is considered. They include: what are 'climate' and 'climatic change'? is there evidence for climatic change in terms of specific 'changes' in temperature, rainfall, etc.? if so how is the climate changing? what are the possible 'causal' mechanisms? can these mechanisms be modified by man? are the effects of a climatic change predictable, and at what degree of accuracy? what are the economic, social and political implications of climatic change?

Although several terms are in common use when describing 'climatic change', such as 'variations', 'trend', 'oscillation', 'periodicity', as well as 'change' itself, there appears to be little uniformity in their use and acceptance. For this reason, it is perhaps desirable to suggest that climate or 'climatic state'* is 'the totality of "weather" conditions existing in a given area over a specific period of time'. If this is accepted, it *may* be considered that the concept of 'climatic change' becomes (or should become) important only when a relatively 'long' time period is considered. But, such a viewpoint is far too restrictive. Indeed because of the finite resources available in the world, the importance of monitoring, analysing, understanding, and forecasting 'variations in the available atmospheric resources' over time periods of months, seasons, years and decades is of far greater importance, whatever the 'true' meaning of climatic change may be.

The *impact* of short-term variations in the available atmospheric resources will continue to increase in importance because of the growing demand for food. Thus any change for the worse in longer-term climate could result in food shortages in many areas that will be far more critical than would occur if the climate remained 'normal'. In any discussion of climatic change, it is important therefore to understand the increasing impact climatic change will have on economic, social and political activities.

Obviously, appropriate meteorological planning must be evolved if we are to live within the limit of our atmospheric resources. But, if this is to be accomplished, the politician and the planner must become more

---

* 'Climatic state' is a term used in a report of the (US) National Academy of Sciences, in their 1975 publication *Understanding Climatic Change* (see Section 4.1 above).

'weather' orientated, for only then will optimum use be made of the climatic resources of the 1980s and the 1990s. Central to this meteorological planning is the need for a much more comprehensive monitoring and analysis of the world's climate, both to detect and predict changes, and to understand the consequences of such changes. The need for advice about the weather therefore offers a challenge to the meteorological community, and a necessary first step is to educate both meteorologists and potential users of meteorological products, in the specific application of meteorological and climatological information to economic, social and political problems.

There is an increasing awareness that weather and climate, and information about them, can play a very important role in the decision-making processes associated with the management of weather-sensitive activities and enterprises, such as dairy production (Maunder, 1974) and agriculture generally (Collis, 1975; World Meteorological Organization, 1975 b), retail trade (Maunder, 1973), power generation (Maunder, 1971) and the construction industry (Russo, 1966). At the international level, the problem is considerably more complex, but it is clear that global and national economics – whether these are monitored by UN organisations, national governments, or private companies – are sensitive to weather and climatic variations.

It could be argued that weather and climatic information cannot really be used in relation to economic, social and political planning, until reasonably accurate climatic forecasts are available. While it might well be assumed that such forecasts will become available eventually, a proper analysis and interpretation of past climatological information is still of considerable value, even if accurate climatic forecasts prove to be impossible. It should be realised, however, that since decisions involving weather-sensitive activities must be made, they will be made irrespective of whether appropriate weather/climatic forecasts are available. Thus, if qualified meteorologists and climatologists do not provide such guidance, then decision-makers will either ignore the possible impact of weather and climatic events on their activity, or obtain 'guidance' from those that are less qualified to provide it. For instance, the state of both the New Zealand and Australian economies is dependent to some extent on next year's weather, yet how many top-level decision-makers in these countries take the weather and the expected weather into account? Further, would more top-level decision-makers take the weather and expected weather into account if there were a more direct advisory communication with meteorological experts?

7.4   The effect on mankind

The range and importance of weather and climatic variations

Extreme weather events, such as hurricanes or severe drought, obviously create headlines in newspapers and concern amongst many people, particularly governments and international agencies. But it is pertinent to consider whether it is not the 'near average' climatic events which really have the more important overall social, economic and political impact. For example, a cooling in Europe in the summer by 1 °C, a decrease in cloud amount over North America in the winter by 5%, or an increase in the temperature of the South Pacific Ocean in the summer by 0.5 °C, particularly if sustained for more than one season, could have social, economic and political consequences which in total could be much greater than the destructiveness – however severe – of hurricanes, severe rain storms, or tornadoes (see e.g. Alexander, 1974; Maunder, 1975; Robertson, 1974; Taylor, 1974).

Moreover, since it is the relatively short-term variations from 'near average' weather that decision-makers of weather-sensitive enterprises, including governments and international agencies, have to contend with (see e.g. McQuigg, 1974) the recognition that such variations are very important – economically, socially and politically – could help to avoid some of the problems which arose in 1972 and again in 1975, when it appears that wheat was traded between a number of countries to an unnecessary degree.

The real value that can be placed on short-term variations in the climate requires the identification of those activities directly or indirectly affected by such variations, and a quantitative analysis of the consequent effects these changes have on such activities. Specific aspects of this problem have been considered by a number of people, but most have been restricted either in time or space. For example, assessing the real value of a 10% reduction in wheat production in North America and Australia over the next decade, or the impact of a 2 °C decrease in winter temperature over North America and Europe over the next decade – both of which could result from relatively small changes in the 'climate', are questions which until very recently have received little attention.

The world's population is growing exponentially, and there are now no longer exploitable virgin territories just 'over the mountains' because the world is at least, in a practical sense, finite. Of major concern is the fact that the crop production areas of the earth are limited in extent and yield. McQuigg (1974) has pointed out that this new world situation leads to climate becoming more and more a significant factor in grain production. As grain reserves continue to fall, climatic variability becomes more important as a sensitive controlling parameter in determining future supply. Thus climate plays an ever more important role in international

330

politics. The United States, for example, provides 'approximately one half of all the grain that moves across international boundaries', according to US Secretary of Agriculture, Butz (US News and World Report, 1976), and so by withholding grain, can exert political pressure. Business Week (1975) also comments on the power that the agricultural capacity of the United States can wield, and suggests that 'there is a growing consensus that the US should be as tough in using food power to achieve national objectives as it is in employing other economic capabilities'.

While some meteorologists seem happy to remain in ignorance of the value of their work, the fact remains that the current and expected future expenditure on meteorology must provide appropriate returns – social, economic and political. It is relevant, therefore, to emphasise, most strongly, that since complex sensing, communications, and data-processing systems are capable of producing a virtually unlimited flow of meteorological data, much more serious attempts should be made by meteorologists and climatologists not only to make better use of this information, but also to evaluate its national and/or global economic, social and political value.

However, while budgetary considerations are important, a more fundamental reason exists for being interested in the impact of weather information, notably that in the near future mankind will undoubtedly try to apply the rapidly improving flow of weather information in a much more deliberate manner, in order to monitor and control significant weather-sensitive aspects of a national economy.

For instance, the deliberate analysis of the weather of one country by another, for purposes other than international aviation and shipping or traditional weather forecasting, such as to forecast agricultural production, is a reality. The deliberate 'control' or 'modification' of the weather of one country by another could well become a reality. In this latter case, it is clear that such control or modification would – at least in the foreseeable future – be limited in area, but its effect on certain aspects of agriculture could be very considerable. Monitoring and control are therefore of fundamental importance, and are matters which will have economic, social and political consequences at the very top level.

Now, few if any models of the physical atmosphere go on to consider the economic and social aspects of the variability of the atmosphere, and generally only token consideration is given to the end products of the meteorological chain. But since weather and climatic information has social, economic and political value it is most important that such information be presented to the decision-maker in much the same way as any consumer good is packaged for efficient and effective marketing, and further that such information is correctly used. But if the improved weather package is to achieve its purpose, the appropriate national or

331

7.4   The effect on mankind

international meteorological authorities and their controlling governments must be convinced that there *are* important customers who need the information that could be provided. The meteorological community therefore has a very important and continuing task of providing ways to establish direct communication between top-level decision-makers and the meteorological system, for only then will the optimum marketing of meteorological information be achieved.

Commodities, weather information and politics

Few stock market experts would consider that there is any proven link between weather variations and the prices on the stock exchange (although this could be challenged); nevertheless, many investors do realise that droughts or floods often do have subsequent effects on the prices of particular shares. However, in another kind of market – the commodities market – prices do vary with the weather, and in this market it is not only the actual weather that is important, but also the weather that is reported to have occurred, as well as the weather that is expected to occur.

The commodities market is the exchange where commodities such as wheat, corn, soybeans, sugar, cocoa and frozen orange juice are bought and sold for delivery at a stated time in the future. The prices for the commodity 'futures' are related to a number of things, but of greatest importance is the anticipated demand for, and the anticipated supply of, the particular commodity at some future date. Consequently, there is a close relationship between the actual, reported and expected weather conditions in the various producing areas, and the price at which 'futures' in a commodity will be bought and sold. Reports of the weather that has occurred, and forecasts of the expected weather, therefore, have a real 'value' as far as the commodity markets are concerned, and consequently an effect on the prices that the consumer will eventually pay for these commodities.

An associated aspect of the commodities market is the role of the meteorological community in supplying weather and climatic information which may or may not affect the market. Indeed, on a wider international level, as already noted, the implications of being able to monitor and predict the weather of other countries, and therefore their potential crop production, are wide-ranging and involve high-level political and even military implications at both national and international levels. In theory, all countries could use internationally exchanged weather information for such purposes, but in practice only a few have the expertise or desire to do so. The United States is one such country, and during 1973/4 a Center for Climatic and Environmental Assessment was established in Columbia, Missouri, under the directorship of Dr J. D. McQuigg (American

332

Meteorological Society, 1974). This Centre, which is part of the National Oceanic and Atmospheric Administration of the US Department of Commerce, has a number of tasks, but one of the principal ones is to monitor and assess crop production both in the United States and elsewhere, based in part on weather information. In addition, the Food and Agriculture Organization, in collaboration with the World Meteorological Organization (WMO), is closely monitoring and, where possible, predicting crop production in many areas of the world, as part of a 'global information and early warning system on food and agriculture'.

The international implications of providing more weather information are therefore considerable, and a proposal for an Agrometeorological Programme in Aid of Food Production was submitted by the Secretary-General of WMO to the Seventh Congress of WMO in May 1975. The programme included the assessment of regional and global crop conditions using current meteorological data, and a number of meteorological centres were to be assigned international responsibility for preparing and disseminating predicted crop-yield assessments on a global basis, with the proviso that assessments be disseminated only if the country concerned did not explicitly oppose this. It was considered, specifically, from the viewpoint of food management and reserves that all nations, developed or developing, would benefit from the free exchange of an objective, weather-based advance estimate of expected crop yields on a regional and global scale, since such predictions could be used to provide more lead-time for planning and decision making.

However, this assessment and prediction scheme was considered to be 'premature' by many Members of the Seventh WMO Congress. Such opinions in the main were based on the stated view that agrometeorological expertise is not yet sufficiently well developed. Nevertheless, perhaps the more fundamental reason is the far-reaching economic, social, political and even military aspects of such information if it is used 'incorrectly'. There will always be competitive advantages to those who know how to use the information, and especially to those who get it first. In certain circumstances, therefore, weather information can become almost too powerful a tool. It is nonetheless clear that there is a need for such information, and one wonders how long a system such as that described for predicting agricultural production from the already available weather data of the World Weather Watch system can remain 'suppressed'. Conversely though, perhaps the time is not too far distant when the *present* free exchange of the world's weather information will be openly questioned, for there is little doubt that the use of weather information, particularly in relation to its use as an early predictor of production, has many political overtones.

Finally, it is pertinent to re-emphasise two important questions which

333

are relevant to the study of climatic change: first – is the climate changing, and if so in what way? second – what is the cause of these changes, and can they be forecast? But there is an even more important question which must also be answered and that is 'what are the likely economic, social and political impacts of such climatic changes?' In this very difficult area, the traditional role of the atmospheric scientist has been one of extreme caution, but the time has now come when the meteorological community must not only give a strong lead in this area, but must also make very sure that decision-makers at the top level are able to use such information for the good of all mankind.

7.5   Mankind, climate and doomsday science
M. CHARLESWORTH

One striking feature of the present age is what is sometimes called the doomsday mentality. It would be of very great sociological interest to know exactly why this curious kind of scientific pessimism has come to have such an intellectual fashion and vogue. The evidence on the basis of which the various kinds of doom are so confidently predicted is almost completely controversial and is so ambiguous that it can be read almost any way. Why then do so many scientists choose to read the evidence in a unilaterally pessimistic way, and why has this doomsday interpretation caught on so readily with ordinary people? It would tell us a great deal about our cultural *psyche* if we could understand the 'unconscious' sociological motivations that underlie this doomsday mood and mentality.

Recent literature about climatic change, in particular Nigel Calder's book *The Weather Machine* (1974), exhibits the same kind of pessimism that has been so successfully purveyed in other fields by Ehrlich (1968), Ehrlich & Ehrlich (1970), Meadows *et al.* (1971) and others. Not only are we to suffer an inescapable doom by overpopulation, and by scarcity of resources and energy; we are, it seems, also doomed by the weather in the shape of a new ice age. The only consolation provided by this latter threat, so we are told, is that it looks as though it will favour those countries which are at present economically disadvantaged and 'strike hardest at the present citadels of the white man's economic power' (Calder, 1974). Some general reflections on the larger issues involved in this whole question are presented here, for there *are* very fundamental human, social and moral issues involved, despite the attempt by the doomsters to present their predictions as being purely scientific and 'value-free'.

In many ways, the present scientific pessimism is the mirror-image of nineteenth-century scientific optimism with its belief in the inexorable progress of mankind towards various kinds of utopia. Darwinian evolution, though random and blind, had ensured that the 'fittest' should survive and had duly produced *homo sapiens*; Adam Smith's 'hidden hand of God' had so adjusted economic forces that private economic self-interest would of necessity bring about the good of all; again, according to Hegel, the dialectic of history meant that man was becoming more and more rational and more and more free; whilst Marx believed that absolutely indisputable socio-economic laws would inevitably bring about the collapse of capitalism and its replacement by a 'classless society' in which people would no longer be economically, socially or psychologically 'alienated'. Clearly, every day in every way, mankind was becoming biologically, economically, socially, psychologically, humanly, better. And these utopian results would come about without any need for effort on our part; regardless of what we may want or choose, the hidden hand of God, or the iron principle of natural selection, or the laws of history, or deterministic socio-economic mechanisms, would bring about the utopia of economic abundance, biological fitness, social harmony, the reign of reason and freedom, the 'classless society'. In other words, the future of mankind was strictly determined but chance had determined it in our favour so that there was a happy coincidence between the necessities of history, or the inexorable laws that govern socio-economic phenomena, and human happiness.

The belief that scientific inquiry and technological invention must inevitably be, in the long run, for the good of mankind, is also part of this world view. For many scientists the possibility that there might *not* be a nice coincidence between scientific inquiry and human welfare is unthinkable, since then, for example, some piece of scientific research may need to be discontinued on the grounds that it would probably bring about more human unhappiness than happiness. Like Adam Smith, many scientists implicitly believe that the 'hidden hand of God' has so nicely arranged things that the progress of scientific inquiry and technological invention will automatically conduce to human happiness. This is a curious residue of nineteenth-century scientific optimism which has persisted even in those scientists who are concerned now to point out pessimistically that the iron laws of history or nature or economics in fact work *against* man and will inexorably lead not to human happiness but to human misery, not to utopia but to doom.

It now appears quite strange and unreal to us that nineteenth-century thinkers should have so naively interpreted the facts before them in this way and accepted without question that the evidence of human history and of science pointed unequivocally towards utopia. From the vantage point that we occupy we can now see how much their history and science

335

was used for what may broadly be called 'ideological' purposes; in other words, they believed what they wanted to believe and they interpreted the historical and scientific evidence to fit in with their preconceptions.

And yet we imagine that things are different with us and that *our* doomsday prophecies are not ideologically motivated. We claim that the evidence of human history and of economics and demography and biology points directly and unequivocally towards doomsday and that we are simply facing up realistically to the brute facts as they present themselves to us. In actual fact, however, there is, as I have already suggested, just as much ideology in contemporary doomsday science as there was in nineteenth-century scientific utopianism. (Recent studies by philosophers of science such as Kuhn (1970), and Feyerabend (1975), have shown in general terms to what a degree ideology enters into science and to what an extent the scientist's method is dictated by extra-scientific motives.) I cannot here enter into a detailed discussion of this question, but it can be said at least that an important part of the ideology which lies behind nineteenth-century scientific utopianism and twentieth-century scientific anti-utopianism is the belief that man's salvation lies in science. In a sense modern science took over the functions of religion as the power and influence of traditional religion in the West declined. And as one of the main functions of religious world-views is to provide an eschatology (that is a futurological 'scenario' of the final culminating state of the universe and of human history) so modern science has also been concerned to provide its own account of the 'last things'.

In the religious view it has always been a problem to see how the divinely preordained future of human history can be reconciled with the fact of man's freedom to either save or damn himself. And in the same way in scientific futurology it is difficult to see how these predictions of utopia or doom can be reconciled with human freedom. Thus at the bottom of both nineteenth-century utopianism and contemporary doomsday science is a naive kind of determinism which effectively denies the place and value of human freedom – the power that mankind has to stand apart from the material factors which condition its existence, and to understand and control them, or at least bend and divert or 'sublimate' them to serve human purposes. And yet, of course, if human beings do not have this power then moral values in general cease to have any real point. Why should we think that human life is something morally valuable, and why should we bother to promote human happiness and alleviate human misery, if human beings are nothing but the product of their biological make-up or their socio-economic environments or their psychological conditioning? If the new ice age will inexorably overtake our successors why should we have any moral concern for them? We may regret the future passing or the deterioration of an interesting species – *homo sapiens* – but that kind of zoological regret is about all the scientific determinist has a right to.

Fortunately, both for them and us, the doomsday experts are not consistent, for along with their naive deterministic assumptions they are fiercely concerned with human values and about the future happiness of humankind. That is, in fact, a paradoxical feature of the history of modern science. Pre-Copernican man thought that by virtue of his reason and free will he occupied a special place in the universe and could not be explained, so to speak, in the same way as we explain the rest of material nature. After Copernicus, however, the earth becomes a relatively insignificant planet of the sun and, equally, mankind is seen as a relatively insignificant part of nature – one particular instance of the great general laws that govern all natural phenomena. And yet, curiously, parallel with this whole tendency to reduce mankind to material terms and to deflate human pretensions to occupy a special place in the world, there is at the same time in the moral and social sphere an extraordinary emphasis upon the values of freedom and autonomy.

There is this same ambivalence in the doomsday scientists, for on the one hand they talk in terms of quasi-deterministic processes, and then on the other hand they speak in tones of prophetic moral concern and appeal to our free will, that is our ability to understand and to 'sublimate' the material factors which condition our future. But once again, we cannot have our cake and eat it: if our final doom is strictly predictable and determined then there is no point in writing books and reports calling on us to *do* something about it; on the other hand, if we can *do* something about it then the future cannot be completely determined and predictable.

At one level, no doubt, the doomsters seem to be saying no more than that if present trends continue, and nothing is done to arrest them, we will end in catastrophe. But they oscillate between this more or less innocent position and a quite different, and far from innocent, position of quasi-determinism according to which mankind's doom is wholly predictable and inescapable. They speak in deterministic tones in order to give their futurological predictions a dramatic air of scientific necessity and inevitability.

The truth is that as human beings we always find ourselves in situations which at once set limits to the possibilities for action and also provide the 'material' for action. Human freedom is never absolutely 'free': we have to make do with what the biological, economic, social and climatic conditions of our life provide and we have to respect the limits set by those conditions. *All* human action, and not just political action, is the 'art of the possible'. Nevertheless, small and limited as the area of human freedom may often be, it is real and important; it is in fact the difference between distinctively human life and non-human life. My conclusion then

337

7.5   The effect on mankind

– and it is hardly a novel or exciting one – is that what we need above all at the present time is neither optimism nor pessimism, neither nineteenth-century scientific utopianism nor twentieth-century scientific doomsday thinking, but a firm commitment to the value of human freedom in all its dimensions. The Gods have *not* so arranged the world that either utopia or doomsday will come about quasi-deterministically, nevertheless the moral value of human beings attempting to understand and care for the fate of humankind remains.

8. Progress and prospect

8.1 The problem of short-term climatic forecasting

R. A. S. RATCLIFFE

The principal parameters of the earth–ocean–atmosphere system have considerable variability about their means, which raises questions about the stability and predictability of climate.

Should a small change in the initial environmental state lead to gross changes in climate, the climate is said to be unstable: if large changes in the initial state do not change the climate, then it is stable. And if two distinct stable climates may exist for the same initial conditions, the climate is said to be intransitive (Lorenz, 1968).

Intransitivity, if existing, will make long-term forecasting exceedingly difficult, if not impossible. On a day-to-day time scale Lorenz (1969, 1973) showed the limit of predictability to be about 12 days; a figure which is substantiated by our work in Britain where analogous synoptic situations have occasionally been found to remain similar for 7–10 days. Theoretical studies indicate a 3-day doubling time of observational error in any initial field. How predictable, then, is climate between these two extremes – that is, on the scale of one month to a year?

The atmosphere itself operates on various time scales from the small scales of turbulence and convection measured in minutes or hours, through intense thunderstorms and the synoptic scale of depressions and anti-cyclones (about 5 days or so), to the long planetary waves which often persist in recognisable forms for a month or more.

Most of the energy in the temperate latitude atmosphere is carried by the longer waves of 10^3 or 10^4 km wavelength, but it is uncertain to what extent the different scales of motion interact. On the whole, however, long-term variations of the large-scale circulation are not thought to be attributable to individual irregularities in the small-scale processes. Although small-scale eddies occur in the oceans also, the sign of sea-surface temperature anomalies over large areas tends to remain constant over quite long periods. Variations of sea temperature, by varying the amount of heat and moisture available to the atmosphere, have been shown by many writers (e.g. Namias, 1973b; Bjerknes, 1969; Ratcliffe & Murray, 1970) to be an important factor in determining monthly and seasonal weather anomalies. It is by studying such interactions that we in the British

339

8.1 Progress and prospect

Meteorological Service have managed to produce over the last 12 years or so some 300 monthly forecasts of which about 75% have been more successful than one would expect by chance. It therefore seems beyond doubt that there is some useful predictability on the monthly time scale.

The economic importance of reliable forecasts over periods between 3 months and a year or more would be very high as they would allow of more reliable planning and allocation of resources. In many temperate-latitude countries, for instance, crops such as tomatoes are marginal; in a good summer, good crops and a high economic yield is possible, while in a poor summer the crop would not be worth growing. However, it is necessary to know the overall character of the summer with some confidence some 5 or 6 months before harvest time if the best decisions are to be made. Similarly it is possible to select the most suitable variety of grain to produce the optimum yield under different growing conditions; some seed does well in a wet year, while other varieties produce their best results in dry seasons.

The demand for many industrial products is also very weather sensitive. For instance, the building industry may almost be halted by a severe winter. Wide variations in demand for building materials may therefore result from variations in winter weather and this example could be extended many times to include, for instance, variations in demand for clothing, fuel and water, etc.

Climatic forecasting on a time scale of decades, centuries or longer, poses difficult problems depending to some extent on man himself. I believe there is more hope of extending present long-range forecasting methods to give general guidance for 3 months to a year ahead.

Short-term climatic forecasting in Britain

One of the most promising fields of research geared to this end involves eigenvector methods (for which see e.g. Barry & Perry, 1973). Using 500 mb data from the period 1949–74, various eigenvector analyses have been carried out over a grid covering 30° to 70° N and 110° W to 30° E. In this work the year has been divided into 6 seasons rather than the more usual 4 because we have found that the general circulation evolution round the year tends to fall into 6 rather distinct periods. If the eigenanalysis is carried out on these 'seasonal' charts it is found that the first seven eigenvectors account for between 85% and 92% of the variance (a minimum in spring and a maximum in pre-winter and winter). Furthermore, the variation of the pattern of the first eigenvector (which accounts for the highest percentage of the variance) from season to season supports the decision to divide the annual progression into 6 natural seasons.

Correlation coefficients have been computed between each eigenvector

340

coefficient in a given season and all eigenvector coefficients in the previous 5 seasons (i.e. up to 10 months previous). The number of correlations found at the various significant levels exceeded the chance level by about 50%, implying that some at least of the apparent correlations are real. It is particularly interesting to note that some of the highest correlations occur several seasons apart, implying the presence of long-period rhythms in the circulation evolution rather than a marked dependence on the immediately preceding season. The distribution of correlations with respect to the nth season have been found, and a forecast procedure based on an analysis of variance technique is being developed. The highest correlations are used as an indication of which season is the best to use as a predictor for a particular seasonal eigenvector.

Our first attempts to do this were carried out before the analysis of variance technique had been developed. We therefore simply took those years in which the coefficients of the highly correlated precursor season were within 10% of the total range of variance compared to this year and then meaned the coefficients for those years in the season to be predicted. The answer was then taken as the probable coefficient in the current year and after repeating the procedure for all correlations and eigenvectors, the forecast anomaly chart was produced.

Figures 8.1.1(a) and 8.1.1(b) show the forecast height anomaly pattern computed for winter 1974/5 and the actual, while Figs. 8.1.2(a) and 8.1.2(b) show similar charts for spring 1975. It is clear from these examples that the method would have correctly predicted both the mild winter and the cold spring of 1975 in Western Europe. An even more impressive forecast was produced for the pre-winter in 1975.

Another approach to the problem of forecasting more than 3 months ahead has resulted from a study of the quasi-biennial oscillation (QBO) in equatorial/stratospheric winds (Ebdon, 1975). The QBO has been recognised by meteorologists for at least 15 years. Since then a large number of papers have been written describing its behaviour: Veryard & Ebdon (1961), Reed (1965); its possible causes: Shapiro & Ward (1962), Holton & Lindzen (1972); and its role as a part of the general circulation: Kac (1964). Long before stratospheric data became available, various workers had detected periodicities of a biennial nature in the tropospheric data and, in recent years, it has been suggested that these may be correlated with the QBO.

If such a relationship did exist it would be important for those modelling the general circulation. In addition, since it is usually a relatively easy matter to forecast the phase of the QBO up to about 12 months ahead, it could be a useful parameter for use in seasonal forecasting.

In our work we selected Januarys and Julys when the QBO was near the maximum easterly phase and compared them with Januarys and Julys

8.1 Progress and prospect

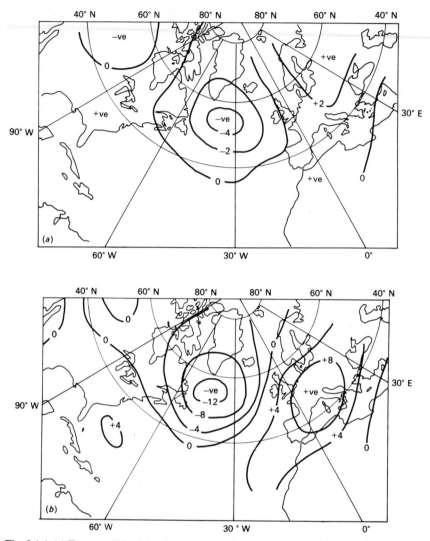

Fig. 8.1.1. (a) Forecast 500 mb height anomalies for winter 1974/5 derived from eigenvector correlations with previous seasons. Units: dam. (b) Actual 500 mb height anomalies for winter 1974/5. Units: dam.

near the maximum westerly phase. Although the data used was from Canton Island or Gan, Veryard & Ebdon (1961) have shown that the QBO appears to be in the same phase all round the equator, so the results are generally applicable.

During the 20 years for which data is available there have been five Januarys when the zonal component of the 30 mb wind near the equator

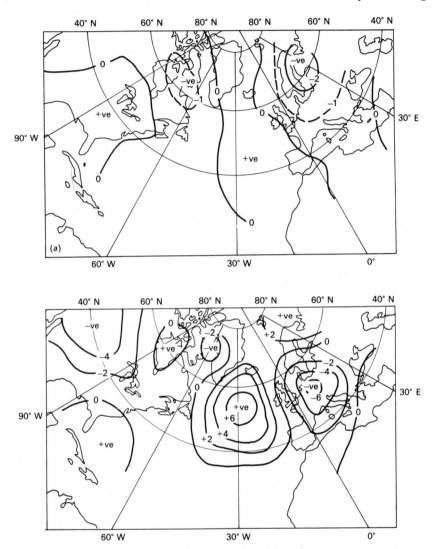

Fig. 8.1.2. (*a*) Forecast 500 mb height anomalies for spring 1975 derived from eigenvector correlations with previous seasons. Units: dam. (*b*) Actual 500 mb height anomalies for spring 1975. Units: dam.

exceeded 10.3 m s^{-1} (20 knots) easterly and five Januarys when the zonal component of the 30 mb wind exceeded 6.2 m s^{-1} (12 knots) westerly.

The mean pressure anomalies for much of the Northern Hemisphere for these two sets of years show a marked difference: in the easterly phase, pressure is anomalously high in the polar regions and low in middle latitudes; in the westerly phase the reverse is roughly true. The difference

343

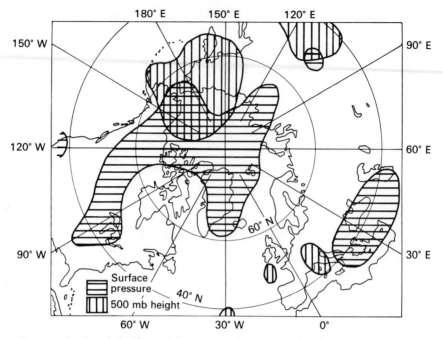

Fig. 8.1.3. Areas of significant difference between means of easterly and westerly phase QBOs in January.

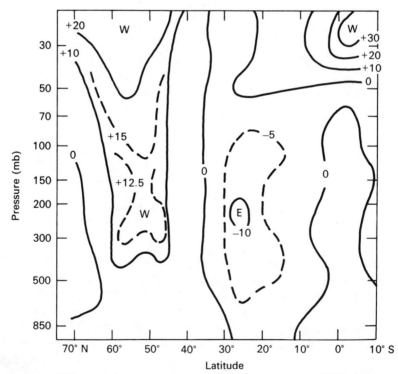

Fig. 8.1.4. Differences between average zonal wind components at 80° W during westerly and easterly phases of the QBO at 30 mb. January 1967, 1972 and 1973 for westerly phase. January 1966, 1968 and 1970 for easterly phase. Speeds in m s⁻¹.

Table 8.1.1. *Relationship between phase of the QBO in July and Central England temperature and sunshine*

	JULY Central England temperatures quintiles				
	Very cool	Cool	Average	Warm	Very warm
Phase of QBO at 30 mb	1	2	3	4	5
Easterly	5	2	2	0	0
Westerly	1	1	1	3	3

	England and Wales sunshine terciles		
	Dull	Average	Sunny
Phase of QBO at 30 mb	1	2	3
Easterly	9	0	0
Westerly	4	0	5

between the two sets of January charts was tested for significance using the method of Ratcliffe (1974). The areas in which the two sets of January charts are significantly different at the 95% confidence level or better are in polar regions and in mid-latitudes (see Fig. 8.1.3). The mean cross-sections of the zonal wind components of the westerly-phase Januarys for which data was available was compared with that for the easterly-phase Januarys along 80° W and the result is shown at Fig. 8.1.4. There is considerable evidence that, above 50 mb, westerly winds in high latitudes are stronger in the westerly phase of the QBO, but also there is evidence of a secondary west-wind maximum in the troposphere between 200 and 300 mb and between 50° and 60° N which, when coupled with an indicated decrease of west wind at this level at about 25°–30° N, implies that the tropospheric wind maximum may be farther north at the time of maximum westerly phase of the QBO.

This enhanced westerly flow was also apparent on the corresponding 500 mb charts. Mean wind speeds were obtained, from grid point 500 mb data, around latitude circles between 30° and 60° N for the entire Northern Hemisphere and for the Atlantic sector separately. In both cases the results were similar and clearly showed, in the mean, the strongest flow at 500 mb to be farther south in the easterly phase.

The tropospheric flow patterns in the nine westerly-phase Julys were then compared with those in the nine easterly-phase Julys (up to 1973). The same result is apparent as for January: namely that the strongest flow is farther south in the easterly-phase years. In Britain, summer weather is very dependent on the latitude of the strongest tropospheric flow: a northward displacement tends to give good summer weather while a southward displacement usually results in cooler and duller summers. Table 8.1.1 shows how the mean temperature and sunshine amounts in July differed between the cases of easterly and westerly QBO.

In fact these results seem to be generally true for much of winter and summer. Six months (January, February, July, August, September and November) showed mean tropospheric flow somewhat farther south in easterly-phase years compared with westerly phase. In the remaining six months no significant difference could be detected.

Another useful method aimed at producing a forecast for the winter at the end of summer has been that due to Folland (1975). He has shown that cool summers in an even year in Britain have been followed by either warm or cold winters but no average ones over the past 100 years. He developed a method of distinguishing between the two likely sequels based on pressure anomalies in the Northern Hemisphere during the summer and indicated how the likely outcome might be confirmed by studying sea-temperature anomalies and pressure anomalies in the autumn.

From these three examples of research work presently continuing, it is clear that there are prospects of making increasingly reliable forecasts of weather or 'climate' in general terms over periods between one month and a year or so ahead which would be of enormous economic benefit.

Relevance to the Southern Hemisphere

An attempt has been made to interpret the QBO in the Southern Hemisphere. Using Australian data together with various Northern Hemisphere data, mean cross-sections were produced along 140° E from 65° N to 45°

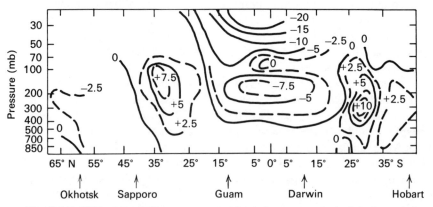

Fig. 8.1.5. Differences between mean zonal wind components in July in easterly and westerly phases of the QBO (4 cases of each). Units: m s⁻¹.

Table 8.1.2. *Latitude (deg S) of maximum 200 mb zonal wind*
component at 140° E for July

Easterly Julys		Westerly Julys	
1956	30.3	1957	27.3
1960	29.6	1959	25.7
1965	28.0	1961	27.3
1968	29.0	1966	27.5
1970	30.0	1969	27.0
1972	27.5	1971	27.5
1974	31.0	1973	25.5
Mean 29.3		Mean 26.8	

S for Julys in easterly- and westerly-phase QBO years. Unfortunately data were only available for 4 years for each class of July. Figure 8.1.5 shows in cross-section form (along 140° E), the differences between the mean zonal wind components in the four cases of westerly QBO compared with the four cases of easterly QBO in July. Positive signs indicate more westerly in the easterly phase (or less easterly) and vice versa. The most notable feature is a tendency for the axis of the strongest flow to move southward in both hemispheres in the easterly phase of the QBO. Most of the tropics between 20° N and 20° S appear to have an increased easterly component during an easterly-phase QBO July, suggesting a direct link with the phase of the QBO. This, however, is not apparent below 500 mb.

An estimate of the latitude of the maximum zonal component at 200 mb in the Southern Hemisphere was made from the cross-sections for individual Julys (140° E). The years 1956–61, for which Muffatti (1963) published the latitude of the subtropical jet over Australia in July, were also considered. Table 8.1.2 compares the latitude of the maximum 200 mb flow for seven easterly- and seven westerly-phase QBO years.

It is noteworthy that not only are the means 2½° of latitude different, but no westerly-phase year has the jet axis farther south than the most northerly year of the easterly-phase years (1971 and 1972 W and E respectively have the same latitude). This result is found to be significant at the 0.1% chance probability level.

These facts so far refer to longitude 140° E. For the whole of the Australasian sector the picture is more complicated, though it is still basically a coherent one. There is general enhancement of flow in the easterly phase compared with the westerly phase but the axis of strongest flow tends to be displaced northwards (easterly phase) in southwest Australia with an actual reduction of 200 mb wind there in the easterly phase. A statistical test of difference between the mean zonal wind

347

8.1   Progress and prospect

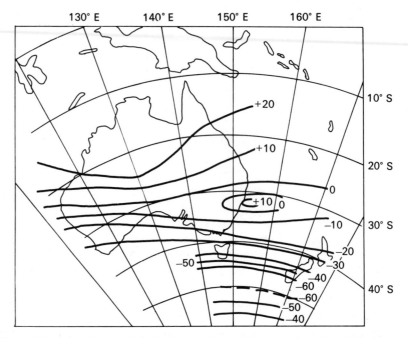

Fig. 8.1.6. Differences of mean 200 mb heights in July between easterly and westerly phases of the QBO (7 cases of each). Units: geopotentialmetres.

components in the easterly and westerly phases was carried out on a station basis. At ten of the thirty stations in the area a difference at the 90% confidence level or better was found and seven stations were significant at the 95% level or better. Mostly these figures represented stronger westerlies in the easterly phase but at Lae and Darwin the significant difference was due to stronger easterlies in the easterly phase.

The difference between the two sets of Julys on a 200 mb height basis (Fig. 8.1.6) shows a similar pattern, generally enhanced flow in the easterly-phase Julys centred near 30° S but the enhanced gradient being somewhat further north in Western Australia. There is also a curious area near 33° S in eastern Australia where 200 mb heights are greater than might be expected (easterly phase).

Two indices relating to surface pressure gradients between 40° and 55° S at 70° W (South American westerlies, SAMNW) and between 40° and 45° S at 172° E (New Zealand westerlies, NEWZW) are maintained in the British Meteorological Office. Mean anomalies of these indices for the seven easterly- and seven westerly-phase Julys since 1953 are shown in Table 8.1.3.

In both areas the surface westerlies appear generally stronger in the easterly phase of the QBO than in the westerly phase though there are

348

Table 8.1.3. *Anomalies of indices of South American and New Zealand westerlies, in mb*

	SAMNW	NEWZW
Easterly Julys	+3.1	+0.3
Westerly Julys	−2.8	−0.6

Overall mean values of SAMNW and NEWZW are +15.7 and +1.4 mb respectively.

some exceptions, perhaps due to the inadequacy of the data or, in the case of the New Zealand westerlies, the narrowness of the latitude band. The difference between the means is significant at the 1% level for the South American westerlies but not significant over New Zealand. These facts are consistent with the general arguments I have put forward and with the cross-section in Fig. 8.1.5.

An analysis of rainfall variation between Julys in the easterly and westerly QBO phases bears out the conclusions drawn so far. The areas of negative 200 mb height anomalies (Fig. 8.1.6) in southwest Australia, the Southern Ocean and New Zealand are areas of increased rainfall while the apparently anomalously high 200 mb heights near 33° S in east Australia are found to be more usually dry in the easterly-phase QBO Julys.

It is clear therefore that the phase of the QBO has an important bearing on the general circulation in both hemispheres. Some remarks on the possibility of forecasting the phase some months in advance may be useful. The tropical stratospheric wind record at 30 mb since about 1954 can be divided into four phases, westerly maximum to west–east changeover: west–east changeover to easterly maximum, easterly maximum to east–west changeover, and back to the westerly maximum. Scatter diagrams can be plotted relating the length of each of these phases with the month of occurrence of the preceding event or for the whole cycle and for half-cycles. These diagrams can be used to estimate the probable date of the next event. For example, the date of the next westerly-phase maximum could be estimated from the date of the previous westerly maximum, the date of the west–east changeover, the date of the following easterly maximum and the date of the east–west changeover. Bearing in mind also that there are preferred times of year for stratospheric events to occur at a particular level it is possible to make useful estimates of probable dates up to 2 years ahead.

8.2   Progress and prospect

8.2   Possible future climatic trends: a panel discussion

This is a highly edited version of a panel discussion on the subject of possible future climatic trends. The panel (Fig. 8.2.1) consisted of: Professor R. A. Bryson of the Institute of Environmental Studies, University of Wisconsin; Dr W. F. Budd of the Antarctic Division, Department of Science, Melbourne; Dr J. Chappell of the Geography Department, Australian National University; Professor H. Flohn of the Meteorological Institute, University of Bonn; and Dr W. J. Gibbs of the Bureau of Meteorology, Melbourne. The panel was chaired by Dr C. H. B. Priestley, Chairman of the CSIRO Environmental Physics Research Laboratories, Aspendale.

The present version is a condensation and adaptation of the recorded transcript. Much good material, including audience participation, has been omitted for the sake of brevity and the development of coherent themes. Wherever possible speakers, including participants from the audience, have had a chance to see and correct the edited version.

The editors believe this discussion, which reveals differences of opinion and emphasis, provides a useful insight into the state of knowledge on the subject of climatic change. They hope that it will help readers to assess the whole subject critically and to focus on some of the areas which need to be further explored.

PROFESSOR FLOHN: I said yesterday that the probability of a transition to the type of climate leading to an ice age in the next 100 years is less than 1%, but that the possibility of large-scale human interference with the climatic system, consisting of ocean, atmosphere, ice and so on, is much larger. It is a question of how much man contributes to the state of the climate we have. If one deals with that matter from the viewpoint of energetics one comes to the result that the sum of man-made interference with this system is of the order of 10% of the energy which is converted in the climatic fluctuations we have experienced in the last 100 or 200 years. The strange fact is that the sum of these man-made effects (perhaps this is a controversial matter) should tend, generally speaking, to warming of the atmosphere, while natural effects result in both warming and cooling, giving rise to non-periodic fluctuations.

Now if we allow man's interference with climate to increase exponentially as it has done in recent years, we sooner or later come to a state where this 10% rises to 100%, resulting in continuous warming made by man superimposed on these natural fluctuations of cooling and warming. This would be a really dangerous situation in that in the Northern Hemisphere we have this extremely sensitive area of the Arctic sea-ice. The few people who have dealt with models of the sea-ice have the feeling

350

Fig. 8.2.1. Panel discussion on 'Possible Future Climatic Trends', Monash University, 11 December 1975. Left to right: Professor R. A. Bryson, Professor H. Flohn, Dr J. Chappell, Dr W. J. Gibbs, Dr W. F. Budd and Dr C. H. B. Priestley (Chairman).

that this is in fact an extremely sensitive system which will reflect very early and very substantially any sizeable warming of the Northern Hemisphere. The lifetime of individual ice floes is 5 or 10 years, certainly not more than 10 years, and once the ice is removed the present situation would not allow the reforming of permanent ice cover as we have it today.

My feeling is that if man's interference with the climatic system is uncontrolled for some decades, together with uncontrolled growth of energy use, sooner or later during the next century the warming will overwhelm natural factors which usually produce cooling. Then the Arctic sea-ice could disappear rather rapidly, some models say in a period of 10 years or less. This could cause the meridional temperature gradients between pole and equator to diminish greatly, to be necessarily followed by a shift of the climatic zones of the whole Northern Hemisphere and perhaps extending a bit beyond the equator. This is not an immediate danger, since certainly in the next few decades we cannot reach this level and man's interference is more or less on a local or regional scale, rather than a global scale. But what will happen in 50 or 80 years is uncertain and depends on the intensity of economic development, and on the options of energy production and energy use. In the future we have to make critical choices.

351

8.2 Progress and prospect

DR BUDD: I would like to speak for the stability of climate over the time scales of many hundreds of years, and for high variability or fluctuation in climatic elements over scales of several years or several seasons. For example, in the last 100 years at the most southerly Antarctic station where we have long-term records, the sea-ice remained on one occasion the whole year, and on another occasion, it didn't come there at all. These kinds of high variability, corresponding to shifts in mean annual temperature of 7 °C over a period of several years, lead me to think that this is a local variation and certainly not large scale. In other places, the sea-ice and temperature do seem to be related on the large scale; more sea-ice cover corresponds in time to colder temperatures, but usually in one spot more so, in another spot less so. Again we do not have evidence over the last, say, 1,000 years of a great deal of change in extent, nor of effects catastrophically spreading to the equator.

Finally, in the longer term I think there will be possible large changes. The longer the term, perhaps the bigger the change we can expect. However, if we are able to monitor and keep up with the climatic scale elements in the same way as we are doing with the current weather on a global scale, then I think we can easily keep track of these long-term events, which I think will come slowly, and perhaps steadily.

DR SHACKLETON (from audience): If Professor A. T. Wilson were here, he would surely ask you how likely you think it is that the Antarctic ice sheet would surge in the foreseeable future. On his behalf may I ask you that question?

DR BUDD: The East Antarctic ice sheet could surge and may have surged about 12,000 years ago on a large scale, and perhaps on a minor scale a few thousand years ago. For Western Antarctica we can expect many thousands of years for it to build up to a surge. I don't think there is anywhere in the Antarctic or for that matter elsewhere where a large ice sheet is currently near a surge position. I might add that with surges, time scales are very slow, the surge might last several hundred years, and the motions of the order of a few kilometres per year are not much different from those of the very fast outlet glaciers.

DR CHAPPELL: I can speak only about the largest scale of climatic change because it's only the geological records with which I am familiar, and these fail to indicate very much about the short-lived events which have human consequences, such as the 'Little Ice Age'. However, turning to that event of low probability but very likely catastrophic consequences should it occur – that is a great ice age – one can examine the geologic record in order to test hypotheses for causes. Some hypotheses are very difficult to examine in the geologic records and turning the principle of Occam's Razor on its side, let's use it not as a test of hypotheses but as a way of directing our work.

The Milankovitch hypothesis of climatic change argues that the orbital perturbations conspired together to produce in the Northern Hemisphere a cold summer–warm winter condition. This seems to occur in the records of the last 700,000 years almost entirely in association with the advance of glaciation there. The alternative condition of warm summer–cold winter seems to be associated with the decline of glaciation. Thus, although there are undoubtedly other preconditions – sea-surface temperature being one and a modulation due to Antarctic ice variations being another – nevertheless, the Milankovitch factors are quite important as far as the records indicate. An argument that has been levelled against these kind of hypotheses is that they sometimes seem to be not on the cards because of some reason of dynamics or energetics. It's most important to test them in an independent way, and to determine whether numerical modelling of some kind or another can in fact satisfactorily test the hypothesis for major glaciation as it appears from the geologic records.

In regard to future climatic trends, as long as other particularly important factors such as North Atlantic temperature, and also the factors of pollutants and dust components of the atmosphere, remain about as they are, the Milankovitch kind of prediction is that glaciation is not on in the next 70,000 years, because we are moving into a period of low orbital eccentricity of the kind we have not seen for about 700,000 years.

DR SHACKLETON: I would like to dispute your conclusion to the extent that at no time in the last 3 million years has any interglacial lasted significantly longer than 10 or 12 thousand years.

DR CHAPPELL: Knowing the records I agree absolutely, and that's why I prefixed my 'prediction' with a qualifying clause. Contributing causes are not equal and are not going to remain so. The North Atlantic temperatures were very much higher than present at the time of initiation of the last glaciation. Whether they were for previous glacial initiations has to be resolved but for the one at 220,000 years ago, the North Atlantic seems to have been no different than at present. There are thus other preconditions than the Milankovitch factor, but I insist that the modulation of the glacial record so closely parallels the orbital processes that it must play an important part.

PROFESSOR FAIRBRIDGE (from audience): The relatively brief interglacials in the geological record of the last couple of million years average out to about 10,000 years in length. If the present one started 10,000 years ago we have just had it. What happens next?

If you look at the marine record of a 100,000-year glacial cycle you will see that there are relatively brief departures to very cool conditions but they soon return to fairly mild conditions almost to the interglacial level, and in the first half of the last glacial cycle that condition persisted for about 50,000 years. We do not therefore predict a true ice age during the

353

next 10,000 years or so. But following the Milankovitch curve, we are definitely well down on the slippery slope. However, a large number of feedback mechanisms have to come into play.

Very brief accelerations in the cooling can take place to a point where they are buffered, and according to my eustatic records, I can see that during the present interglacial we have had departures that have involved more than 5 metres drop in sea level. This involves an enormous temporary increase in ice volume, bearing in mind that 1 millimetre of sea-level change is equivalent to 380×10^9 cubic metres of ice! So we have very good evidence that sea level has changed during the last 5,000 years at times within periods of the order of 100 or 200 years, through more than 5 metres. That means a tremendous but brief climate change, an oscillation that could be catastrophic for many farmers as well as for the world in general, and we are not speaking about ice ages at all.

DR BUDD: It's worth considering that these changes could be due not to climate change, but to ice discharge which can occur independently of climate change, although no doubt affected by or interacting with it. For example, the large production of icebergs in a certain area can affect the isotopic composition of sea water in the local area quite strongly, without being a global climate change.

If you have several large ice basins surging periodically but independently of each other, you can have combinations of, for example, two, each having variations of about 20,000 years which come into phase at about 100,000 year intervals.

PROFESSOR BRYSON: We have had some very, very fine discussion of sea-bottom sediment data, and how it parallels the Milankovitch curve. I am convinced that the best answer is that the Milankovitch variations are explaining what we see, but I am profoundly disturbed by the 'hand-waving' kind of discussion. As a young man, I often caught hell for doing it myself. You see, the curves look alike, and therefore probably one is causing the other; it's not likely that the earth's climate is changing the sun. We get an admirable model of glaciers and glacier surging, but the connection between the model and reality is disturbing. If you will pardon my putting it in very rough terms, the fact that the model hiccups periodically, does not mean that this behaviour is roughly on the scale of the climate changing. In order to relate the Milankovitch variations to the climate, to relate the glacier pulsations (which are real) to the climate, you have to have a model of how they interrelate, and that gentlemen, is where we have fallen down. We have got the data, we have got an idea, but nobody has made that model which tells you how the variations in the solar radiation as given by the Milankovitch calculations will actually drive the climate to do what it is presumed to have done.

DR BUDD: Professor Bryson has pointed out that we have not developed

354

a model that includes the ocean, not only the surface but also the deep ocean, and the ice cap. This is quite true. We have recently got models to operate on ordinary glaciers, but only in the last few months have we got them to operate on surging glaciers and the Antarctic ice sheet. He has put forward a very good idea for the future, and it will not be an impossible task to do, but it's a very difficult task because of various stages of development of all these models.

DR HUNT (from audience): As a modeller, I might try to bring the discussion perhaps up to date or into the future, as the most important question is, can we forecast the future, interesting as the past may be? The most important time scale is the next 5 to 50 years, and one doesn't need to worry about Milankovitch, surging glaciers, etc., in that time scale. But if we are heading towards a 'Little Ice Age', which is the worst thing that could happen in this time scale, it's very important that we should know. I am not sure as a modeller that one can forecast climate. I would like to hear other people comment on whether they think it is possible to forecast climatic change.

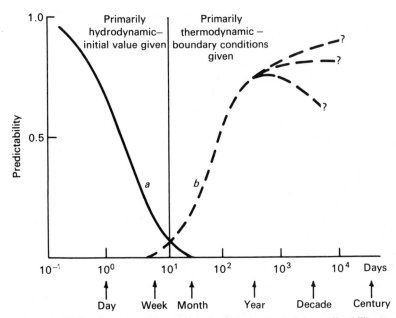

Fig. 8.2.2. Highly schematic diagram drawn by Bryson of possible predictability at various time scales (for example, 1-day forecast of conditions over a 2-hour interval, 1-year forecast of conditions over a month interval). From theory and practice we know that the predictability at short scales, primarily hydrodynamic and with initial values given, decays to small values after about 2 weeks, as in curve *a*. Is it possible that 'climate', defined over longer intervals, is primarily thermodynamic, and that the predictability might be quite high if the boundary conditions were known (as in curve *b*)?

355

8.2 Progress and prospect

PROFESSOR BRYSON: Let me draw a schematic diagram here. How predictable is the future on a scale of a day, a week, an hour? The work of Lorenz and other people says that the weather predictability drops off rather rapidly, something like curve 'a' (Fig. 8.2.2). Because, very simply, the approach is an initial-value problem, and in two weeks the atmosphere has forgotten where it started, the original pattern has been forgotten. But climate in the longer range is not an initial-value problem, but a boundary-condition problem, and my opinion is that the predictability goes something like curve 'b' in the diagram (Fig. 8.2.2). This refers not to our ability, but to the potential predictability. Further, it is a predominantly hydrodynamic problem in the short-term prediction area, where very careful treatment of the hydrodynamics is necessary, but you can get away with crude parameterisation of the thermodynamics. In long-range prediction it is almost entirely a thermodynamic problem and you can get away with the crudest parameterisation of the hydrodynamics. Now, when you extend the global circulation models to predicting the future for the next 5 to 10 years, what you are trying to do is push the initial value forecast out there, and I say, 'Fellows, you ain't going to get there.' Let's instead try a new kind of model that focuses on boundary conditions and thermodynamics and really start modelling climate rather than 90 consecutive January 15ths (which has never happened in the history of man, or the history of the earth). If we can then say that boundary conditions have changed, the thermodynamic inputs have changed, then we can say that climate has changed.

DR GREEN (from audience): General circulation models also include boundary conditions and everything that simple-minded people like you and I put into climate prediction models.

PROFESSOR BRYSON: You are right, but they also include an enormous amount of noise, which when integrated, hides the climate, and my pitch here is to leave out this noise, so that the experts develop a nice, quiet, global circulation model.

DR WEBSTER (from audience): There is one difficult thing you have to do to make your thermodynamic prediction or your boundary value prediction – you have to forecast the boundary values.

DR ZILLMAN (from audience): Is everything to the right of your vertical line your definition of climate?

PROFESSOR BRYSON: Yes. Climate, to me, is dominantly thermodynamic. Weather is dominantly hydrodynamic, but we must have both in.

DR GIBBS: Although I am in a game where one must be a pragmatist and do the best one can with the instruments or the equipment available, I am not a knocker of the palaeoclimatologists. I am glad they have been here, because they give one a correct time perspective. But I believe it's obvious that they can't help us with the question of what the climate will be in the next 25 years. I think you will agree that so far this afternoon we have not had an answer to this question.

Apart from diurnal and annual oscillations, there is no clear periodicity in climate unless perhaps in the Milankovitch cycle. But it's obvious that there are many other agencies at work; I suspect that the association of circulation indices with sunspots is not an accident; I am inclined to think volcanic dust might have some influence on climate; I am convinced the Walker oscillation is real, but none of these fluctuations appears to be periodic. I am sure the ocean must cause climatic fluctuations but I can't believe that its effect is simple. We must be in the data-gathering business and particularly with respect to the ocean; our ignorance of the ocean is incredible, and it's probably more important to increase observation of the ocean than it is to get more observations of the atmosphere. Until we have more and better observations, we will not be able to answer questions relating to climatic change and variability.

We have been asked what the climate will be like in the next 25 years? I consider the answer to this question lies in the observations of good quality which have been made over the last 100 years. From these records we can estimate, for instance, how many droughts we are likely to get in Australia and in the other grain-producing areas of the world during this period and whether they are likely to be synchronous; a strategy for combating drought can be developed. We might find that some of these processes are not completely random, although randomness does not mean that there is no cause–effect relationship. It may mean that the cause occurs randomly.

DR PATERSON (from audience): A discussion of food supply and climate is something which we should perhaps give more consideration to. One problem we have got to face is, can we afford to be too successful with climatic forecasting for food supply? We must recognise that the loss of world food supply is only one element in the world system, and only one of the destabilising elements. The effects of climatic change are to produce stress, here, but not there, now but not a little bit later, and in fact if we rely purely on our traditional knowledge of meteorology, we will find the crises which we face, with famine, will be crises which are much more isolated and of smaller extent and sooner than if we become successful at forecasting climate and make this a basis for organising world agriculture. In the latter case we will find we are facing a system which is much more highly tuned to our technical abilities, and a system which we have to be continually more successful at if we are to avoid a more serious crisis later.

DR LA MARCHE (from audience): I disagree that one can neglect long-term climatic variability in anticipating the immediate climatic future. For

357

example, there is a surprising amount of agreement in the records of temperature variations in northern temperate latitudes over the last thousand years. Scientists are unanimous in saying that the past 100 years have been unusually warm and that such warmth has occurred only once before for a comparable length of time in the last 1,000 years. I have been told the Australians are a sporting race, so I will give you 5:1 odds that the next 25 years will be different from the past 100, and I will even let you choose your own variable.

PROFESSOR FLOHN: I should perhaps comment on Bryson's diagram (Fig. 8.2.2). The atmosphere has a memory of the order of a few days, the ocean has a memory of the order of a few months, the ice has a memory of the order of a few years. If we wish to predict for the period of the order of one year, then we should not insist on thinking in terms of atmospheric physics alone. We have to deal with the interaction between these three subsystems of the climatic system, and to take into account the many interactions with other subsystems, like the soil and its moisture. This is the challenge for our modellers, to think more in broad terms of the climatic system as a whole. The present approach of general circulation models cannot lead, at least not in the foreseeable future, to that purpose.

DR CHAPPELL: It is entirely on the scale of records which Dr La Marche has just indicated – that of the Holocene, the last few thousand years – where geologists and earth scientists can in fact with some precision supply histories of some of the boundary conditions for some of the parameters as they varied. It's still just not on to predict the way any boundary condition will vary from our historical geologic records as they stand at the present, as far as I can see. I feel one can commend geophysical work which aims at this because one of the most profoundly influential factors generated within the solid earth, that is volcanic dust, develops at several scales. This is an important factor; we are now beginning to know something about the kind of connections and relationships within the solid earth processes, which a few years ago, we thought just did not exist. We for our part can concentrate on trying to specify if not a prediction of volcanic events, at least predictors of volcanic processes.

DR GIBBS: I don't believe we should make forecasts if we cannot do it with skill. There is a peculiar psychological desire on the part of the community for long-range forecasts and it is quite clear that the community will accept forecasts having no skill. But in the absence of predictive skill it is better to give probability estimates based on previous records.

There is a finite probability that a climatic shift will occur in the next 25 years. I believe that probability is very low and that it is more likely to result from man's activities than from natural causes. With respect to man-made effects, we might go along with Professor Flohn that if man's activities are to produce a shift in temperature it's more likely to be up than down. But with respect to major climatic shifts within the next 25 years I will take up Dr La Marche's challenge. I won't go so far as to say that for evidence of a climatic shift Port Phillip Bay must be frozen every winter – I think that would be cheating – but I will wager that snow will not lie on the ground for more than a day in Melbourne every winter for the next 25 years.

DR BUDD: I think Professor Bryson's diagram is somewhat confusing in the sense that it assumes that the general circulation models are concerned with the weather over that time scale. That's okay, but for climate they can equally well use variable boundary conditions, as for any other model. In fact, there's no reason why they can't use all the thermodynamics as proposed by that kind of model, as well. Some of us in different time scales will be collecting boundary conditions so that the models can be tested with past data as well as present data and so that we can be more reliant on them for predicting the future.

PROFESSOR BRYSON: We were asked what the next few years are going to be like and there was a surprising lack of forecasts. I have worked enough with my own simple model to have some confidence that it works for its purpose. Now, the variables in that model are the transmissivity of the atmosphere, due to volcanic dust, due to man and due to carbon dioxide. We have heard forecasts in this meeting of what the carbon dioxide will do, very confident forecasts that it will go up, and so I have used that kind of forecast in my model. We have also no reason to believe that the population of the earth in the next few years is going to change its ways, or that the population of the earth will change its course.

I can make an estimate of the man-made input into the atmosphere from the increase of population for the next 5 or 10 years. Volcanoes – I assume that volcanoes would at least stay the way they are, and if they change we might know that they change, and then can change the forecasts. On this basis I can predict that the general climate of the next 5 years will be similar to the last 5 years. That is a very gloomy forecast, because at the beginning of the last 5 years, there were 400 million less people than now. At the end of the next 5 years there will be 400 million more. If we have the next 5 years like the last 5, another 1972, another 1974, and assume nothing about going up or down, but just staying put, that's a gloomy forecast.

8.3   Reflections on climatological research

*A climatologist's view**
G. MANLEY

For some of us climatology has emerged rather rapidly from being among the dullest subjects in the world to the most entrancing. You now find people declaring in our scientific journals that it is the next really big field of research, linked as it is with the increasing anxieties of a world approaching overpopulation, that is, an ever-increasing demand for food and energy.

Climatology has long suffered from the treatment it was given, for many, in our schooldays. Climate was something we took for granted as a rather depressing part of school geography – a routine knowledge of Trade Winds and monsoons and doldrums, monthly mean temperatures for January and July, seasons of rainfall and continental climates with a very little about anticyclones and depressions that was presented in a way that did little to stimulate inquiry. For a few there was a little interest – the daily weather map in the newspapers – but the endless flow of charts brought on fatigue and even they could not really cope. Perhaps we can epitomise the reaction as a result of 'too many observations'. But this was all said by a Sheffield doctor (Thomas Short) in 1749; 'Many gentlemen start taking daily readings of their glasses and recording their weather, but it being a dry subject, soon weary of it.'

Fifty years ago climatology tended to be the static end-product of those routine observations that every meteorological service had to maintain as a 'public memory of the weather'; something that came to rest in the room at the end of the corridor. Most meteorologists were in the nature of things dynamic men, lovers of movement; eternally bothered by the behaviour of this queerly mobile mixture of gases at the bottom of which they have to live. They might, when depressed, feel rather like butterfly-hunters, chasing about expending a lot of energy, with inadequate results in a way that amused the public.

Our earliest scientists were quick to realise what was needed. In 1699 Edmond Halley sailed down into the cold South Atlantic on what we can recognise as the world's first geophysical expedition and made observations of temperature on the way. Soon afterwards, William Derham from the Royal Society was corresponding in Latin with Scheuchzer at Zurich about his barometric readings and the time of arrival of changes in wind and weather; but how far could they go when it took two months to get the answer to his letter?

---

* This is based on notes which were to be used by Professor Manley at the conference: however, he was unable to attend.

Climate as we know makes an emotional topic; subjective views linger; and throughout Victorian times the conflict of classicists and romantics went on, developing with the advent of Darwinism into those environmental–determinist arguments from which we are still by no means free. Huntington gave civilisations the highest marks in places where there was some cold, snow and discomfort to do people good. How fortunate for the inhabitants of Sydney that they too had snow one memorable day in June 1836!

Far too much emotion has ranged around climate in the absence of firm evidence, until it began to be gathered from quite other sciences, notably geologists and botanists. The incontrovertible evidence of glaciation, of changes in the character of prevailing natural vegetation, and of varying spread of sea-ice gradually accumulated and a meteorological explanation began to be sought; but there was still the question whether events on the geological time scale could in any way be related to events in historical time. Moreover, within historical time there began to be evidence, especially on the desert margins, of large and lasting changes in rainfall.

If we are to set a time when recognition that the greater oscillations of climate appeared to be of the same kind as the lesser oscillations that were beginning to be espied within the instrumental record, I should put it about 1910. That was when the Swedish varve-chronology gave us, at last, a dating of the events of Late and Post-glacial time. About the same decade, tree rings had begun to be counted and were attracting notice. On the borders of historical geography and archaeology, there was the bold leap forward into the climatic determination of history that we associate with Huntington. Pollen analysis then emerged and post-glacial fluctuations in northwest Europe became better understood. Then in the 1920s we saw the emergence of knowledge of the relationship between glacier behaviour and climate, followed by renewal of interest and further extension of the instrumental observations that had indeed begun much earlier, about 1890, with Buchan, Renou, Brückner and Voyeykov. Fluctuations on the desert margin and the polar margin began to be integrated; fluctuations of climate began to emerge as fluctuations of the general circulation of the atmosphere which might arise from solar or terrestrial variations or both; this we are still trying to find out.

Five-sevenths of the earth's surface, however, is water, most of it mobile with a complicated density-structure. We are still only beginning to find out the extent to which surface temperature of the oceans has varied in the past, with the aid of deep-ocean core-samples largely beginning in the 1930s. From 1948 we can put the beginnings of radiocarbon dating, and in the last decade or so, isotopic ratios in the ice layers of the great ice caps and, most recently, in tree rings within the historic period of the past few centuries.

8.3 Progress and prospect

Climatic fluctuations, regarded as the result of varied behaviour of the general circulation, are not going to be explained by the meteorologist until he has enough numbers to put in to his models and, most importantly, has dates and time intervals during which changes of longer duration have been established.

We all know how long-term fluctuations of climate might be agreeably ascribed to periodic variations in the elements of the earth's orbital motion and the tilt of its axis; short-term episodes to some kind of varying receipt of energy at the earth's surface, linked with sunspots. The latter began to be noticeable in the 1860s among economists interested in grain prices and varying harvest yields, and the early solar-physics men. All these tended to encourage interest in periodicities, notably in Australia where the sun is something that you are conscious of. But we are all aware how often periodicity-hunts and claims have become discredited with time. Fifty years ago, we can recall Napier Shaw's memorable list of a large number of the periodicities that had then been put forward; it covered more than four pages in his *Manual of Meteorology* ((1928), vol. 2, pp. 320–4).

Fashions in thought may well be linked with the environment in which research goes on. Scandinavians are much affected by the northern latitude, where the sun is a welcome object for part of the year, but the mountains and their glaciers, the visible limits of vegetation, even the stories from primitive mythology all influence the mind to dwell on the terrestrial field-evidence related to ice ages and vegetation changes. In Britain we incline to preoccupation with wind and ocean, moisture and evaporation. Australia presents a new environment; full of the sun, a subtropical desert margin, yet with nothing between it and the Antarctic.

Mathematical-physical scientists whose traditions incline them to follow the astronomers, still look for periodicities, or for links demonstrable by statistics; they still are capable of basing their efforts on what are really too few observations covering too short a period from too limited areas of the globe.

Then there are the people who incline, as Hoyle said, to tamper with the atmosphere; volcanic dust, man-made dust, the polyatomic gases, damage to the ozone layer, and so on. They too lack adequate observations. There are others who reckon in terms of statistical investigation of existing records to establish links between various parts of the world; are they always certain that their data are adequately representative?

There are those who think that our knowledge of the vicissitudes of ocean surface temperature, linked with the density-structure of the upper water-layers and with the prevailing meteorology, is too scanty for comfort.

This is really a plea for more observations, better planned and designed, and more purposeful. To this we might add the more careful devising of 'numerical models' and, always, the more precise establishment in *time* of those climatic fluctuations for which we have good evidence. The movement of considerable areas of ocean-surface water differing markedly in temperature is a most effective way of changing the heat balance in particular parts of the world. Do we now need to know more about the possible effects off Tasmania, and the signs of short-term oscillations provided by glacier behaviour in New Zealand? Might not further detailed field work and dating in Kerguelen benefit climatology?

From the standpoint of present-day requirements, we need to know more about the small fluctuations which have economic effect, especially on those desert margins that Australians know so well: the 'groups of years' that appear to be characteristic of the historic period in many parts of the world. Even in England, however, 300 years is about the limit for instrumental material. The question is how to confirm what we have and extend it. Isotopic analysis of the elements in tree rings as well as in glacier layers and sedimentary deposits looks as if it is just coming over the horizon. Perhaps it will be for you in Australia to get at some other annual variable, such as weathering-rates. With world population increasing so rapidly we urgently need further knowledge of these short-term vicissitudes over the past 1,000 years, which have already provided us with cause for anxiety and will do so again, for there are many marginal lands beside Russia and the Sahel, and many marginal seas besides those waters around Iceland that have given us our most recent problems.

A geographer's view
J. OLIVER

We have learned a great deal about climatic change, but perhaps controversies indicate that there is a long way to go. I am going to look at this as a geographer. Progress in and knowledge of climatic change has moved ahead in this century rather selectively, in specific fields of interest for particular periods of time or for particular areas. There is no longer any doubt that climates can and do change. We have been able to improve very significantly the chronology of events in different parts of the world and we have seen some major advances in setting up integrated circulation models of atmospheric energy, momentum and moisture transfer both on a global basis and in quantitative terms. The previously very disparate though numerous fragments of the jigsaw of past climatic experiences in different areas have been provided with an improved physical framework which has permitted a more critical eye to be cast over the causal speculations and the interpretations that have been indulged in at least over this century. We are still speculating on the causes although focusing more closely on certain aspects such as solar factors, atmospheric changes due to volcanic dust, and man-initiated pollution. But we are now more highly conscious of the inescapable need to assemble our much increased information on past climates in a more coherent form. There is the problem of distinguishing between climatic change and climatic variability and this relates closely to the problems of time and space scales of study. It could be argued that there is no such thing as a 'normal' climate except within the general constraints imposed by the solar–terrestrial system.

There seem to be an innumerable set of basic questions both about the mechanism and the behaviour of the climate in the past in time scales of more than a decade or a few decades. A number of these questions were apparent in one form or other thirty or fifty years ago; for example, in C. E. P. Brooks' publication *Climate Through the Ages* (1926). Many uncertainties arose then in his interpretations when trying to integrate the piecemeal evidence that had already been assembled indicating the existence of climatic changes or variability. There has been undoubted progress since then. But some of the present discussions are still reminiscent of the same sort of question which appeared in that work and in other discussions at the time.

The disappointment that has been expressed about our continuing inability to make long-range forecasts is perhaps an indicator of the extent to which we have limitations in our knowledge about climatic change. There is a very great gap despite significant improvements in data collection and widening of the sources of information on the character and dating of past climates over the globe or over large parts of it. Frequently we have detailed information on small areas and we then sometimes try to make correlations without filling in the gaps between those areas. Our knowledge of the working of the overall climatic system is thus based on various subsystems.

It seems that our best guides to the future are going to be those largely based on statistical assessments of probabilities and relationships derived from an improving assemblage of past recorded experiences. Perhaps we admit this by referring so often to our climatic change studies as stochastic. Reconstructions of former climates often are empirically based and very dependent on our data bank on past climatic events. In this respect there has been a major improvement with the more concerted action of national meteorological organisations and world-wide data-collecting networks, giving us far more precise information over a wider range of meteorological parameters. There has been a concurrent major improvement in the many increasingly refined techniques of interpretation and data measurement used in the field by natural scientists. Great advances have occurred in radiometric dating, such as carbon-14, or oxygen isotope determination,

364

and with the exciting new tool for improving our knowledge of the physical system provided for us through the medium of satellite cloud photography.

The extent of our present knowledge can be briefly reviewed under four headings. First, what do we know about the evidence for climatic change and what are our abilities in interpreting this evidence? This task involves knowledge of other natural systems with which climate interacts. There are many dangers in looking at local unique areas and in ignoring the disturbing factors, obviously the anthropogenic disturbances are a clear example.

Secondly, we need to be able to set out in precise form, described mathematically and cartographically, what the patterns of climatic change are. There is a great deal of divergence of view still on this, on details such as extent, synchronism and phase or lag relationships, teleconnections, intensities of change, persistence, periodicities, suddenness or progressiveness of change, scale of behaviour and so on.

Thirdly, when we look at the application of this knowledge then we are perhaps, so far as meteorologists and climatologists *per se* are concerned, in greater difficulties, because application requires as much knowledge of the processes to which the climatic factors are to be related as of the climatic factors and the climatic changes themselves. The task is a challenging one and there is clear justification for the development of interdisciplinary groupings of scientists in the attempt to co-ordinate environmental and resource management studies. There are considerable gaps in communication between the expert climatologists and meteorologists, who are the specialists in climatic change, on the one hand, and the environmental scientists on the other.

My fourth point is that we clearly require most of all an improved understanding of the causal mechanism of climatic change. For example, until we can appreciate the causes and the characteristics of the Southern Oscillation, its relationship, if any, to quasi-biennial oscillations, to shifts in the Australian anticyclone, to changes in the amplitude and location of the planetary long waves and their present interactions including related feedback situations, we cannot really integrate our knowledge of climatic change in the Australian and adjacent ocean regions. Nor can we decide how that knowledge fits into the global pattern of energy transfers and related weather. The literature on the causes of climatic change is frequently characterised by special pleading for a particular explanation.

As I see it, the most fruitful future probably lies in numerical and physical modelling of the atmospheric system and its regional subsystems. These require accurate data input as well as improved precise physical understanding. We have some way to go, perhaps a very long way to go in this, although it can be fairly claimed that in the last thirty years a major advance has been achieved. Studies of climatic change are no longer in danger of being regarded as science fiction.

An applied mathematician's view

B. R. MORTON

None of us doubts the existence of climatic change, though we have difficulties in agreeing on objective definitions for *climate* and for *change*. We accept that past changes in climate have brought ice cover to large areas of land now free from permanent ice, and have been responsible for ice-sculpting mountains, such as those of Tasmania, where glaciers have been unknown for recent millennia. But what are we to make of the apparently contrary situation at Wyperfeld in northwest Victoria where during the first six months of 1976, in a period of extensive and serious regional drought, a group of small lakes previously dry for 50 years has been refilling? These are fed by the Wimmera River which rises in the Grampian Mountains but has for most of this century terminated 20 km 'upstream' in Lake Albacutya (area ~ 50 km^2), which has itself intermittently dried and filled on a time scale of order a decade. Albacutya filled 2 years ago in spite of increasing diversion of water for dry farming, and the Wimmera River is again extending slowly northwards towards the River Murray. During this century the water table in the Murray Basin has risen, in places almost to ground level, and there has been a consequent increase in surface runoff. The reasons for this aquifer recharge are difficult to establish with certainty, but include deforestation, over-irrigation and climatic change. Thus the filling of drought-cracked lake beds at Wyperfeld is a complicated phenomenon which provides an example of overlapping climatic and man-induced effects with different time scales, the local drought of order 1 year, the partial filling of Lake Albacutya of order 10 years, and the aquifer recharge perhaps of order 10^2 years. Climatic phenomena often arise from a variety of causes, including effects due to man, and exhibit complicated patterns of change with time.

The division between *weather* (an atmospheric event or state) and *climate* (an average of atmospheric states over a selected period) is subjective and can be determined only in the context of particular situations. So too is the distinction between *climatic fluctuations* (variations about the mean climate, usually random but possibly periodic or quasi-periodic with period small relative to that chosen for climatic averaging) and *climatic change* (temporal variations of the climatic mean). Time scales clearly play an important role, and what is a climatic change on one averaging period may become a climatic fluctuation over a longer period. The sequence of cyclone and anticyclone in temperate latitudes is a weather phenomenon; whereas an unusually dry season is part of climate, although whether of a continuing change in mean climate or a random fluctuation about the climatic mean can be judged only by comparison with a recent sequence of the same seasons. We cannot know until afterwards just when a fluctuation becomes a trend.

Variations in climate may result from widespread systematic changes in weather or from displacement of weather patterns, and *length scales* are important in determining the geographical extent of climatic regions and the separation distances of regions of opposite climatic anomaly.

An analogy from turbulence

We can gain limited but useful insight into the relationship between weather and climate by considering properties of a turbulent flow in homogeneous fluid, such as the turbulent boundary layer. A turbulent velocity field can, for a selected averaging period, be divided into mean flow and fluctuations about the mean which are largely random (but may retain a limited imprint of the forcing system). Although boundary-layer turbulence has certain characteristic features, these are difficult to measure by direct spectral analysis of the velocity field because of phase incoherence in the eddy structure. The energy field, however, can be analysed without difficulty. It is found that the energy is fed initially into the larger component wavelengths, experiences progressive inertial transfer to smaller wavelengths with progressive loss of information about the larger scales, and is ultimately dissipated at the smallest scales. Measurements taken in the lowest three metres of the atmospheric boundary layer serve to determine the detailed eddy structure for wavelengths up to some three metres together with the mean velocity profile in that region, but they cannot in themselves supply information about the structure of the larger eddies, or about the mean flow in the overlying kilometre of air.

A more thorough study of the physics of atmospheric boundary layers reveals that these are generally self-similar flows in the sense that mean velocity profiles in a homogeneous atmosphere are of standard profile shape but have scales which are determined by the principal features of the forcing system, including the geostrophic wind above the boundary layer and the surface roughness. Mean eddy properties are part of this self-similarity structure, and thus it is possible to relate the mean behaviour of large and small eddies when values for the main flow parameters are known. Suppose, however, that we did not understand the physics, but had happened to notice a relationship between the mean properties for eddies of different sizes. We might still be able to 'predict' with limited success the behaviour of large eddies from observations on small eddies and a knowledge of the relevant gross flow parameters. These predictions

367

would appear as expectations of particular types of behaviour, and might be termed empirical or *pragmatic predictions*. They depend not on direct causal relationships, but on a far more complex net of weak causal connections. Similar predictions might be possible on an empirical basis without direct knowledge of the determining parameters in systems where these vary little.

The mean structure of a turbulent boundary layer at the lower boundary of a density-stratified fluid exhibits a range of profiles which depends on the parametric relationship between mechanical and buoyant effects. In this case the structure is more complex, but provided that we know both the parameter value and the principal features of the forcing, it should again be possible to inter-relate mean local and larger-scale quantities through a knowledge of the gross dynamics.

Sources of climatic variation

The atmosphere is a vastly more complicated dynamical system than a turbulent boundary layer. It responds to the influx of radiative energy from the sun, and we should expect this response to exhibit a well-defined mean state ('climate') together with a wide range of random, and especially randomly-phased departures from the mean ('weather'), very much like turbulence. This motion, however, is affected strongly by the rotation of the earth, by density stratification, by inhomogeneity of the boundaries, and by the anisotropy and periodic rhythms of the solar flux. The effects of rotation are particularly strong at larger scales and combine with the effects of stratification to produce a multiparametric system which is not strictly self-similar like the homogeneous turbulent boundary layer but may still be expected to exhibit quasi-similarity over limited parametric ranges and hence to show identifiable inter-relationships between scales.

The dynamical time scale for the troposphere is that for single cyclonic events and is a few days; smaller events have shorter time scales, for example, half an hour for a fine-weather cumulus cloud. Thus reasonably detailed weather forecasts for individual cyclonic systems can be given with good accuracy for one or two days, with much less accuracy for three or four days, and not usefully beyond seven days. A major complicating factor in estimating characteristic time scales for systematic changes in weather systems (i.e. fluctuations in climate) is that the atmosphere is closely coupled with the oceans, which play a significant role in the storage and transport of energy, with time scales ranging from a few months for the persistence of sea-surface temperature patterns to a few years for more general dynamical interactions. However, the oceans display the same combination of mean behaviour with random fluctuations, and even if the solar influx were uniform would themselves exert

368

random secondary forcing on the atmosphere with time scales ranging from a few months to a few years. This is one or two orders of magnitude greater than the time scale for a cyclone, and oceanic forcing must surely relate to climatic fluctuations, although we do not yet understand the detailed mechanisms. Ocean forcing of the atmosphere is complicated further by differences in length scales for oceanic and atmospheric disturbance systems.

Polar ice provides both an energy and a water reservoir, and changes in the volume of polar ice may provide another source of random secondary forcing for both oceanic and atmospheric systems, with interaction time scales that range from a few decades to centuries or millennia. Changes in sea-ice cover introduce an important seasonal scale that affects surface transports of heat and water vapour, and albedo. We lack knowledge of the relative importance of these and other influences, and whether any of them may trigger larger effects; and in the long term we need to strengthen our understanding of the physical mechanisms by which they may induce climatic change.

Climatic variations arise also from changes in solar emission and hence in the energy flux reaching the earth, and from changes in the earth's albedo which affect the proportion and distribution of incident radiation retained. We cannot be sure about past changes in solar emission, although these may have been responsible for substantial changes in climate. Changes in albedo and in the spatial distribution of the absorption of radiation result at the ground from deforestation, drought, the spread of desert, or reforestation and the spread of other vegetation; changes in cloud cover may be caused by variations in droplet nuclei or freezing nuclei, possibly from volcanic eruptions, or by changes in available water vapour; while changes in the distribution of ozone, carbon dioxide, water vapour and haze all affect radiation. Finally, mountain building, variations in sea level and continental drift have in the long term produced substantial changes in the configuration and albedo of the earth's surface. This very mixed bag of perturbations of the flux of solar energy to the atmosphere covers a tremendous range of time scales, from 10^{-1} to 10^{8} years. All may be regarded as direct causes of climatic change, although each will set up associated patterns of fluctuations.

Thus the study of climate and especially of climatic fluctuations and climatic change presents a mixed deterministic/stochastic problem, with a bewildering variety of sources of variation acting over an extremely wide range of time scales. In such problems a clear understanding of the underlying physics is an essential goal if we are to judge which properties relate most directly to the time scales of interest, what we should observe, and how we should analyse our observations. By emphasising the importance of energy we may avoid difficulties in analysing systems lacking

369

phase coherence, and at the same time direct emphasis to dynamically important issues. And although long-term climatic changes may be of overwhelming importance to future inhabitants of the earth, our own time scale is short: indeed, if I am persuaded that this winter will be unusually cold I may go out and buy a warm coat; and if I believe that we are entering a cold decade I may consider building a warmer house; but if told that next century will be a 'Little Ice Age' I shall tell my grandchildren.

The climatic record

Our knowledge of past and present climates in all parts of the world constitutes the *climatic record*, and the most exciting aspect of current work on the record is the assembling of climatic evidence from an astonishing range of scientific and technical studies of the surrogate, ancient and historic records. This is a tremendous achievement; and yet the reconstruction is like trying to assemble a jigsaw when the picture is unknown, more than half the pieces are missing, and others are chipped about the edges. Although excellent work is going on, our data cover on climate and climatic change is still extremely sparse both in time and space, and the quantitative relationship of some surrogate data to past climate is uncertain. This makes interpretation difficult, and I imagine that as our understanding of climate deepens we shall have to reinterpret substantial parts of the record.

There has been a major improvement in both measurement techniques and areal cover in meteorological networks, and improvements are planned for oceanographic networks which at present lag well behind. The resulting flow of data is already very large and will grow, and difficult questions will arise on the content and management of data banks. Partial networks may fail to detect effects due to pattern displacement, and are likely to distort the record by emphasising observations from land stations at the expense of those over oceans. Even where relatively long meteorological records exist there may be inhomogeneities of data that have been caused by relocation of stations, change of instruments, or modification of operating methods. Thus careful checking of historic data and detailed quality control of current data are essential parts of upgrading the climatic record. So too is calibration of surrogate data by intercomparison and where possible by comparison with historic records. In spite of recent developments, our most urgent need is still to extend and consolidate our knowledge of the climatic record. This will allow in the short term a search for pattern inter-relationships that may provide a basis for pragmatic prediction of climatic change; it will provide also the data base against which all climatic models must be tested, particularly those relating to long-term changes. It should be emphasised, however, that it is not enough merely to possess the climatic record, and we must seek continually to understand the dynamical processes by which it has been fashioned. Even if we were to have a perfect record of past climate, we should be unable to predict future climate with any real assurance until we have a better understanding of those processes.

Modelling changes in climate

Climate studies are at the stage where the extent and quality of the record is improving rapidly, but where our understanding of mechanisms causing climatic change, of relative importance and of interactions are at a primitive level. Seasonal climatic fluctuations and longer-term variations extending to perhaps a decade are significant factors in agricultural planning and production, with large potential economic consequences; and seasonal forecasts of moderate accuracy could already be of real benefit. In these circumstances *climate modelling* has an important role and requires active stimulation.

There are two quite different approaches to the modelling of weather and climate: (i) *operational models*, in which the physical content is in a sense incidental and the sole emphasis is on making useful predictions; and (ii) *fundamental models*, where the primary emphasis is on exploring the physical nature of the environmental system, although prediction remains a desirable secondary objective. Actual models tend to lie between these extremes, although both aspects of modelling are important. The two approaches should, ideally, converge in a general model which includes only the relevant physics for reasons that are understood, and which is effective in prediction. However, the physical complexity of atmosphere and ocean behaviour is such that an operational emphasis is likely to be necessary for many generations of models yet. This need is reinforced by the fact that the most urgent task in climatology is to 'predict', at least on a probability basis, regionally significant anomalies of duration a few months to a few years. In this task we are concerned primarily with climatic fluctuations.

A good example of a strongly *operational model* is provided by the simple one-dimensional convective cloud models used in early cumulus cloud-seeding programmes in the United States. These models visibly ignored much of the seemingly relevant physics, were based on an assumed steady source below cloud base, and excluded any possibility of downward transmission of information. However, when 'tuned' to a particular locality they appeared successful, although a tuned American model proved unsuccessful in the rather different Australian conditions. The

371

use of climatological averages to incorporate cloud cover in short-range weather models provides a good example of an operational element in a model.

Fundamental models provide the basic method for exploring physical significance in the physically complex problems involving weather and climate. They allow the study of simplified systems and the testing of physical concepts and assumptions, but if they are not to become a substitute for reality they must always be tested against real situations, no matter how partially. If modelling is to play its full role in climate studies, a variety of models should be developed and compared, including extensions of explicit general circulation models to find early responses to unusual forcing situations, statistical-dynamical models to find mean fluctuations, and as wide a range as possible of simplified models, especially those emphasising energy relationships, mean fluxes of energy and momentum, and effects of cloud distribution on energy inflow. There is great scope for ingenuity in the search for fresh approaches.

Many aspects of physical behaviour may usefully be explored with models, both for the insight that may be gained into the mechanisms and for the development of modelling experience. For example, in view of the relatively small changes in atmospheric patterns of behaviour over shorter times, what causes are likely to produce significant climatic changes over tens of thousands of years? What is the role of cloud cover in controlling albedo, and how can sufficiently accurate deductions of cloud cover be obtained without undue parameterisation? What role is played by the sea-ice in polar regions? How critical is the location of the Antarctic Convergence and other major ocean-current features? Is the flux in the Circumpolar Current important? What effects might result from a drop in sea level that would expose the 'Arafura Plain' (at present the shallow Arafura Sea north of Australia)? Might critical changes in the atmospheric radiation windows be produced in any likely way, and what effects would follow from the changed distribution of atmospheric heating? What changes in water-vapour budgets would result from specified changes in mean atmospheric temperature? In view of the large ratio of atmospheric to ocean velocity does the ocean play a passive role in air–sea momentum exchange; and is the magnitude of air–sea drag (or other quantities often parameterised) critical? What effects are likely to be caused by various concentrations of pollutants in specified layers? And can the roles of cumulus, cumulo-nimbus and other mesoscale systems in atmospheric transport be better resolved?

The exchange of water and water vapour between the ocean and the atmosphere is not well understood and needs further attention. Because of its high latent heat, water vapour evaporated over one ocean region may be advected downstream as a carrier of potential thermal energy and may produce a significant atmospheric disturbance far downstream. This may provide one mechanism through which sea-surface temperature anomalies can influence motion in the atmosphere.

Validation of climate models

Much greater emphasis is needed on the *validation* of models. There are still modellers who are pleased to obtain a reasonably plausible result from a model and believe that detailed validation is a job that can be left to others. One speaker at the Conference suggested (tongue in cheek) that the results of climate models are still so bad that validation is not yet a problem; but another claimed a high degree of accuracy in predicting the corn harvest for the Midwest United States, as though the quantitative assessment of model predictions were already an exact science. Validation *is* a problem already, and may be the *major* problem in climate modelling, and the more complex the models become the more difficult it will be. We must beware of the temptations to adjust a model to fit a particular set of observations, even though these should be judged characteristic, as there are sufficient degrees of freedom left in most models to fit a particular pattern, especially if the fit is to be assessed from a few gross parameters. Fitting a solution to a single set of observations is a process analogous to 'curve fitting' and may be devoid of physical meaning, whereas the real purpose of comparing model solution and observations is to ensure the physical significance of the model. Again, good physical understanding may help to expose high correlations which have arisen by chance or for spurious reasons, and would yield poor predictability. In this respect it is important that we accumulate data sets that are large enough so that hypotheses can be generated from one subset of the data and tests carried out against the remainder (and not the original subset!).

In some cases it will prove difficult to find appropriate tests for models of climatic change. Thus, if a particular model situation produces a 'Little Ice Age', it is not a sufficient test that such phenomena have been known in the past, because we are unable to judge whether the situations are comparable, since the best-known historic examples are beyond the span of accurate instrumental records. In such situations all that remains is a thorough exploration of the parametric ranges of the solution in a study of the effects of various assumptions on the production of cold periods.

Energy and energy fluxes, from the net influx of radiant energy to long-wave re-emission or ultimate dissipation, are responsible for driving motion in the fluid envelope, and hence are the fundamental variables in the system. They share with momentum fluxes the added advantage that they do not exhibit the phase incoherence of the velocity field, and thus provide a more suitable base for analysis.

373

8.3   Progress and prospect

Climate modellers are at a very great disadvantage in relation to weather modellers because of the long delay before their predictions are tested. A weather forecast may be publicly called in question within twenty-four hours, and even if this is not the same as an objective validation it acts as a constraint on the forecaster. Thirty-day extended-range forecasts are again subject to critical treatment before passing from the memory of either forecaster or public. But predictions of climatic change over decades may not be tested in practice during the working lifetime of the forecasters! Thus, if we are to develop viable climatic models we shall have to do so using past observations from the climatic record in an attempt to predict the present or recent past. This kind of *retrospective modelling* can be carried further as the climatic record improves, so that we may seek to consolidate our understanding of the record by relating past conditions to earlier states. Retrospective validation using the past record will be difficult, as this record is still very imperfect, and there will be temptations to arrange successful predictions of what are, after all, known states. This is a field that merits further study. It is time also that climate modellers began to define their objectives more clearly so that they can ask practical meteorologists, oceanographers, and those studying the palaeo-record for data in the form that may be most useful or, if possible, critical in their models.

Climatic prediction

What are the prospects for the prediction of climatic fluctuations and climatic change? Strong disappointment was expressed during the Conference at our failure to give a clear lead on *climatic prediction*. There was, however, disagreement about immediate priorities and about what might prove to be an optimal route to long-term forecasting, and there was pessimism over the magnitude of the task involved. Although a start has been made in a number of Northern Hemisphere countries with thirty-day extended-range forecasts, we simply do not know enough at present about climate or its variations to embark on more ambitious schemes for prediction without a great deal of preparatory work. This work has started, but is confined largely to the Northern Hemisphere, in spite of indications that atmosphere–ocean–ice interactions in the zone south of 50° S play a significant and perhaps even a principal role in the modulation of climate. Australia has a special opportunity and a special responsibility to contribute to these studies. She is already active in meteorological research and in Antarctic ice research, has efficient meteorological services and maintains scientific stations on Antarctica and Macquarie Island. She has, however, an inadequate effort in physical oceanography and has

374

yet to bring together the component parts of a viable programme on climate research.

In the short term, with relatively limited improvements in our understanding of climatic processes, we should see a more general introduction of thirty-day forecasts and then of seasonal forecasts (which may, in fact, be the immediate objective for Australia) and predictions of annual climatic trends. These are likely to develop initially as pragmatic predictions in the form of probability estimates of seasonal indices, and will depend heavily on our recent climatic record. Almost the only echo of this pragmatic approach to be heard at the Conference was the account of United Kingdom experience with twelve years of thirty-day forecasts. These have shown better than random success, and it seems probable that meteorological correlations can be shown to exist in the record extending back through a number of years. There appears to be at least a measure of predictability for climate in the short term, although we need a better understanding of the reasons for these correlations to use them more effectively.

I imagine that to predict when seasonal and annual forecasts will form a useful part of our meteorological services is itself a long-range forecast, and at present beyond our skills. However, the recognition of patterns in data is a normal part of the development of any science, even if it is only the first step. The next steps call for consolidation of the data record, past and present, for much more careful statistical analyses of data, and for a wide range of numerical experiments to improve models and to build up experience in their use. But essential though each of the observational, statistical and numerical tasks may be, climate is too important a problem to leave solely to data engineers, statisticians and numerical modellers, and the central and long-range goal must be to integrate these various contributions into an improved physical understanding of the responses of the fluid envelope of the earth. Although we have already made considerable progress, a great deal remains to be done. We cannot at present assess the potential rewards with certainty, but the ability to predict variations in climate is likely to play an increasing role in the management of human affairs. A role, indeed, that will place a new responsibility on climatologists to ensure that their skills are used for good and not for harm.

375

References

Ackley, S. F. & Keliher, T. E. (1975). Antarctic sea ice dynamics and its possible climatic effects. *Technical Note, Corps of Engineers*, US Army Cold Regions Research and Engineering Laboratory, Hanover, New Hampshire, 16 pp.

Adam, D. P. (1969). Ice ages and the thermal equilibrium of the Earth. *Research Report 15, Department of Geochronology*, University of Arizona, 26 pp.

Adam, D. P. (1973). Ice ages and the thermal equilibrium of the Earth. *Journal of Research*, US Geological Survey, i, 587–96.

Adam, D. P. (1975). Ice ages and the thermal equilibrium of the Earth, *Quaternary Research*, 5, 161–71.

Addicott, W. O. (1969). Tertiary climatic change in the marginal northeastern Pacific Ocean. *Science*, 165, 583–6.

Adem, J. (1975). A critical appraisal of simple climatic models. In *The Physical Basis of Climate and Climate Modelling*, GARP Publications Series, No. 16, ICSU-World Meteorological Organization, pp. 163–70.

Adler, R. F. (1975). A comparison of the general circulations of the Northern and Southern Hemispheres based on satellite, multi-channel radiance data. *Monthly Weather Review*, 103, 52–60.

Ahlmann, H. W. (1948). *Glaciological Research on the North Atlantic Coasts*. Royal Geographical Society, London, 83 pp.

Albrecht, F. (1960). Jahreskarten des Wärme-und Wasserhaushaltes der Ozeane, *Bertrage Deutscher Wetterdienstes*, 9 (66), 3–19.

Albrecht, F. (1962). Paläoklimatologie und Wärmehaushalt. *Meteorologische Rundschau*, 15, 38–44.

Alexander, T. (1974). Ominous changes in the world's weather. *Fortune*, February 1974, pp. 90–5, 142, 146, 150, 152.

Ambe, Y. (1967). Secular variation of aridity in the world. *Japanese Journal of Geology and Geography, Transactions*, 38, 43–61.

American Meteorological Society (1974). NOAA Laboratory for Environmental Assessment. *Bulletin American Meteorological Society*, 55, 1049–50.

Anderson, J. B. (1972). *The Marine Geology of the Weddell Sea*. Report 36. Florida State University, Sedimentology Research Laboratory, 222 pp.

Anderson, J. R. (1970). Rainfall correlations in the pastoral zone of eastern Australia. *Australian Meteorological Magazine*, 18, 94–101.

Andrews, J. T. (1973). The Wisconsin Laurentide ice sheet: dispersal centers, problems of rates of retreat, and climatic implications, *Arctic and Alpine Research*, **5**, 185–99.

Andrews, J. T., Barry, R. G., Davis, P. T., Dyke, A. S., Mahaffy, M., Williams, L. D. & Wright, C. (1975). The Laurentide ice sheet: Problems of the mode and speed of inception. In *WMO/IAMAP Symposium on Long-Term Climatic Fluctuations* (WMO-No. 421), World Meteorological Organization, Geneva, pp. 87–94.

Andrews, J. T. & Miller, G. H. (1972). Quaternary history of northern Cumberland peninsula, Baffin Island, NWT, Canada; part 4: maps of the present glaciation limits and lowest equilibrium line altitudes for north and south Baffin Island. *Arctic and Alpine Research*, **4**, 45–59.

Angell, J. K. & Korshover, J. (1975). Estimate of the global change in tropospheric temperature between 1958 and 1973. *Monthly Weather Review*, **103**, 1007–12.

Arakawa, H. (1955). Meteorological conditions of the great famines in the last half of the Tokugawa period, Japan. *Papers in Meteorology and Geophysics*, **6**, 101–15.

Arakawa, H., Fujita, T., Itoo, H., Masuda, Y., Matsumoto, S., Murakami, T., Ozawa, T., Suzuki, E., Takiuchi, M. & Tomatsu, K. (1955). Climatic abnormalities as related to the explosions of volcano and hydrogen-bomb. *Geophysical Magazine, Tokyo*, **26**, 231–55.

Arnold, G. W. & Bennett, D. (1975). The problem of finding the optimum solution. In *Study of Agricultural Systems*, ed. G. E. Dalton. Applied Science Publishers Ltd, London, pp. 129–73.

Arnold, G. W. & Campbell, N. A. (1972). A model of a ley farming system, with particular reference to a sub-model for animal production. *Proceedings Australian Society of Animal Production*, **9**, 23–30.

Arnold, G. W., Carbon, B. A., Galbraith, K. A. & Biddiscombe, E. F. (1974). Use of a simulation model to assess the effects of a grazing management on pasture and animal production. In *Proceedings of the 12th International Grasslands Congress*, Moscow. Sectional papers 'Grassland Utilisation', pp. 47–52.

Arnold, G. W. & Galbraith, K. A. (1974). Predicting the value of lupins for sheep and cattle in cropping and pastoral farming systems. *Proceedings Australian Society of Animal Production*, **10**, 383–6.

Arnold, P. W. (1954). Losses of nitrous oxide from soil. *Journal of Soil Science*, **5**, 116–28.

Asakura, T. (1972). *Unusual Weather and Environmental Pollution*. Tokyo, Kyoritsu Shuppan.

Ashkanasy, N. N. & Weeks, W. D. (1975). Flood frequency distribution in a catchment subject to two storm rainfall producing mechanisms. *Institute of Engineers, Australia, Hydrological Symposium, Armidale, National Conferences Publication* No. 75/3, pp. 153–7.

Ashton, T. S. (1959). *Economic fluctuations in England 1700–1800*. Oxford University Press, Oxford, 199 pp.

Atkinson, B. W. (1971). The effect of an urban area on the precipitation of a moving thunderstorm. *Journal of Applied Meteorology*, **10**, 47–55.

References

Auer, A. H. (1975). The production of cloud and Aitken nuclei by the St Louis metropolitan area (Project Metromex). *Journal de recherches atmosphériques*, **9**, 11–22.

Australian Academy of Sciences (1976). *Report of a Committee on Climatic Change*, Report No. 21, 92 pp.

Australian Bureau of Meteorology (1961). *Selected Tables of Australian Rainfall and Related Data*. Bureau of Meteorology, Melbourne, 146 pp.

Australian Bureau of Meteorology (1968). Review of Australia's water resources. *Monthly Rainfall and Evaporation*. Bureau of Meteorology, Melbourne, 2 volumes, variously paginated.

Bacastow, R. & Keeling, C. D. (1973). Atmospheric carbon dioxide and radiocarbon in the natural carbon cycle: II Changes from AD 1700 to 2070 as deduced from a geochemical model. *Brookhaven Symposium Biology* (AEC–CONF–720510), **25**, 86–135.

Baerreis, D. A. & Bryson, R. A. (1965). Climatic episodes and the dating of the Mississippian cultures. *Wisconsin Archaeologist*, **46**, 203–20.

Baerreis, D. A. & Bryson, R. A. (1967). Mississippian cultural developments in the light of historical climatology. Paper presented at 32nd annual meeting, Society for American Archaeology, Ann Arbor, Michigan, 4–6 May.

Baerreis, D. A. & Bryson, R. A. (1968). Climatic change and the Mill Creek culture of Iowa. *Iowa Archaeological Society Newsletter*, **15**, 1–34.

Bagehot, W. (1915). *Lombard Street*, 5th ed. King, London, 359 pp. Referred to by Jones (1964).

Ballantyne, E. R. (1975 a). Climatic design data and the effect of climate on indoor environment. *Australian Refrigeration, Air Conditioning and Heating Journal*, **29**, 20–30.

Ballantyne, E. R. (1975 b). Energy costs of dwellings. *Proceedings of the 5th Australian Building Research Congress*, Melbourne, July 1975.

Bandy, O. L., Butler, E. A. & Wright, R. C. (1969). Alaskan Upper Miocene marine glacial deposits and the *Turborotalia pachyderma* datum plane, *Science*, **166**, 607–8.

Bandy, O. L., Casey, R. E. & Wright, R. C. (1971). Late Neogene planktonic zonation, magnetic reversals and radiometric dates, Antarctic to the tropics. In *Antarctic oceanology I*, ed. J. L. Reid, Antarctic Research Series, vol. 15. Washington, DC. American Geophysical Union, pp. 1–26.

Barker, P. F., Dalziel, I. W. D. *et al.*, (1974). Southwestern Atlantic, Leg 36. *Geotimes*, **19**, 16–18.

Barkov, N. I., Gordienko, F. G., Korotkevitch, E. S. & Kotlyakov, V. M. (1975). Isotope oxygen studies of 500 m ice core taken from drill hole at Vostok station. *Bulletin of the Soviet Antarctic Expedition*, No. 90, pp. 39–49.

Barrett, E. C. (1974). *Climatology from Satellites*. Methuen, London, 418 pp.

Barry, R. G. (1973). Conditions favouring glacierisation and deglacierisation in North America from a climatological viewpoint. *Arctic and Alpine Research*, **5**, 171–84.

Barry, R. G., Andrews, J. T. & Mahaffy, M. A. (1975). The growth of continental ice sheets. *Science*, **190**, 979–81.

Barry, R. G. & Perry, A. H. (1973). *Synoptic Climatology: Methods & Applications*. Methuen, London, 555 pp.

Basinski, J. J. (1960). Water resources of the Yass Valley and their present and potential use. *CSIRO, Australia, Division of Land Research and Regional Survey Technical Paper* No. 7.

Bates, D. R. & Hays, P. B. (1967). Atmospheric nitrous oxide. *Planetary Space Science*, **15**, 189–97.

Batten, E. S. (1964). A model of the seasonal and latitudinal variation of zonal winds and temperatures in the stratosphere above 30 km. *Rand Corporation, Santa Monica, California, Memorandum* RM-4144-PR, 28 pp.

Batten, E. S. (1974). The atmospheric response to a stratospheric dust cloud as simulated by a general circulation model. *Rand Corporation, Santa Monica, California, RAND Report* R-1324-ARPA, 17 pp.

Berger, A. L. (1975). The astronomical theory of paleoclimates: a cascade of inaccuracy. In *Proceedings of the WMO/IAMAP Symposium on Long-Term Climatic Fluctuations*, Norwich (WMO-No. 421). World Meteorological Organization, Geneva, pp. 65–72.

Berger, W. H. (1972). Deep sea carbonates: dissolution facies and age–depth constancy. *Nature*, **246**, 392–5.

Berggren, R. (1975). Economic benefits of climatological services. *World Meteorological Organization, Technical Note* No. 145, Geneva, 43 pp.

Berggren, W. A. (1972*a*). A Cenozoic time-scale – some implications for regional geology and paleobiogeography. *Lethaia*, **5**, 195–215.

Berggren, W. A. (1972*b*). Late Pliocene–Pleistocene glaciation. In *Initial Reports of the Deep Sea Drilling Project*, vol. 12, Washington, US Government Printing Office, pp. 953–63.

Bergthorsson, P. (1969). An estimate of drift ice and temperature in Iceland in 1,000 years. *Jökull*, **19**, 94–101.

Berkner, L. V. & Marshall, L. C. (1967). The rise of oxygen in the earth's atmosphere with notes on the Martian atmosphere. In *Advances in Geophysics*, vol. 12. Academic Press, New York, pp. 309–31.

Berlage, H. P. (1961). Variations in the general atmospheric and hydro-spheric circulation of periods of a few years duration affected by variations of solar activity. *Annals New York Academy of Sciences*, **95**, 354–67.

Berry, F. A., Bollay, E. & Beers, N. R. (1945). *Handbook of Meteorology*. McGraw-Hill, New York and London, 1068 pp.

Bigg, E. K. (1975). Stratospheric particles. *Journal of the Atmospheric Sciences*, **32**, 910–17.

Birrer, W. M. (1974). Some critical remarks on trend analysis of total ozone data. *Pure and Applied Geophysics*, **112**, 523–32.

References

Bischof, W. (1975). The influence of the carrier gas on the infrared gas analysis of atmospheric CO_2. *Tellus*, **27**, 59–61.

Bjerknes, J. (1969). Atmospheric teleconnections from the equatorial Pacific. *Monthly Weather Review*, **97**, 163–72.

Blasing, T. J. (1975). Methods for analyzing climatic variations in the North Pacific sector and western North America for the last few centuries. Unpublished Ph.D. Thesis, University of Wisconsin, Madison, Department of Meteorology.

Blifford, I. H. (1970). Tropospheric Aerosols. *Journal of Geophysical Research*, **75**, 3099–103.

Bloom, A. L., Broecker, W. S., Chappell, J. M. A., Matthews, R. K. & Mesolella, K. J. (1974). Quaternary sea level fluctuations on a tectonic coast: new $^{230}Th/^{234}U$ dates from the Huon peninsula, New Guinea. *Quaternary Research*, **4**, 185–205.

Bobek, H. (1937). Die Rolle der Eiszeit in Nordwestiran. *Zeitschrift für Gletscherkunde*, **25**, 130–83.

Boer, G. J. & Kyle, A. C. (1974). Cloudiness, precipitation and vertical motion. In *The General Circulation of the Tropical Atmosphere*, eds. R. E. Newell *et al.* MIT Press, Cambridge, Massachusetts, pp. 143–78.

Boland, F. M. & Hamon, B. V. (1970). The East Australian Current, 1965–1968. *Deep-Sea Research*, **17**, 777–94.

Bolin, B. & Bischof, W. (1970). Variations of the carbon dioxide content of the atmosphere in the northern hemisphere. *Tellus*, **22**, 431–42.

Bolin, B. & Keeling, C. D. (1963). Large scale atmospheric mixing as deduced from seasonal and meridional variations of carbon dioxide. *Journal of Geophysical Research*, **68**, 3899–920.

Bowler, J. M. (1970). Late Quaternary environments: a study of lakes and associated sediments in south-eastern Australia. Unpublished Ph.d. Thesis, Australian National University, Canberra, Department of Biogeography and Geomorphology.

Bowler, J. M. (1971). Pleistocene salinities and climatic change: evidence from lakes and lunettes in south-eastern Australia. In *Aboriginal Man and Environment in Australia*, eds. D. J. Mulvaney & J. Golson. Australian National University Press, Canberra, pp. 47–65.

Bowler, J. M. Recent developments in reconstructing late Quaternary environments in Australia. In *The Biological Origin of the Australians*, eds. R. L. Kirk & A. G. Thorne. Australian Institute of Aboriginal Studies, Canberra, in press.

Bowler, J. M. & Hamada, T. (1971). Late Quaternary stratigraphy and radiocarbon climatology of water level fluctuations in Lake Keilambete, Victoria. *Nature*, **232**, 330–2.

Bowler, J. M., Hope, G. S., Jennings, J. N., Singh, G. & Walker, D. (1976). Late Quaternary climates in Australia and New Guinea. *Quaternary Research*, **6**, 359–94.

Bradley, R. S. (1973). Seasonal climatic fluctuations on Baffin Island during the period of instrumental records. *Arctic*, **26**, 230–43.

Bradley, R. S. (1976). *Precipitation History of the Rocky Mountain States.* Westview Press, Boulder, Colorado, 334 pp.

Braslau, N. & Dave, J. V. (1973). Effects of aerosols on the transfer of solar energy through realistic model atmospheres. Part I: Non absorbing aerosols. *Journal of Applied Meteorology,* 12, 601–15.

Bray, J. R. (1959). An analysis of the possible recent change in atmospheric carbon dioxide concentration. *Tellus,* 11, 220–30.

Bray, J. R. (1971). Vegetational distribution, tree growth and crop success in relation to recent climatic change. *Advances in Ecological Research,* 7, 177–233.

Bray, J. R. (1974 a). Glacial advance relative to volcanic activity since 1500 AD. *Nature,* 248, 42–3.

Bray, J. R. (1974 b). Volcanism and glaciation during the past 40 millennia. *Nature,* 252, 679–80.

Bray, J. R. & Struik, G. J. (1963). Forest growth and glacial chronology in Eastern British Columbia and their relation to recent climatic trends. *Canadian Journal of Botany,* 41, 1245–71.

Breeding, R. J., Lodge, J. P. Jr. *et al.,* (1973). Background trace gas concentrations in the Central United States. *Journal of Geophysical Research,* 78, 7057–64.

Broadbent, F. E. & Clark, F. (1965). Denitrification. In *Soil Nitrogen,* eds. W. V. Bartholomew & F. E. Clark, *Agronomy Monograph* No. 10, American Society of Agronomy, Madison, Wisconsin, pp. 344–59.

Broecker, W. S. (1974). *Chemical oceanography.* Harcourt Brace Jovanovich. New York, 214 pp.

Broecker, W. S., Li, Y-H., & Peng, T-H. (1969). Carbon dioxide – man's unseen artifact. In *Impingement of Man on the Oceans,* ed. D. W. Hood. Wiley-Interscience, New York, pp. 287–324.

Broecker, W. S., Thurber, D. L. Goddard, J., Ku, T-L., Matthews, R. K. & Mesolella, K. J. (1968). Milankovitch hypothesis is supported by precise dating of coral reefs and deep-sea sediments. *Science,* 159, 297–300.

Broecker, W. S. & van Donk, J. (1970). Insolation changes, ice volumes and the ^{18}O record in deep sea cores. *Reviews of Geophysics and Space Physics,* 8, 169–98.

Brookfield, H. C. & Hart, D. (1966). Rainfall in the tropical southwest Pacific. *Research School of Pacific Studies Department of Geography, Australian National University, Publication* G/3, 90 pp.

Brooks, C. E. P. (1926). *Climate through the ages.* Ernest Benn, London. Rev. edn. 1949, 395 pp.

Bryan, K. & Cox, M. D. (1968 a). A nonlinear model of an ocean driven by wind and differential heating: Part I. Description of the three-dimensional velocity and density fields. *Journal of the Atmospheric Sciences,* 25, 945–67.

Bryan, K. & Cox, M. D. (1968 b). A nonlinear model of an ocean driven by wind and differential heating: Part II. An analysis of the heat, vorticity and energy balance. *Journal of the Atmospheric Sciences*, **25**, 968–78.

Bryan, K. & Gill, A. E. (1971). Effects of geometry on circulation in a three-dimensional southern hemisphere ocean model. *Deep-Sea Research*, **18**, 685–721.

Bryson, R. A. (1966). Airmasses, streamlines, and the Boreal Forest, ONR technical report no. 24, Nonr. 1202(07). *Geographical Bulletin (Canada)*, **8**, 228–69.

Bryson, R. A. (1973 a). Drought in Sahelia, who or what is to blame? *Ecologist*, **3**, 366–71.

Bryson, R. A. (1973 b). World food prospects and climatic change. Testimony before joint meeting of the US Senate Subcommittees on foreign agricultural policy and agricultural production, marketing, and stabilization of prices, 18 October 1973.

Bryson, R. A. (1974 a). A perspective on climatic change. *Science*, **184**, 753–60.

Bryson, R. A. (1974 b). World climate and world food systems III: the lessons of climatic history. *Institute for Environmental Studies Report 27*, University of Wisconsin-Madison, 17 pp.

Bryson, R. A. & Baerreis, D. A. (1967). Possibilities of major climatic modification and their implications: Northwest India, a case for study. *Bulletin American Meteorological Society*, **48**, 136–42.

Bryson, R. A., Baerreis, D. A. & Wendland, W. M. (1970). The character of late-glacial and post-glacial climatic changes. In *Pleistocene and recent environments of the Central Great Plains*, eds. W. Dort, Jr., J. Jones & J. Knox, Jr. University Press of Kansas, pp. 53–74.

Bryson, R. A. & Julian, P. R. (1962). *Proceedings of the Conference on the Climate of the Eleventh and Sixteenth Centuries, Aspen, Colorado. NCAR Technical Notes* 63–1. National Center for Atmospheric Research, Boulder, Colorado, 102 pp.

Bryson, R. A., Lamb, H. H. & Donley, D. L. (1974). Drought and the decline of the Mycenae. *Antiquity*, **48**, 2–6.

Bryson, R. A. & Wendland, W. M. (1967). Tentative climatic patterns for some late glacial and postglacial episodes in central North America. In *Life, Land and Water, Occasional Paper* No. 1, ed. W. J. Mayer-Oakes. Anthropology Department, University of Manitoba, Winnipeg, pp. 271–98.

Bryson, R. A. & Wendland, W. M. (1970). Climate effects of atmospheric pollution. In *Global Effects of Environmental Pollution*, ed. F. S. Singer. Reidel, Holland, pp. 130–8.

Buch, H. S. (1954). Hemispheric wind conditions during the year 1950. *General Circulation Project, Final Report*, Part 2. Contract No. AH19-122-153. Department of Meteorology, Massachusetts Institute of Technology, Massachusetts, 126 pp.

Budd, W. F. (1969). The Dynamics of ice masses. *ANARE Scientific Reports Series A (IV), Glaciology Publication* No. 108. Antarctic Division, Department of Science, Melbourne, 216 pp.

Budd, W. F. (1975 *a*). A first simple model for periodically surging glaciers. *Journal of Glaciology*, **14**, 3–21.

Budd, W. F. (1975 *b*). The evidence of long term climate change from $^{16}O/^{18}O$ isotope ratios in deep sea cores. Australian Academy of Sciences, interim Report on Climatic Change (unpublished).

Budd, W. F. (1975 *c*). Antarctic sea-ice variations from satellite sensing in relaton to climate. *Journal of Glaciology*, **15**, 417–27.

Budd, W. F. & Jenssen, D. (1975). Numerical modelling of glacier systems. In *Snow and Ice Symposium* (*Proceedings of the Moscow Symposium, August 1971*). *International Association of Hydrological Sciences Publication* No. 104, pp. 251–91.

Budd, W. F., Jenssen, D. & Radok, U. (1970). The extent of basal melting in Antarctica. *Polarforschung*, **6**, 293–306.

Budd, W. F., Jenssen, D. & Radok, U. (1971). Derived physical characteristics of the Antarctic ice sheet. ANARE Interim Reports Series A (IV) *Glaciology Publication* No. 120. Antarctic Division, Department of Science, Melbourne, 178 pp.

Budd, W. F., Landon-Smith, I. H. & Wishart, E. R. (1967). The Amery ice shelf. In *Physics of snow and ice*, ed. Hirobumi Ôura, vol. 1, Part I, Institute of Low Temperature Science, Hokkaido University, pp. 447–67.

Budd, W. F. & McInnes, B. (1975). Modelling periodically surging glaciers. *Science*, **186**, 925–7.

Budd, W. F. & Morgan, V. (1973). Isotope measurements as indications of ice flow and palaeo-climate. In *Palaeoclimatology of Africa and of the Surrounding Islands and Antarctica*, ed. van Zinderen Bakker Sr. Balkema, Cape Town, pp. 5–22.

Budd, W. F. & Radok, U. (1971). Glaciers and large ice masses. *Reports on Progress in Physics*, **34**, No. 1, 70 pp.

Budyko, M. I. (ed.) (1963). Atlas teplovogo balansa. *Gidrometeorologichesko Izdatel'skos*. Leningrad,

Budyko, M. I. (1969). Climatic change. *Soviet Geography*, **10**, No. 8, pp. 429–51.

Budyko, M. I. (1974). *Climate and life*. Academic Press, New York, 508 pp.

Bull, C. (1971). Snow accumulation in the Antarctic. In *Research in the Antarctic*, ed. L. Quam. American Association for the Advancement of Science, Washington, DC, pp. 367–421.

Bunting, A. H., Dennett, M. D., Elston, J. & Milford, J. R. (1976). Rainfall trends in the West African Sahel. *Quarterly Journal Royal Meteorological Society*, **102**, 59–64.

Burdecki, F. (1969). The climate of SANAE, Part 1. *Notos*, **18**, 3–60.

Burrows, C. J. (1973). Studies on some glacial moraines in New Zealand II. The ages of moraines of the Mueller, Hooker and Tasman glaciers. *New Zealand Journal of Geology and Geophysics*, **16**, 831–56.

Burrows, C. J. (1975). Late pleistocene and holocene moraines of the Cameron Valley, Arrowsmith Range, Canterbury, New Zealand. *Arctic and Alpine Research*, **7**, 125–40.

References

Burrows, C. J. (1976). Icebergs in the southern ocean. *New Zealand Geographer*, **32**, 127–38.

Business Week (1975). US food power, ultimate weapon in world politics? *Business Week*, 15 December, 54–60.

Butzer, K. W., Isaac, G. L., Richardson, J. L. & Washbourn-Kamau, C. (1972). Radiocarbon dating of East African lake levels. *Science*, **175**, 1069–76.

Byerly, T. C. (1970). Nitrogen compounds used in crop production. In *Global Effects of Environmental Pollution*, ed. S. F. Singer. Reidel Holland, pp. 104–9.

Cadle, R. D. (1973). Particulate matter in the lower atmosphere. In *Chemistry of the Lower Atmosphere*, ed. S. I. Rasool. Plenum Press, New York, pp. 69–120.

Cadle, R. D. (1975). Volcanic emissions of halides and sulfur compounds to the troposphere and stratosphere. *Journal of Geophysical Research*, **80**, 1650–2.

Caine, N. & Jennings, J. N. (1968). Some blockstreams of the Toolong Range, Kosciusko State Park, New South Wales. *Journal Proceedings Royal Society New South Wales*, **101**, 93–103.

Calder, N. (1974). *The Weather Machine and the Threat of Ice*. British Broadcasting Corporation, London, 143 pp.

Canterford, R. & Pierrehumbert, C. (1975). Unpublished Study. Bureau of Meteorology, Australia.

Capurro, L. R. A. (1973). USNS Eltanin's 55 cruises – scientific accomplishments. *Antarctic Journal of the United States*, **8**, 57–60.

Carpenter, R. (1966). *Discontinuity in Greek civilization*. Cambridge University Press, Cambridge, 80 pp.

Castleman, A. W., Munkelwitz, H. R. & Manowitz, B. (1973). Contribution of volcanic sulphur compounds to the stratospheric aerosol layer. *Nature*, **244**, 345–6.

Caviedes, C. N. (1973). Sêcas and El Niño: two simultaneous climatical hazards in South America. *Proceedings of the Association of American Geographers*, **5**, 44–9.

Ceplecha, V. J. (1971). The distribution of the main components of the water balance in Australia. *The Australian Geographer*, **11** (5), 455–62.

Changnon, S. A. Jr. (1968). The La Porte weather anomaly – fact or fiction? *Bulletin American Meteorological Society*, **49**, 4–11.

Changnon, S. A. Jr. (1970). Reply. *Bulletin American Meteorological Society*, **51**, 337–42.

Changnon, S. A. Jr. (1973 a). Inadvertent weather and precipitation modification by urbanization. *Journal of Irrigation and Drainage Division, American Society of Civil Engineers*, March, 27–41.

Changnon, S. A. Jr. (1973 b). Urban-industrial effects on clouds and precipitation. In *Proceedings of the Inadvertent Weather Modification Workshop*, Logan, August 1973. Utah State University, pp. 111–39.

Changnon, S. A. Jr. & Jones, D. M. A. (1972). Review of the influence of the Great Lakes on weather. *Water Resources Research*, **8**, 360–71.

384

Chapman, S. (1930). A theory of upper atmospheric ozone. *Memoirs of the Royal Meteorological Society*, **3**, 103–25.

Chappell, J. (1973). Astronomical theory of climatic change: status and problem. *Quaternary Research*, **3**, 221–36.

Chappell, J. (1974 a). Geology of coral terraces, Huon Peninsula, New Guinea: a study of Quaternary tectonic movements and sea level changes. *Geological Society of America, Bulletin*, **85**, 553–70.

Chappell, J. (1974 b). Relationships between sealevels, ^{18}O variations and orbital perturbations, during the past 250,000 years. *Nature*, **252**, 199–202.

Chappell, J. (1975). On possible relationships between Upper Quaternary glaciation, geomagnetism, and vulcanism. *Earth Planetary Science Letters*, **26**, 370–6.

Chappell, J. (in press). Strong factors influencing upper pleistocene ice ages. *Quaternary Research*.

Chappell, J. & Veeh, H. H. (1977). Quaternary uplift and sea level changes at Portuguese Timor and Atauro Island. *Geological Society of America, Bulletin* (in press).

Charlson, R. J., Vanderpol, A. H., Covert, D. S., Waggoner, A. P. & Ahlquist, N. C. (1974). $H_2SO_4/(NH_4)_2SO_4$ background aerosol: optical detection in St Louis region. *Atmospheric Environment*, **8**, 1257–67.

Charney, J. G. (1975). Dynamics of deserts and drought in the Sahel. *Quarterly Journal Royal Meteorological Society*, **101**, 193–202.

Charney, J. G., Stone, P. H. & Quirk, W. J. (1975). Drought in the Sahara: a biogeophysical feedback mechanism. *Science*, **187**, 434–5.

Chervin, R. M. & Schneider, S. H. (1976). On determining the statistical significance of climate experiments with general circulation models. *Journal of the Atmospheric Sciences*, **33**, 405–12.

Chriss, T. & Frakes, L. A. (1972). Glacial marine sedimentation in the Ross Sea: In *Antarctic geology and geophysics*, ed. R. J. Adie. Oslo (Universitets-forlaget), pp. 747–62.

Chuan, G. K. & Lockwood, J. G. (1974). An assessment of topographical controls on the distribution of rainfall in the Central Pennines. *Meteorological Magazine*, **103**, 275–87.

Chu Ko-Chen (1973). *Scientia Sinica*, **16**, 226–56.

Chýlek, P. & Coakley, J. A., Jr. (1974). Aerosols and climate. *Science*, **183**, 75–7.

Ciesielski, P. F. & Weaver, F. M. (1974). Early Pliocene temperature changes in the Antarctic seas. *Geology*, **2**, 511–15.

Claiborne, R. (1970). *Climate, man, and history*. W. W. Norton and Co., New York., 444 pp.

Clapp, P. F. (1970). Parameterization of macro-scale transient heat transport for use in a mean-motion model of the general circulation. *Journal of Applied Meteorology*, **9**, 554–63.

Clark, E. C. & Whitby, T. W. (1975). Measurements of aerosols produced by the photochemical oxidation of SO_2 in air. *Journal of Colloid and Interface Science*, **51**, 477–90.

385

References

CLIMAP Project Members (1976). The surface of ice age earth. *Science*, **191**, 1131–7.

Climate/Food Research Group (1975). A detailed model of the production and consumption of spring wheat in the United States. Institute for Environmental Studies, University of Wisconsin-Madison. *IES Report* No. 49, 103 pp.

Colhoun, E. A. (1976). The glaciation of the lower Forth Valley, north-western Tasmania. *Australian Geographical Studies*, **14**, 83–102.

Collis, R. T. H. (1975). Weather and world food. *Bulletin American Meteorological Society*, **56**, 1078–83.

Coote, G. G. & Cornish, E. A. (1958). The correlation of monthly rainfall with position and altitude of observing stations in South Australia. *CSIRO, Division of Mathematical Statistics, Technical Paper* No. 4, Melbourne.

Cornish, E. A. & Evans, M. J. (1964). An analysis of daily temperature at Adelaide, South Australia. *CSIRO, Division of Mathematical Statistics, Technical Paper* No. 17, Melbourne.

Costin, A. B. (1971). Vegetation, soils and climate in late-quaternary southeastern Australia. In *Aboriginal Man and Environment in Australia*, eds. D. J. Mulvaney & J. Golson. Australian National University Press, Canberra, pp. 26–37.

Costin, A. B. (1972). Carbon-14 dates from the Snowy Mountains area, south-eastern Australia and their interpretation. *Quaternary Research*, **2**, 579–90.

Costin, A. B., Thom, B. G., Wimbush, D. W. & Stuiver, M. (1967). Non-sorted steps in the Mt Kosciusko area, Australia. *Bulletin Geological Society America*, **78**, 979–92.

Court, A. & Salmela, H. A. (1963). Improbable weather extremes and measurement needs. *Bulletin American Meteorological Society*, **44**, 571–5.

Coventry, R. J. (1976). Abandoned shorelines and the Late Quaternary history of Lake George, New South Wales. *Journal of the Geological Society of Australia*, **23**, 249–73.

Cramer, J. & Myers, A. L. (1972). Rate of increase of atmospheric carbon dioxide. *Atmospheric Environment*, **6**, 563–73.

Crary, A. P. (1961). Glaciological studies at Little America Station, Antarctica, 1957 and 1958. *IGY Glaciological Report Series*. New York, IGY World Data Center A, Glaciology, No. 5, 197 pp.

Crowe, B. L. (1969). The tragedy of the Commons revisited. *Science*, **166**, 1103–7.

Crutzen, P. J. (1974). Photochemical reactions initiated by and influencing ozone in the troposphere. *Tellus*, **26**, 47–57.

CSIRO (1963). *Oceanographic Cruise Report* No. 6. Oceanographic Observations in the Pacific Ocean in 1960. HMAS Gascoyne cruise G 3/60. Commonwealth Scientific and Industrial Research Organization, Melbourne.

Currey, D. T. (1970). Lake systems in Western Victoria. *Australian Society for Limnology Bulletin*, **3**, 1–13.

Curry, L. (1962). Climatic change as a random series. *Association of American Geographers, Annals*, **52**, 21–31.

Dalby, D. & Harrison-Church, R. J. (eds.) (1973). *Drought in Africa*. Centre for African Studies, University of London.

Dansgaard, W. & Johnsen, S. J. (1969). A flow model and a time scale for the ice core from Camp Century, Greenland. *Journal of Glaciology*, **8**, 215–23.

Dansgaard, W., Johnsen, S. J., Clausen, H. B. & Langway, C. C. Jr. (1971). Climatic record revealed by the Camp Century ice core. In *The Late Cenozoic Glacial Ages*, ed. K. K. Turekian. Yale University Press, New Haven, pp. 37–56.

Dansgaard, W., Johnsen, S. J., Clausen, H. B. & Langway, C. C. Jr. (1972). Speculations about the next glaciation. *Quaternary Research*, **2**, 396–8.

Dansgaard, W., Johnsen, S. J., Moller, J. & Langway, C. C. Jr. (1969). One thousand centuries of climatic record from Camp Century on the Greenland Ice Sheet. *Science*, **166**, 377–81.

Dansgaard, W., Johnsen, S. J., Reeh, N., Gundestrup, N., Clausen, H. B. & Hammer, C. U. (1975). Climatic changes, Norsemen and modern man. *Nature*, **255**, 24–8.

Darbyshire, J. (1964). A hydrological investigation of the Agulhas Current area. *Deep-Sea Research*, **11**, 781–815.

Darmstadter, J. & Schurr, S. H. (1974). World energy resources and demand. *Philosophical Transactions Royal Society, London*, **A276**, 413–30.

Davies, J. L. (ed.) (1965). *Atlas of Tasmania*. Lands and Surveys Department, Hobart, 128 pp.

Davies, J. L. (1967). Tasmanian landforms and Quaternary climates. In *Landform Studies from Australia and New Guinea*, eds. J. N. Jennings & J. A. Mabbutt. Australian National University Press, Canberra. pp. 1–25.

Davis, R. (1973). *The Rise of the Atlantic Economies*. Weidenfeld & Nicholson, London, 380 pp.

Deacon, E. L. (1953). Climatic change in Australia since 1880. *Australian Journal of Physics*, **6**, 209–18.

Deacon, G. E. R. (1937). The hydrology of the Southern Ocean. *Discovery Reports*, **15**, 1–124.

Deacon, G. E. R. (1963). The Southern Ocean. In *The Sea*, ed. M. N. Hill, vol. 2. Interscience Publishers, London, pp. 281–96.

Deacon, G. E. R. (1966). Subtropical convergence. In *The Encyclopedia of Oceanography*, ed. R. W. Fairbridge. Reinhold, New York, pp. 884–5.

Deane, P. & Cole, W. A. (1967). *British economic growth 1688–1959, trends and structure*. Cambridge University Press, Cambridge, 350 pp.

Degens, E. T. & Hecky, R. E. (1974). Paleoclimatic reconstruction of late Pleistocene and Holocene based on biogenic sediments from the Black

References

Sea and a tropical African lake. *Colloquium Internationale Centre Nationale Recherche Scientifique*, No. 219, pp. 13–24.

Deirmendjian, D. (1969). *Electromagnetic Scattering on Spherical Polydispersions*. American Elsevier, New York, 290 pp.

Deirmendjian, D. (1973). On volcanic and other particulate turbidity anomalies. *Advances in Geophysics*, vol. 16. Academic Press, New York, pp. 267–96.

Delany, A. C., Parkin, D. W., Griffin, J. J., Goldberg, E. D. & Reiman, B. E. F. (1967). Airborne dust collected at Barbados. *Geochimica et Cosmochimica Acta*, **31**, 885–909.

Delany, A. C., Shedlovsky, J. P. & Pollock, W. H. (1975). Stratospheric aerosol. The contribution from the troposphere. *Journal Geophysical Research*, **79**, 5646–50.

de Lisle, J. F. (1957). Fluctuations of seasonal pressure patterns in the vicinity of New Zealand. *New Zealand Journal of Science and Technology*, **38B**, 400–15.

Denman, K. L. & Miyake, M. (1973). Upper-layer modification at ocean station Papa, observations and simulation. *Journal of Physical Oceanography*, **3**, 185–96.

Denmead, O. T., Simpson, J. R. & Freney, J. R. (1974). Ammonia flux into the atmosphere from a grazed pasture. *Science*, **185**, 609–10.

Denton, G. H. (1974). Ice-age ice sheets: their global distribution 18,000 years ago and subsequent disintegration. (Abstract), *Eos*, **55**, p. 258 only.

Denton, G. H. & Armstrong, R. L. (1969). Miocene-Pliocene glaciations in southern Alaska. *American Journal of Science*, **267**, 1121–42.

Denton, G. H., Armstrong, R. L. & Stuiver, M. (1971). The late Cenozoic glacial history of Antarctica. In *The Late Cenozoic Glacial Ages*, ed. K. K. Turekian. Yale University Press, New Haven, pp. 267–306.

Denton, G. H. & Karlén, W. (1973). Holocene climatic changes, their pattern and possible cause. *Quaternary Research*, **3**, 155–205.

Derbyshire, E. (1972). Pleistocene glaciation of Tasmania; review and speculations. *Australian Geographical Studies*, **10**, 79–94.

Devereux, I. (1967). Oxygen isotope palaeotemperatures on New Zealand tertiary fossils. *New Zealand Journal of Science*, **10**, 988–1011.

Devereux, I., Hendy, C. H. & Vella, P. (1970). Pliocene and early Pleistocene sea temperature fluctuations, Mangaopari Stream, New Zealand. *Earth and Planetary Science Letters*, **8**, 163–8.

Dickson, R. R., Lamb, H. H., Malmberg, S. A. & Colebrook, J. H. (1975). Climatic reversal in the northern North Atlantic. *Nature*, **256**, 479–82.

Dinger, J. E., Howell, H. B. & Wojciechowski, T. A. (1970). On the source and composition of cloud nuclei in a subsident air mass over the North Atlantic. *Journal of the Atmospheric Sciences*, **27**, 791–7.

Dobson, G. M. B., Harrison, D. N. & Lawrence, J. (1927). Measurements of the amount of ozone in the earth's atmosphere and it relation to other geophysical conditions. *Proceedings of the Royal Society, London,* **236A**, 187–93.

Dodson, J. R. (1974a). Vegetation history and water fluctuations at Lake Leake, south-eastern South Australia. I. 10,000 BP to Present. *Australian Journal of Botany,* **22**, 719–41.

Dodson, J. R. (1974b). Vegetation and climatic history near Lake Keilambete, western Victoria. *Australian Journal of Botany,* **22**, 709–17.

Dodson, J. R. (1974c). Calcium carbonate formation by *Enteromorpha nana* algae in a hypersaline volcanic crater lake. *Hydrobiologie,* **44** (2), 247–53.

Dodson, J. R. (1975). Vegetation history and water fluctuations at Lake Leake, south-eastern South Australia. II. 50,000 to 10,000 BP. *Australian Journal of Botany,* **23**, 815–31.

Donn, W. L. & Ewing, M. (1968). The theory of an ice-free Arctic Ocean. *Meteorological Monographs,* **8**, No. 30, pp. 100–5.

Donn, W. L. & Shaw, D. M. (1966). Heat budgets of an ice-free and ice-covered Arctic Ocean. *Journal of Geophysical Research,* **17**, 1087–93.

Donn, W. L. & Shaw, D. M. (1968). Maintenance of an ice-free Arctic Ocean. *Progress in Oceanography,* Oxford, **4**, 105–13.

Dordick, I. L. (1953). Climate and work in Australian New Guinea. *Acta Tropica,* **10**, 233–50.

Dorman, F. H. (1966). Australian Tertiary palaeotemperatures. *Journal of Geology,* **74**, 49–61.

Douglas, R. G. & Savin, S. M. (1971). Isotope analyses of planktonic foraminifera from the Cenozoic of the northwest Pacific. In *Initial Reports of the Deep Sea Drilling Project,* vol. 6. Washington, DC, US Government Printing Office, pp. 1123–7.

Dreimanis, A. & Karrow, P. F. (1972). Glacial history of the Great Lakes–St Lawrence Region, the classification of the Wisconsin stage, and its correlatives. *International Geological Congress,* 24th Session, Montreal, Section 12, pp. 5–15.

Dronia, H. (1967). The influence of cities on the world-wide temperature trend (in German). *Meteorologische Abhandlungen,* **74** (4), 1–65.

Dronia, A. H. (1974). Uber temperäturänderung der freien Atmosphäre auf der Nordhalbkügel in den letzten 25 Jahren. *Meteorologische Rundshau,* Berlin, **27**, 166–74.

Düing, W. & Szekielda, K. H. (1971). Monsoonal response in the Western Indian Ocean. *Journal of Geophysical Research,* **76**, 4181–7.

Dunbar, M. (1973). Increasing severity of ice conditions in Baffin Bay and Davis Strait and its effect on the extreme limits of ice. In *Sea Ice,* ed. T. Karlsson. National Research Council, Reykjavik, pp. 87–93.

Duplessy, J. C., Labeyrie, J., Lalou, C. & Nguyen, H. V. (1971). La mesure des variations climatiques continentales. *Quaternary Research,* **1**, 162–74.

Durham, J. W. (1950). Cenozoic marine climates of the Pacific coast. *Geological Society of America Bulletin,* **61**, 1243–64.

References

Dutton, J. A. & Johnson, D. R. (1967). The theory of available potential energy and a variational approach to atmospheric energetics. *Advances in Geophysics*, vol. 12. Academic Press, New York, pp. 333–436.

Dzerdzeevski, B. L. (1963). Fluctuations of general circulation of the atmosphere and climate in the twentieth century. In *Changes of Climate, Arid Zone Research*, vol. 20. UNESCO, Paris, pp. 285–95.

Dzerdzeevski, B. L. (1969). Circulation epochs in the 20th century and some comments on analysis of past climates. In *Quaternary Geology and Climate*, ed. H. E. Wright, Jr, *INQUA* vol. 16, National Academy Science Publication 1701, Washington, DC, pp. 49–60.

East, W. G. (1962). *An Historical Geography of Europe*, 5th edn. (1st edn. 1935). Methuen, London.

Ebdon, R. A. (1975). The quasi-biennial oscillation and its association with tropospheric circulation patterns. *Meteorological Magazine*, **104**, 282–97.

Ehrlich, P. R. (1968). *The population bomb*. Ballantine Books, New York, 223 pp.

Ehrlich, P. R. & Ehrlich, A. H. (1970). *Population, Resources, Environment: Issues in Human Ecology*. Freeman, San Francisco, 383 pp.

Eichenlaub, V. (1971). Further comments on the climate of the mid-nineteenth century United States compared to current normals. *Monthly Weather Review*, **99**, 847–50.

Ekdahl, C. A. & Keeling, C. D. (1973). Atmospheric carbon dioxide and radiocarbon in the natural carbon cycle: 1. Quantitative deductions from records at Manua Loa Observatory and at the South Pole. *Brookhaven Symposium Biology* (AEC-CONF-720510), **25**, 51–85.

Emery, K. O. (1969). The continental shelves. *Scientific American*, **221** (3), 106–22.

Emiliani, C. (1966a). Isotopic paleotemperatures. *Science*, **154**, 851–7.

Emiliani, C. (1966b). Paleotemperature analysis of Caribbean cores P 6304–8 and P 6304–9 and a generalised temperature curve for the past 425,000 years. *Journal of Geology*, **74**, 109–26.

Emiliani, C. (1971). The last interglacial: paleotemperatures and chronology. *Science*, **171**, 571–3.

Emiliani, C. & Shackleton, N. J. (1974). The Brunhes epoch: isotopic paleotemperatures and geochronology. *Science*, **183**, 511–14.

Ericson, D. B. & Wollin, G. (1968). Pleistocene climates and chronology in deep-sea sediments. *Science*, **162**, 1227–34.

Eriksson, E. (1959). The yearly circulation of chloride and sulphur in nature; meteorological, geochemical and pedological implications Part I. *Tellus*, **11**, 375–403.

Faegri, K. (1948). On the variations of Western Norwegian glaciers during the last 200 years. *Association Internationale d'Hydrologie Scientifique, Assemblée Générale d'Oslo 1948*, vol. 2. Commission de la Neige et des Glaciers, pp. 293–303.

Fairbridge, R. W. (1961). Eustatic changes in sea level. In *Physics and Chemistry of the Earth*, vol. 4. Pergamon Press, London, pp. 99–185.

Fairbridge, R. W. (1962). World sea level and climatic changes. *Quaternaria*, **6**, 111–34.

Fairbridge, R. W. (ed.) (1967). *The Encyclopedia of Atmospheric Sciences and Astrogeology*. Reinhold, New York, 1200 pp., see in particular, pp. 454–74.

Fairbridge, R. W. (1976). Shellfish-eating preceramic Indians in Coastal Brazil. *Science*, **191**, 353–9.

Fairhall, A. W. (1973). Accumulation of fossil CO_2 in the atmosphere and the sea. *Nature*, **245**, 20–3.

Farmer, J. G. & Baxter, M. S. (1974). Atmospheric carbon dioxide levels as indicated by the stable isotope record in wood. *Nature*, **247**, 273–5.

Feyerabend, P. K. (1975). *Against method: outline of an anarchistic theory of knowledge*. New Left Books, London, 339 pp.

Fletcher, J. O. (1969). Ice extent on the Southern Ocean and its relation to world climate. *Rand Corporation, Santa Monica, California, Memorandum* RM-5793-NSF, 111 pp.

Flint, R. F. (1971). *Glacial and Quaternary geology*. Wiley, New York, 892 pp.

Flint, R. F. & Gale, W. A. (1958). Stratigraphy and radiocarbon dates at Searles Lake, California. *American Journal of Science*, **256**, 689–714.

Flohn, H. (1964). Grundfragen der Paläoklimatologie im Lichte einer theoretischen Klimatologie. *Geologische Rundschau*, **54**, 504–51.

Flohn, H. (1969). *Climate and Weather*. Weidenfeld & Nicolson, London, 253 pp.

Flohn, H. (1971 a). Investigations on the climatic conditions of the advancement of the Tunisian Sahara. (WMO-No. 279). *World Meteorological Organization Technical Note* No. 116, Geneva, 31 pp.

Flohn, H. (1971 b). Tropical circulation pattern. *Bonner Meteorologische Abhandlungen*, **15**, 55 pp.

Flohn, H. (1973). Globale energiebilanz und klimaschwankungen. *Bonner Meteorologische Abhandlungen*, **19**, 75–117.

Flohn, H. (1974 a). Background of a geophysical model of the initiation of the next glaciation. *Quaternary Research*, **4**, 385–404.

Flohn, H. (1974 b). Instabilität und anthropogene Modifikation des Klimas. *Annalen der Meteorologie*, NF, **9**, 25–31.

Foley, J. C. (1957). Droughts in Australia. *Commonwealth Bureau of Meteorology Bulletin* No. 43. Melbourne, 281 pp.

Folland, C. K. (1975). A relationship between cool summers in central England and the temperature of the following winter for summers occurring in an even year. *Weather*, **30**, 348–58.

Frakes, L. A. (1975). Oceanographic and biological influences on sediment distribution in the Southern Ocean. *Search*, **6**, 339–41.

Frakes, L. A. & Kemp, E. M. (1972). Influence of continental positions on early Tertiary climates. *Nature*, **240**, 97–100.

Frakes, L. A. & Kemp, E. M. (1973). Palaeogene continental positions and evolution of climate. In *Implications of Continental Drift to the Earth Sciences*, vol. 1, eds. D. H. Tarling & S. K. Runcorn. Academic Press, New York, pp. 539–58.

References

Frenzel, B. (1973). *Climatic fluctuations of the ice age.* Case Western Reserve University Press, Cleveland, 306 pp.

Frenzel, B. (1975). The distribution pattern of Holocene climatic change in the northern hemisphere. *Proceedings of the WMO/IAMAP Symposium on Long-Term Climatic Fluctuations*, Norwich, 18–23 August, 1975 (WMO-No. 421). World Meteorological Organization, Geneva, pp. 105–18.

Freyer, H. D. & Wiesberg, L. (1973). ^{13}C decrease in modern wood due to the large-scale combustion of fossil fuels. *Naturwissenschaften*, **60**, 517–18.

Freyer, H. D. & Wiesberg, L. (1976). Review on different attempts of tracking back the increasing atmospheric CO_2 level to pre-industrial times in *Proceedings of the 2nd International Conference on Stable Isotopes*, Oak Brook, October 1975. ERDA, 768 pp.

Fridriksson, S. (1973). Crop production in Iceland. *International Journal Biometeorology*, **17**, 359–62.

Fritts, H. C. (1971). Dendroclimatology and dendroecology. *Quaternary Research*, **1**, 419–49.

Fritts, H. C., Blasing, T. J., Hayden, B. P. & Kutzbach, J. E. (1971). Multivariate techniques for specifying tree-growth and climate relationships for reconstructing anomalies in paleoclimate. *Journal of Applied Meteorology*, **10**, 845–64.

Fuller, J. F. C. (1970). *The Decisive Battles of the Western World: and their Influence on History*, vol. 2. Eyre and Spottiswoode, London, 561 pp.

Gaffney, D. O. (1975). Rainfall deficiency and evaporation in relation to drought in Australia. Paper presented at 46th ANZAAS Congress, Canberra. (Available from Bureau of Meteorology, Melbourne.)

Galbally, I. E. (1971). Ozone profiles and ozone fluxes in the atmospheric surface layer. *Quarterly Journal of Royal Meteorological Society*, **97**, 18–29.

Galbally, I. E. (1975). Emission of oxides of nitrogen (NO_x) and ammonia from the earth's surface. *Tellus*, **27**, 67–70.

Galloway, R. W. (1963). Glaciation in the Snowy Mountains: a reappraisal. *Proceedings Linnean Society New South Wales*, **88**, 180–98.

Galloway, R. W. (1965). Late Quaternary climates in Australia. *Journal of Geology*, **73**, 603–18.

Galloway, R. W. (1970). The full-glacial climate in the southwestern United States. *Annals Association of American Geographers*, **60**, 245–56.

Galloway, R. W., Hope, G. S., Loffler, E. & Peterson, J. A. (1973). Late Quaternary glaciation and periglacial phenomena in Australia and New Guinea. In *Palaeoecology of Africa and of the Surrounding Islands and Antarctica*, ed. E. M. Van Zinderen Bakker, vol. 8. Balkema, Cape Town, pp. 125–38.

Gani, J. (1975). The use of statistics in climatological research. *Search*, **6**, 504–8.

Garratt, J. R. & Pearman, G. I. (1973). Large-scale CO_2 fluxes in the southern hemisphere troposphere. *Nature (Physical Science)*, **242**, 54–6.

Gates, W. L. (1976). Modelling the ice age climate. *Science*, **191**, 1138–44.

Gates, W. L. & Imbrie, J. (1975). Climatic Change. *Reviews of Geophysics and Space Physics*, **13**, 726–31.

Gebhart, R. (1967). On the significance of the shortwave CO_2 absorption in investigations concerning the CO_2 theory of climatic change. *Archiv für Meteorologie, Geophysik und Bioklimatologie*, **B15**, 52–61.

Geitzenauer, K. R. (1972). The Pleistocene calcareous nannoplankton of the subantarctic Pacific Ocean. *Deep-Sea Research*, **19**, 45–60.

Geitzenauer, K. R., Margolis, S. V. & Edwards, D. (1968). Evidence consistent with Eocene glaciation in a South Pacific deep sea sedimentary core. *Earth and Planetary Science Letters*, **4**, 173–7.

Gentilli, J. (1967). A history of meteorological and climatological studies in Australia. *Western Australia University Studies in History*, **5**, 88 pp.

Gentilli, J. (1971). Climatic fluctuations. In *Climates of Australia and New Zealand*, ed. J. Gentilli, World Survey of Climatology, vol. 13. Elsevier, Amsterdam, pp. 189–211.

Geyh, M. A. & Jäkel, D. (1974). Late Glacial and Holocene climatic history of the Sahara Desert derived from a statistical assay of carbon-14 dates. *Palaeogeography, Palaeoclimatology and Palaeoecology*, **15**, 205–8.

Gibbs, W. J. (1975). Drought – its definition, delineation and effects. In *Special Environmental Report* No. 5. *World Meteorological Organization*, Geneva, pp. 1–39.

Gibbs, W. J. & Maher, J. V. (1967). Rainfall deciles as drought indicators, *Commonwealth Bureau of Meteorology Bulletin* No. 48, Melbourne, not paginated.

Gilchrist, A., Corby, G. A. & Newson, R. L. (1973). A numerical experiment using a general circulation model of the atmosphere. *Quarterly Journal of the Royal Meteorological Society*, **99**, 2–34.

Gill, A. E. & Clarke, A. J. (1974). Wind-induced upwelling, coastal currents and sea level changes. *Deep-Sea Research*, **21**, 325–45.

Gill, A. E., Green, J. S. A. & Simmons, A. (1974). Energy partition in the large-scale ocean circulation and the production of mid-ocean eddies. *Deep-Sea Research*, **21**, 499–528.

Gill, A. E. & Niiler, P. P. (1973). The theory of seasonal variability in the ocean. *Deep-Sea Research*, **20**, 141–77.

Gillette, D. A. (1974). On the production of soil wind erosion aerosols having the potential for long range transport. *Journal de Recherches Atmosphériques*, **8**, 735–44.

Gillette, D. A., Blifford, I. H. & Fenster, C. R. (1972). Measurements of aerosol size distributions and vertical fluxes of aerosols on land subject to wind erosion. *Journal of Applied Meteorology*, **11**, 977–87.

Giovinetto, M. B. (1964). The drainage systems of Antarctica: accumu-

References

lation. In *Antarctic Snow and Ice studies*, Antarctic Research Series, vol. 2, Washington DC, American Geophysical Union, pp. 127–55.

Giovinetto, M. B. (1970). The Antarctic ice sheet and its bimodal response to climate. In *International Symposium on Antarctic Glaciological Exploration (ISAGE)*, IASH, Commission of Snow and Ice, Publication No. 86, pp. 347–58.

Giovinetto, M. B. & Schwerdtfeger, W. (1966). Analysis of a 200 year snow accumulation series from the South Pole. *Archiv für Meteorologie Geophysik und Bioklimatologie*, **A15**, 227–50.

Godfrey, J. S. (1973 a). On the dynamics of the western boundary current in Bryan & Cox's (1968) numerical model ocean. *Deep-Sea Research*, **20**, 1043–58.

Godfrey, J. S. (1973 b). Comparison of the East Australian current with the western boundary current in Bryan & Cox's (1968) numerical model ocean. *Deep-Sea Research*, **20**, 1059–76.

Godfrey, J. S. (1975). On ocean spindown. I: A linear experiment. *Journal of Physical Oceanography*, **5**, 399–409.

Golson, J. (1972). The remarkable history of Indo-Pacific Man: missing chapters from every world prehistory. *Search*, **3**, 13–21.

Gordon, A. L. (1967). Structure of Antarctic waters between 20° W and 170° W. *Antarctic Map Folio Series*, folio 6, ed. V. C. Bushnell. American Geographical Society, New York.

Gordon, A. L. (1973). USNS Eltanin Cruise 50: physical oceanography of the southeast Indian Ocean. *Antarctic Journal of the United States*, **7**, 38–40.

Gordon, A. L. (1975). General ocean circulation. In *Numerical Models of Ocean Circulation*. Academy of Sciences, Washington, DC, pp. 39–53.

Gordon, A. L. & Bye, J. A. T. (1972). Surface dynamic topography of Antarctic waters. *Journal of Geophysical Research*, **77**, 5993–9.

Gordon, A. L. & Goldberg, R. D. (1970). Circumpolar characteristics of antarctic waters. *Antarctic Map Folio Series*, folio 18, ed. V. Bushnell. American Geographical Society, New York.

Gow, A. J., Ueda, H. T. & Garfield, D. E. (1969). Antarctic ice sheet: preliminary results of first core hole to bedrock. *Science*, **161**, 1011–13.

Gray, B. M. (1975). Japanese and European winter temperatures. *Weather*, **30**, 359–68.

Gray, W. M., Frank, W. M., Corrin, M. L. & Stokes, C. A. (1974). Weather modification by carbon dust absorption of solar energy. *Atmospheric Science Paper* No. 225. Department of Atmospheric Science, Colorado State University, Fort Collins, Colorado, 190 pp.

Green, J. S. A. (1970). Transfer properties of the large-scale eddies and the general circulation of the atmosphere. *Quarterly Journal of Royal Meteorological Society*, **96**, 157–85.

Gribbin, J. (1975). Aerosol and climate: hotter or cooler? *Nature*, **253**, 162.

Griffin, I. J., Windom, H. L. & Goldberg, E. D. (1968). The distribution of clay minerals in the world ocean. *Deep-Sea Research*, **15**, 433–59.

Grobecker, A. J. (1975). *CIAP Monograph* No. 5 The impacts of climatic changes on the biosphere, Part 1, Ultraviolet Radiation Effects, DOT-TST-75-55. Department of Transport, Washington, DC.

Grobecker, A. J., Coroniti, S. C. & Cannon, R. H. (1974). *Effects of Stratospheric Pollution by Aircraft.* DOT-TST-75-50. Department of Transport, Washington, DC.

Grove, J. M. (1966). The little ice age in the massif of Mont Blanc. *Institute of British Geographers, Transactions and Papers,* **40**, 129–43.

Hadley, G. (1735). Concerning the cause of the general Trade Wind. *Philosophical Transactions, Royal Society (London),* **39**, 58–73.

Hahn, J. (1974). The North Atlantic Ocean as a source of atmospheric N_2O. *Tellus,* **26**, 160–8.

Hall, C. A. S., Ekdahl, C. A. & Wartenberg, D. E. (1975). A fifteen-year record of biotic metabolism in the Northern Hemisphere. *Nature,* **255**, 136–8.

Haltiner, G. J. (1971). *Numerical weather prediction.* Wiley, New York, 317 pp.

Hamon, B. V. (1965). The East Australian current, 1960–1964. *Deep-Sea Research,* **12**, 899–921.

Hare, F. K. (1971). Future climates and future environments. *Bulletin of the American Meteorological Society,* **52**, 451–6.

Harman, J. R. & Elton, W. M. (1971). The La Porte, Indiana, precipitation anomaly. *Annals Association of American Geographers,* **61**, 468–80.

Harman, J. R. & Hehr, J. G. (1972). Lake breezes and summer rainfall. *Annals Association of American Geographers,* **62**, 375–87.

Harris, S. (1976). Economic aspects of possible climatic changes. Appendix 3 of Report of a Committee on Climatic Change, *Australian Academy of Science, Report* No. 21, Canberra, pp. 83–91.

Haude, W. (1969). Erfordern die Hochstände des Toten Meeres die Annahme von Pluvial-Zeiten während des Pleistozäns? *Meteorologische Rundschdau,* **22**, 29–40.

Hayes, D. E. & Frakes, L. A. (1975). General synthesis. In *Initial Reports of the Deep Sea Drilling Project,* vol. 28. Washington, DC, US Government Printing Office, pp. 919–42.

Hayes, D. E. & Frakes, L. A. *et al.* (1975). Leg 28 – Antarctic. In *Initial Reports of the Deep Sea Drilling Project,* vol. 28. Washington, DC, US Government Printing Office, 1117 pp.

Hays, J. D., Imbrie, J. & Shackleton, N. J. (in press). Variations in the earth's orbit: pacemaker of the ice ages. *Science,* **194**, 1121–32.

Hays, J. D., Lozano, J., Shackleton, N. & Irving, G. (1976). Reconstruction of the Atlantic Ocean and western Indian Ocean sectors of the 18,000 BP Antarctic Ocean. In *Investigation of Late Quaternary Paleoceanography and Paleoclimatology,* eds. R. M. Cline & J. D. Hays. *Geological Society of America Memoir,* vol. 145, pp. 337–72.

Hays, J. D., Saito, T., Opdyke, N. D. & Burckle, L. H. (1969). Pliocene–Pleistocene sediments of the equatorial Pacific: their paleomagnetic,

395

References

biostratigraphic and climatic record. *Geological Society of America, Bulletin*, **80**, 1481–514.

Healy, T. V., McKay, H. A. C., Pilbean, A. & Scargill, D. (1970). Ammonia and ammonium sulfate in the troposphere over the United Kingdom. *Journal of Geophysical Research*, **75**, 2317–21.

Hellerman, S. (1967). An updated estimate of the wind stress on the world ocean. *Monthly Weather Review*, **95**, (9), 607–26.

Hellerman, S. (1968). Correction. *Monthly Weather Review*, **96** (1), 62–74.

Her Majesty's Stationery Office (1972). *The Meteorological Glossary*. Her Majesty's Stationery Office, London, 319 pp.

Hess, W. N. (ed.) (1974). *Weather and Climate Modification*. Wiley, New York, 842 pp.

Hesstvedt, E. (1964). On the water vapour content in the high atmosphere. *Geofysiske Publikasjoner* (Geophysica Norvegica), **25**, No. 3, 67–87.

Hester, N. E., Stephens, E. R. & Taylor, O. C. (1975). Fluorocarbon air pollutants, measurements in lower stratosphere. *Environmental Science and Technology*, **9**, 875–6.

Hitchcock, D. R. (1975). Dimethyl sulfide emissions to the global atmosphere. *Chemosphere*, **3**, 137–8.

Hobbs, P. V., Radke, L. F. & Shumway, S. E. (1970). Cloud condensation nuclei from industrial sources and their apparent influence on precipitation in Washington State. *Journal of the Atmospheric Sciences*, **27**, 81–9.

Hoffert, M. I. (1974). Global distributions of atmospheric carbon dioxide in the fossil-fuel era. A projection. *Atmospheric Environment*, **8**, 1225–49.

Hoffman, E. J. & Duce, R. A. (1974). The organic carbon content of marine aerosols collected on Bermuda. *Journal of Geophysical Research*, **79**, 4474–7.

Hoffman, G. (1963). Die höchsten und die tiefsten Temperaturen auf der Erde. *Umschau, Frankfurt am Main*, **63** (1), 16–18.

Hollin, J. T. (1965). Wilson's theory of ice ages. *Nature*, **208**, 12–16.

Hollin, J. T. (1969). The Antarctic ice sheet and the Quaternary history of Antarctica. In *Paleoecology of Africa and of the Surrounding Islands and Antarctica*, ed. E. M. Van Zinderen Bakker, vol. 5, Balkema, Cape Town, pp. 109–38.

Hollin, J. T. (1972). Interglacial climates and Antarctic ice surges. *Quaternary Research*, **2**, 401–8.

Hollister, C. D., Craddock, C. *et al.* (1974). Deep drilling in the southeast Pacific basin. *Geotimes*, **19**, 16–19.

Holton, J. R. & Lindzen, R. S. (1972). An updated theory for the quasi-biennial cycle of the tropical stratosphere. *Journal of the Atmospheric Sciences*, **29**, 1076–80.

Holzman, B. G. (1971). The La Porte precipitation fallacy, *Science*. **172**, p. 847 only.

396

Hope, G. S. & Peterson, J. A. (1975). Glaciation and vegetation in the high New Guinea mountains. *Bulletin Royal Society of New Zealand*, No. 13, pp. 155–62.

Hope, G. S., Peterson, J. A., Radok, U. & Allison, I. (eds.) (1976). *The Equatorial Glaciers of New Guinea*. Balkema, Rotterdam. 244 pp.

Hoskins, W. G. (1964). Harvest fluctuations and English economic history, 1480–1619. *Agricultural History Review*, 12, 28–46.

Hoskins, W. G. (1968). Harvest fluctuations and English economic history 1620–1759. *Agricultural History Review*, 16, 15–31.

Hosler, C. L. (1974). Overt weather modification. *Reviews of Geophysics and Space Physics*, 12, 523–7.

Hsü, K. J. (1972). When the Mediterranean dried up. *Scientific American*, 227, No. 6, pp. 26–36.

Huff, F. A. & Changnon, S. A. Jr. (1972). Climatological assessment of urban effects on precipitation at St Louis. *Journal of Applied Meteorology*, 11, 823–42.

Huff, F. A., Changnon, S. A. Jr., & Jones, D. M. A. (1975). Precipitation increases in the low hills of Southern Illinois. Part I. Climatic and network studies. *Monthly Weather Review*, 102, 823–36.

Huff, F. A. & Semonin, R. G. (1975). Potential of precipitation modification in moderate to severe droughts. *Journal of Applied Meteorology*, 14, 974–9.

Hughes, T. (1973). Is the West Antarctic ice sheet disintegrating? *Journal of Geophysical Research*, 78, 7884–910.

Hughes, T. (1975). The West Antarctic ice sheet: instability, disintegration and initiation of ice ages. *Reviews of Geophysics and Space Physics*, 13, 502–26.

Hunkins, K., Be, A. W. H., Opdyke, N. D. & Mathieu, G. (1971). The late Cenozoic history of the Arctic Ocean. In *Late Cenozoic Glacial Ages*, ed. K. K. Turekian. Yale University Press, New Haven, pp. 215–38.

Hunt, B. G. (1969). Experiments with a stratospheric general circulation model. Part III. Large scale diffusion of ozone including photochemistry. *Monthly Weather Review*, 97, 287–306.

Hunt, B. G. (1976a). Experiments with a stratospheric general circulation model. Part IV. Inclusion of the hydrologic cycle. *Monthly Weather Review*, 104, 333–50.

Hunt, B. G. (1976b). A simulation of the possible consequences of a volcanic eruption on the general circulation of the atmosphere. *Monthly Weather Review*, 105, 247–60.

Hunt, B. G. & Manabe, S. (1968). Experiments with a stratospheric general circulation model. Part II. Large-scale diffusion of tracers in the atmosphere. *Monthly Weather Review*, 96, 503–39.

Huntington, E. (1915). *Civilization and climate*. Yale University Press, New Haven, 333 pp.

Huzayyin, S. (1956). Changes in climate, vegetation, and human adjust-

References

ment in the Saharo–Arabian Belt, with special reference to Africa. In *Man's Role in Changing the Face of the Earth*, ed. W. L. Thomas. University of Chicago Press. pp. 304–23.

Imbrie, J. & Kipp, N. G. (1971). A new micropaleontological method for quantitative paleoclimatology: application to a Late Pleistocene Caribbean core. In *The Late Cenozoic Glacial Ages*, ed. K. K. Turekian. Yale University Press, New Haven, pp. 71–181.

Imbrie, J., Van Donk, J. & Kipp, N. G. (1973). Paleoclimatic investigation of a late Pleistocene Caribbean deep-sea core: comparison of isotopic and faunal methods. *Quaternary Research*, **3**, 10–38.

International Federation of Institutes for Advanced Study (1974). *Statement by IFIAS on Climate Change and World Food Production, adopted by IFIAS Board of Trustees*, 3 October 1974. International Federation of Institutes for Advanced Study, Stockholm, 6 pp.

International Federation of Institutes for Advanced Study (1975). The policy implications of food and climate interactions. *Summary of an IFIAS Project Workshop held at Berlin*, 5–7 February, 1975, IFIAS, Stockholm, 36 pp. plus appendices.

Ives, J. D. (1956). Till patterns in central Labrador. *Canadian Geographer*, **8**, 25–33.

Ives, J. D. (1957). Glaciation of the Torngat Mountains, Northern Labrador. *Arctic*, **10**, 67–87.

Ives, J. D., Andrews, J. T. & Barry, R. G. (1975). Growth and decay of the Laurentide ice sheet and comparisons with Fenno-Scandinavia. *Naturwissenschaften*, **62**, 118–25.

Jaenicke, R. (1973). Monitoring of aerosols by measurement of single parameters. In *Stockholm Tropospheric Aerosol Seminar, Report AP-14*, Institute of Meteorology, University of Stockholm.

Jagannathan, P., Arlery, R., ten Kate, H. & Zavaring, M. V. (1967). A note on climatological normals (WMO-No. 208, Technical Paper 108), *World Meteorological Organization*, Note 84, Geneva, 19 pp.

Jeffreys, H. (1926). On the dynamics of geostrophic winds. *Quarterly Journal of Royal Meteorological Society*, **52**, 85–104.

Jenne, R. L. (1975). Data sets for meteorological research. *NCAR Technical Note*, National Center for Atmospheric Research, Boulder, Colorado (in press).

Jennings, J. N., Noakes, L. C. & Burton, G. M. (1964). Notes on the Lake George and Lake Bathurst excursion. In *Geological Excursions, Canberra District*. Commonwealth of Australia, Department of National Development, Canberra, pp. 24–34.

Jenssen, D. (1977*a*). Elevation and climatic changes from gas content and stable isotope measurements. In *Isotopic and Temperature Profiles in Ice Sheets*, ed. G. deQ. Robin. Cambridge University Press, Cambridge (in press).

Jenssen, D. (1977*b*). A three-dimensional polar ice sheet model. Submitted to *Journal of Glaciology*.

Jenssen, D. & Radok, U. (1961). Transient temperature distributions in ice caps and ice shelves. *IASH Publication* No. 55, pp. 112–22.

Jenssen, D. & Radok, U. (1963). Heat conduction in thinning ice sheets. *Journal of Glaciology*, **4**, 387–98.

J. G. (1973). Climatic changes in China over the past 5,000 years. *Nature*, **246**, 375–6.

Johnsen, S. J., Dansgaard, W., Clausen, H. B. & Langway, C. C. Jr (1970). Climatic oscillations 1200–2000 AD. *Nature*, **227**, 482–3.

Johnsen, S. J., Dansgaard, W., Clausen, H. B. & Langway, C. C. Jr. (1972). Oxygen isotope profiles through the Antarctic and Greenland ice sheets. *Nature*, **235**, 429–34.

Johnston, H. S. (1974). Catalytic reduction of stratospheric ozone by nitrogen oxides. *Advances Environmental Science and Technology*, **4**, 263–380.

Joint Organizing Committee for GARP (1975). *The Physical Basis of Climate and Climate Modelling.* GARP Publications Series, No. 16, ICSU-World Meteorological Organization, 265 pp.

Jones, E. L. (1964). *Seasons and Prices. The Role of the Weather in English Agricultural History.* Allen & Unwin, London, 193 pp.

Jones, R. (1975). The Neolithic, Palaeolithic and the Hunting Gardeners: man and land in the Antipodes. In *Quaternary Studies*, eds. R. P. Suggate & M. M. Cresswell. The Royal Society of New Zealand, Wellington, pp. 21–34.

Judson, S. (1968). Erosion of the land – or what's happening to our continents? *American Scientist*, **56**, 356–74.

Junge, C. E. (1972 a). Our knowledge of the physico-chemistry of aerosols in the undisturbed marine environment. *Journal of Geophysical Research*, **77**, 5183–200.

Junge, C. E. (1972 b). The cycle of atmospheric gases – natural and man made. *Quarterly Journal of Royal Meteorological Society*, **98**, 711–29.

Junge, C. E. (1975). The possible influences of aerosols on general circulation and climate and possible approaches for modelling. In *The Physical Basis of Climate and Climate Modelling*, GARP Publications Series, No. 16, ICSU-World Meteorological Organization, pp. 244–51.

Junge, C. E. & Jaenicke, R. (1971). New results in background aerosol studies from the Atlantic expedition of the RV Meteor, Spring 1969. *Aerosol Science*, **2**, 305–14.

Kac, A. L. (1964). The two year periodicity in the equatorial stratosphere and the general circulation of the atmosphere. *Meteorologiya i Gidrologiya*, Leningrad, No. 6, pp. 3–10 (in Russian).

Kachelhoffer, S. J. (1973). Effects of the annual migration of the Antarctic pack-ice border on the cyclonic storm activity over the Southern Ocean. M.Sc. Thesis, Meteorology Department, University of Wisconsin, 49 pp.

Kalnicky, R. A. (1974). Climatic change since 1950. *Annals, Association of American Geographers*, **64**, 100–12.

References

Kaplan, L. D. (1960). The influence of carbon dioxide variations in the atmospheric heat balance. *Tellus*, **12**, 204–8.

Karelsky, S. F. (1956). Classification of the surface circulation in the Australian region. *Bureau of Meteorology, Australia, Meteorological Study* No. 8, 36 pp.

Keany, J. & Kennett, J. P. (1972). Pliocene–early Pleistocene paleo-climatic history recorded in Antarctic–Subantarctic deep-sea cores. *Deep-Sea Research*, **19**, 529–48.

Keeling, C. D. (1973 a). The carbon dioxide cycle: reservoir models to depict the exchange of atmospheric carbon dioxide with the oceans and land plants. In *Chemistry of the Lower Atmosphere*, ed. S. I. Rasool. Plenum Press, New York, pp. 251–329.

Keeling, C. D. (1973 b). Industrial production of carbon dioxide from fossil fuels and limestone. *Tellus*, **25**, 174–98.

Keeling, C. D., Ekdahl, C. A., Guenther, P. R., Waterman, L. S. & Chin, J. F. S. (1972). Atmospheric carbon dioxide variations at Mauna Loa Observatory Hawaii (unpublished mimeo), 61 pp. Scripps Institute of Oceanography, University of California.

Kellogg, W. W. (1973). Climatic feedback mechanism involving the polar regions. In *Climate of the Arctic*, 24th Alaska Science Conference. pp. 111–16.

Kellogg, W. W., Coakley, J. A. Jr. & Grams, G. W. (1975). Effect of anthropogenic aerosols on the global climate. In *Proceedings, WMO/IAMAP Symposium on Long-Term Climatic Fluctuations*, Norwich (WMO-No. 421). World Meteorological Organization, Geneva, pp. 323–30.

Kellogg, W. W. & Schneider, S. H. (1974). Climate stabilization: for better or for worse? *Science*, **186**, 1163–72.

Kemp, E. M. (1975). Palynology of Leg 28 drill sites, deep sea drilling project. In *Initial Reports of the Deep Sea Drilling Project*, vol. 28. Washington, DC, US Government Printing Office, pp. 599–624.

Kemp, E. M., Frakes, L. A. & Hayes, D. E. (1975). Paleoclimatic significance of pelagic biogenic facies. In *Initial Reports of the Deep Sea Drilling Project*, vol. 28. Washington, DC, US Government Printing Office, pp. 909–18.

Kemp, R. A. & Armstrong, J. G. (1972). An investigation of temperature trends at Sydney (Observatory Hill) over all years of record. *Bureau of Meteorology Australia, Working Paper* 154, 19 pp.

Kennett, J. P. & Huddlestun, P. (1972 a). Late Pleistocene paleo-climatology, foraminiferal biostratigraphy and tephrochronology, western Gulf of Mexico. *Quaternary Research*, **2**, 38–69.

Kennett, J. P. & Huddlestun, P. (1972 b). Abrupt climatic change at 90,000 years BP: faunal evidence from Gulf of Mexico cores. *Quaternary Research*, **2**, 384–95.

Kennett, J. P. & Thunell, R. C. (1975). Global increase in Quaternary explosive volcanism. *Science*, **187**, 497–503.

References

Kennett, J. P. & Vella, P. (1975). Late Cenozoic planktonic foraminifera and paleooceanography at DSDP Site 284 in the cool, subtropical South Pacific. In *Initial Reports of the Deep Sea Drilling Project*, vol. 29, Washington, DC, US Government Printing Office, pp. 769–99.

Kennett, J. P. & Watkins, N. D. (1970). Geomagnetic polarity changes, volcanic maxima, and faunal extinctions in the South Pacific. *Nature*, **227**, 930–4.

Kershaw, A. P. (1974). A long continuous pollen sequence from northeastern Australia. *Nature*, **251**, 222–3.

Kershaw, A. P. (1975 a). Stratigraphy and pollen analysis of Bromfield Swamp, Northeastern Queensland, Australia. *New Phytologist*, **75**, 173–91.

Kershaw, A. P. (1975 b). Late Quaternary vegetation and climate in northeastern Australia. *Bulletin Royal Society of New Zealand* No. 13, 181–7.

Kershaw, A. P. (1976). A late-Pleistocene and Holocene pollen diagram from Lynch's crater Northeastern Queensland, Australia. *New Phytologist*, **77** (2), 469–98.

Kevan, S. (1971). Psychology and climate: a brief review of some subjective views. *Climatological Bulletin*, No. 10. Department of Geography, McGill University, pp. 24–30.

Kidson, E. (1925). Some periods in Australian weather. *Bureau of Meteorology, Bulletin* No. 17, Paper 1, Melbourne, 33 pp.

Kidson, J. W. (1975 a). Eigenvector analysis of monthly mean surface data. *Monthly Weather Review*, **103** (3), 177–86.

Kidson, J. W. (1975 b). Tropical Eigenvector Analysis and the Southern Oscillation. *Monthly Weather Review*, **103** (3), 187–96.

Kinzl, H. (1932). Die grössten nacheiszeitlichen Gletschervorstösse in den schweizer Alpen und in der Mont-Blanc Gruppe. *Zeitschrift für Gletscherkunde und Glazialgeologie*, **20**, 269–397.

Kirch, R. (1966). Temperaturverhältnisse in der Arktis während der letzten 50 Jahre. *Meteorologische Abhandlungen*, **69** (3), 102 pp.

Klein, C. (1961). On the fluctuations of the level of the Dead Sea since the beginning of the 19th century. *Hydrological Paper No. 7*. Hydrological Service, Israel.

Knapp, W. W. (1967). Formation, persistence and disappearance of open water channels related to the meteorological conditions along the coast of the Antarctic Continent (WMO-No. 211, TP. 111). *World Meteorological Organization, Technical Note* No. 87, pp. 89–104.

Kondratyev, K. Ya (1969). *Radiation in the Atmosphere*. Academic Press, New York and London, 912 pp.

Kondratyev, K. Ya (1972). *Radiation Processes in the Atmosphere* (WMO-No. 309). World Meteorological Organization, Geneva, 214 pp.

Kondratyev, K. Ya (1973). *The Complete Atmospheric Energetics Experiment*. GARP Publications Series, No. 12. ICSU-World Meteorological Organization, Geneva, 44 pp.

Kondratyev, K. Ya & Nikolsky, G. A. (1970). Solar radiation and solar

References

activity. *Quarterly Journal of the Royal Meteorological Society*, **96**, 509–62.

Kraus, E. B. (1954). Secular changes in the rainfall regime of southeast Australia. *Quarterly Journal of the Royal Meteorological Society*, **80**, 591–601.

Kraus, E. B. (1955 a). Secular changes of tropical rainfall regimes. *Quarterly Journal of the Royal Meteorological Society*, **81**, 198–210.

Kraus, E. B. (1955 b). Secular changes of east-coast rainfall regimes. *Quarterly Journal of the Royal Meteorological Society*, **81**, 430–9.

Kraus, E. B. (1956). Secular changes of the standing circulation. *Quarterly Journal of the Royal Meteorological Society*, **82**, 289–300.

Kraus, E. B. (1958). Recent climatic changes. *Nature*, **181**, 666–8.

Kraus, E. B. (1963). Recent changes of east-coast rainfall regimes. *Quarterly Journal of the Royal Meteorological Society*, **89**, 145–6.

Kraus, E. B. (1972). *Atmosphere–Ocean Interaction*. Clarendon Press, Oxford, 275 pp.

Krey, P. W. & Lagomarsino, R. J. (1975). Stratospheric concentrations of SF_6 and CCl_3F. *Environmental Quarterly*, July 1975, HASL, US Energy Research and Development Administration.

Krinsley, D. B. (1970). *A Geomorphological and Paleoclimatological Study of the Playas of Iran*. US Air Force, Cambridge Research Laboratories, Massachusetts. Contract No. PRO CP 70–800. Final report, 2 volumes.

Krinsley, D. B., Woo, C. C. & Stoertz, G. E. (1968). Geological characteristics of seven Australian playas. In *Playa Surface Morphology*, ed. J. T. Neal. US Air Force, Cambridge Research Laboratories, Massachusetts, pp. 59–103.

Krishnamurti, T. N., Kanimitsu, M., Koss, W. J. & Lee, J. D. (1973). Tropical east–west circulations during the northern winter. *Journal of the Atmospheric Sciences*, **30**, 780–7.

Kruss, D. (1976). A flexible computer model of glaciers. Unpublished B.Sc. (Honours) Thesis, Meteorology Department, Melbourne University, 63 pp.

Ku, T. L., Kimmel, M. A., Easton, W. H. & O'Neil, T. J. (1974). Eustatic sea level 120,000 years ago on Oahu, Hawaii. *Science*, **183**, 959–62.

Kuhn, T. S. (1970). *The Structure of Scientific Revolutions*, 2nd edn. University of Chicago Press, Chicago, 210 pp.

Kukla, G. J. (1975). Milankovitch and climate: the missing link? *Nature*, **253**, 600–8.

Kukla, G. J. & Kukla, H. J. (1974). Increased surface albedo in the northern hemisphere. *Science*, **183**, 709–14.

Kulkarni, R. N. (1973). Ozone trend and haze scattering. *Quarterly Journal of the Royal Meteorological Society*, **99**, 480–9.

Kutzbach, J. E. & Bryson, R. A. (1974). Variance spectrum of Holocene climatic fluctuations in the North Atlantic sector. *Journal of the Atmospheric Sciences*, **31**, 1958–63.

La Marche, V. C., Jr (1973). Holocene climatic variation inferred from treeline fluctuations in the White Mountains, California. *Quaternary Research*, **3**, 632–60.

La Marche, V. C., Jr (1975). Potential of tree rings for reconstruction of past climatic variations in the southern hemisphere. In *Proceedings of the WMO/IAMAP Symposium on Long-Term Climatic Fluctuations*, Norwich. (WMO-No. 421). World Meteorological Organization, Geneva, pp. 21–30.

Lamb, H. H. (1958). The occurrence of very high surface temperatures. *Meteorological Magazine, London*, **87**, 39–43.

Lamb, H. H. (1959). Our changing climate, past and present. *Weather*, **14**, 299–318.

Lamb, H. H. (1963). What can we find out about the trend of our climate? *Weather*, **18**, 194–216.

Lamb, H. H. (1965 a). Frequency of weather types. *Weather*, **20**, 9–12.

Lamb, H. H. (1965 b). The early Medieval warm epoch and its sequel. *Palaeogeography, Palaeoclimatology, Palaeoecology*, **1**, 13–37.

Lamb, H. H. (1966 a). *The changing climate: selected papers*. Methuen, London, 236 pp.

Lamb, H. H. (1966 b). Climate in the 1960's. Changes in the world's wind circulation reflected in prevailing temperatures, rainfall patterns and the levels of African lakes. *Geographical Journal*, **132**, 183–212.

Lamb, H. H. (1967 a). Review of Rhys Carpenter: Discontinuity in Greek Civilization. *Antiquity*, **41**, 233–4.

Lamb, H. H. (1967 b). On climatic variations affecting the Far South. In *Polar Meteorology* (WMO-No. 211, Technical Paper 111). *World Meteorological Organization, Technical Note* No. 87, Geneva, pp. 428–53.

Lamb, H. H. (1968). Climatic changes during the course of early Greek history. *Antiquity*, **42**, 231–3.

Lamb, H. H. (1969). The new look of climatology. *Nature*, **223**, 1209–15.

Lamb, H. H. (1970). Volcanic dust in the atmosphere, with chronology and assessment of its meteorological significance. *Philosophical Transactions of the Royal Society*, **A266**, 425–533.

Lamb, H. H. (1971). Volcanic activity and climate. *Palaeogeography, Palaeoclimatology, Palaeoecology*, **10**, 203–30.

Lamb, H. H. (1972 a). *Climate: Present, Past and Future*, vol. 1. *Fundamentals and Climate Now*. Methuen, London, 613 pp.

Lamb, H. H. (1972 b). British Isles weather types and a register of the daily sequence of circulation patterns, 1861–1971. *Geophysical Memoirs*, vol. 16 (No. 116). Her Majesty's Stationery Office, London, 85 pp.

Lamb, H. H. (1973). Whither climate now? *Nature*, **244**, pp. 395–7.

Lamb, H. H. (1974 a). Is the earth's climate changing? *Ecologist*, **4**, 10–15.

Lamb, H. H. (1974 b). The current trend of world climate – a report on the early 1970s and a perspective, University of East Anglia, *Climatic Research Unit Research Report*, No. 3, Norwich, 27 pp.

Lamb, H. H. & Johnson, A. I. (1966). Secular variations of the atmospheric circulation since 1750. *Geophysical Memoirs*. vol. 14, No. 110. Her Majesty's Stationery Office, London, 125 pp.

Lamb, H. H., Lewis, R. P. W. & Woodroffe, A. (1966). Atmospheric circulation and the main climatic variables between 8000 and 0 BC: meteorological evidence. In *World Climate from 8000 to 0 BC*. Royal Meteorological Society, London, pp. 174–217.

Lamb, H. H. & Woodroffe, A. (1970). Atmospheric circulation during the last ice age. *Quaternary Research*, **1**, 29–58.

Landsberg, H. E. (1953). The origin of the atmosphere. *Scientific American*, offprint No. 824. 8 pp.

Landsberg, H. E. (1970 a). Man-made climatic changes. *Science*, **170**, 1265–74.

Landsberg, H. E. (1970 b). Climates and urban planning. In *Urban Climates* (WMO-No. 254). *World Meteorological Organization, Technical Note* No. 108, Geneva, pp. 364–72.

Landsberg, H. E. (1975). Drought, a recurrent element of climate (WMO-No. 403). *World Meteorological Organization Special Environment Report* No. 5, Geneva, pp. 41–90.

Landsberg, H. E. & Albert, J. M. (1974). The summer of 1816 and volcanism. *Weatherwise*, **27**, 63–6.

Landsberg, H. E., Mitchell, J. M., Jr, Crutcher, H. L. & Quinlan, F. T. (1963). Surface signs of the biennial atmospheric pulse. *Monthly Weather Review*, **91**, 549–56.

Langbein, W. B. (1961). Salinity and hydrology of closed lakes. *United States Geological Survey Professional Paper* 412, 20 pp.

Lawrence, E. N. (1971). Urban climate and day of the week. *Atmospheric Environment*, **5**, 935–48.

Lazarus, A. L. & Gandrud, B. W. (1974). Stratospheric sulphate aerosol. *Journal Geophysical Research*, **79**, 3424–31.

Lee, C. H. (1911). Precipitation and altitude in the Sierra. *Monthly Weather Review*, **39**, 1092–9.

Lee, R. B. (1972). !Kung spatial organization: an ecological and historical perspective. *Human Ecology*, **1**, 125–47.

Leetmaa, A. (1972). The response of the Somali current to the southwest monsoon of 1970. *Deep-Sea Research*, **19**, 319–25.

LeMasurier, W. E. (1972). Volcanic record of Cenozoic glacial history of Marie Byrd Land. In *Antarctic Geology and Geophysics*, ed. R. J. Adie. Oslo (Universitets-forlaget), pp. 251–60.

Le Roy Ladurie, E. L. (1959). History and climate. In *Economy and Society in Early Modern Europe: Essays from Annales*, ed. P. Burke. Routledge and Kegan Paul, London, pp. 134–69.

Le Roy Ladurie, E. L. (1970 and 1971). *Times of Feast, Times of Famine: a History of Climate since the Year 1000*. George Allen & Unwin (London, 1971) and Doubleday (New York, 1970), 428 pp.

Libby, L. M. & Pandolfi, L. J. (1974). Temperature dependence of isotope ratios in tree rings. *Proceedings of the National Academy of Sciences, USA*, **71**, 2482–6.

Lighthill, M. J. (1969). Dynamic response of the Indian Ocean to onset of the southwest monsoon. *Philosophical Transactions Series A, Royal Society of London,* **265**, 45–92.

Lliboutry, L. (1964). *Traité de glaciologie,* vol. 1. Masson, Paris, 427 pp.

Lloyd, J. W. (1973). Climatic variations in north central Chile from 1866 to 1971. *Journal of Hydrology,* **19**, 53–70.

Lockwood, G. A. (1975). Planetary brightness changes: evidence for solar variability. *Science,* **190**, 560–2.

Loewe, F. (1948). Variability of annual rainfall in Australia. Bureau of Meteorology *Bulletin* No. 39, Melbourne, pp. 1–13.

Loewe, F. (1974). Über Änderungen des Luftdrucks Während der Eiszeit. *Zeitschrift für Gletscherkunde und Glazialgeologie,* **9**, No. 1–2, 229–30.

London, J. & Kelley, J. J. (1974). Global trends in total atmospheric ozone. *Science,* **184**, 987–9.

Longley, R. W. (1974). Spatial variation of precipitation over the Canadian Prairies. *Monthly Weather Review,* **102**, 307–12.

Lorenz, E. N. (1955). Available potential energy and the maintenance of the general circulation. *Tellus,* **7**, 157–67.

Lorenz, E. N. (1968). Climatic determinism. In, Causes of Climatic Change, *Meteorological Monographs,* **8**, No. 30, pp.1–3.

Lorenz, E. N. (1969). The predictability of a flow which possesses many scales of motion. *Tellus,* **21**, 289–307.

Lorenz, E. N. (1973). On the existence of extended range predictability. *Journal of Applied Meteorology,* **12**, 543–6.

Lorenz, E. N. (1975). Climatic predictability. In *The Physical Basis of Climate and Climate Modelling,* GARP Publications Series, No. 16, World Meteorological Organization, Geneva, pp. 132–6.

Lovelock, J. E. (1974). Atmospheric halocarbons and stratospheric ozone. *Nature,* **252**, 292–3.

Lovelock, J. E. (1975). Natural halocarbons in the air and in the sea. *Nature,* **256**, 193–4.

Lumb, F. E. (1965). Cycles and trends of rainfall over East Africa. In *Proceedings 3rd Specialist Meeting on Applied Meteorology,* Mguga, Kenya, East African Common Services Organization, Mimeo.

Lysgaard, L. (1963). On the climatic variation. In *Changes of Climate, Arid Zone Research,* vol. 20. UNESCO, Paris, pp. 151–9.

McConnell, J. C. & McElroy, M. B. (1973). Odd nitrogen in the atmosphere. *Journal of the Atmospheric Sciences,* **30**, 1465–80.

McCormick, R. A. (1958). An estimate of the minimum possible surface temperature at the South Pole. *Monthly Weather Review,* **86**, 1–5.

MacCracken, M. C. & Potter, G. L. (1975*a*). Comparative climatic impact of increased stratospheric aerosol loading and decreased solar constant in a zonal climate model. Lawrence Livermore Laboratory. Preprint UCRL-76132, Livermore, California. 8 pp.

MacCracken, M. C. & Potter, G. L. (1975*b*). Comparative impact of increased stratospheric aerosol loading and decreased solar constant in a zonal climate model. In *Proceedings of the WMO/IAMAP Symposium*

References

on *Long-Term Climatic Fluctuations*, Norwich (WMO-No. 421), World Meteorological Organization, Geneva, pp. 415–20.

McElroy, M. B., Wofsy, S. C., Penner, J. E. & McConnell, J. C. (1974). Atmospheric ozone: possible impact of stratospheric aviation. *Journal of the Atmospheric Sciences*, **31**, 287–303.

Machta, L. (1972). The role of the oceans and biosphere in the carbon dioxide cycle. In *The Changing Chemistry of the Oceans*. Proceedings of the Twentieth Nobel Symposium, eds. D. Dyrssen & D. Jagner. Wiley Interscience, New York, and Almqvist & Wiksell, Stockholm, pp. 121–45.

McInnes, B. (1976). Numerical modelling of self-surging glaciers. Unpublished M.Sc. Thesis. Meteorology Department, Melbourne University, 165 pp.

McIntyre, A. (1967). Coccoliths as paleoclimatic indicators of Pleistocene glaciation. *Science*, **158**, 1314–17.

McIntyre, A., Ruddiman, W. F. & Jantzen, R. (1972). Southward penetration of the North Atlantic polar front: faunal and floral evidence of the large-scale surface movements over the last 225,000 years. *Deep-Sea Research*, **19**, 61–77.

Mackintosh, N. A. (1946). The Antarctic convergence and the distribution of surface temperatures in Antarctic waters. *Discovery Reports*, **23**, 177–212.

Mackintosh, N. A. & Herdman, H. F. P. (1940). Distribution of pack ice in the Southern Ocean. *Discovery Reports*, **19**, 285–96.

McMullen, T. B., Faoro, R. B. & Morgan, G. B. (1970). Profile of pollutant fractions in non-urban suspended particulate matter. *Journal of Air Pollution Control Association*, **20**, 369–72.

Macphail, M. K. & Peterson, J. A. (1975). New deglaciation dates from Tasmania. *Search*, **6**, 127–30.

McQuigg, J. D. (1974). World without hunger? The weather factor grows more critical. *Economic Impact*, No. 8, pp. 34–8.

Maksimov, I. V. (1952). On the eighty-year cycle of terrestrial climatic fluctuations. *Doklady Akad. Nauk SSSR*, **86**, 917–20. (Translated Defence Research Board, Canada, T81R, 1953, by E. R. Hope.)

Manabe, S. (1975). The use of comprehensive general circulation modelling for studies of the climate and climate variation. In *The Physical Basis of Climate and Climate Modelling*, GARP Publications Series, No. 16, World Meteorological Organization, Geneva, pp. 148–62.

Manabe, S. & Bryan, K. (1969). Climate calculations with a combined ocean-atmosphere model. *Journal of the Atmospheric Sciences*, **26**, 786–9.

Manabe, S. & Hunt, B. G. (1968). Experiments with a stratospheric general circulation model. *Monthly Weather Review*, **96**, 477–539.

Manabe, S., Smagorinsky, J., Holloway, J. L., Jr & Stone, H. M. (1970). Simulated climatology of a general circulation model with a hydrologic cycle: III. Effects of increased horizontal computational resolution. *Monthly Weather Review*, **98**, 175–212.

Manabe, S. & Strickler, R. F. (1964). Thermal equilibrium of the atmosphere with a convective adjustment. *Journal of the Atmospheric Sciences*, **21**, 361–85.

Manabe, S. & Wetherald, R. T. (1967). Thermal equilibrium of the atmosphere with a given distribution of relative humidity. *Journal of the Atmospheric Sciences*, **24**, 241–59.

Manabe, S. & Wetherald, R. T. (1975). The effects of doubling the CO_2 concentration on the climate of a general circulation model. *Journal of the Atmospheric Sciences*, **32**, 3–15.

Manley, G. (1955). On the occurrence of ice domes and permanently snow-covered summits. *Journal of Glaciology*, **2**, 453–6.

Manley, G. (1957). Climatic fluctuations and fuel requirements. *Scottish Geographical Magazine*, **73**, 19–28.

Manley, G. (1958). The revival of climatic determinism. *Geographical Review*, **48**, 98–105.

Manley, G. (1961). Meteorological factors in the Great Glacier Advance (1690–1720). *International Association of Scientific Hydrology Publication* No. 54, pp. 388–91. (Published in Belgium, 1961.)

Manley, G. (1962). *Climate and the British Scene*. Collins, London, 314 pp.

Manley, G. (1974). Central England temperatures: monthly means 1659 to 1973. *Quarterly Journal of the Royal Meteorological Society*, **100**, 389–405.

Margolis, S. V. & Kennett, J. P. (1971). Antarctic glaciation during the Tertiary recorded in sub-Antarctic deep-sea cores. *Science*, **170**, 1085–7.

Margules, M. (1903). Uber die Energie der Sturme, Jahrb Zentralanst, Vienna, 1–26. English translation: C. Abbe, 1910: *The Mechanics of The Earth's Atmosphere*, 3rd Collection, Washington, Smithsonian Institution, pp. 533–95.

Martin, H. A. (1973). Palynology and historical ecology of some cave excavations in the Australian Nullarbor. *Australian Journal of Botany*, **21**, 283–316.

Mason, B. J. (1966). The role of meteorology in the national economy. *Weather*, **21**, pp. 382–93.

Matthes, F. E. (1942). Glaciers. Chapter V, physics of the earth. *Hydrology*, ed. O. E. Meinzer, vol. 9. Dover, New York, pp. 149–219.

Matthews, R. K. (1973). Relative elevation of late Pleistocene high sea level stands: Barbados uplift rates and their implications. *Quaternary Research*, **3**, 147–53.

Maunder, W. J. (1970). *The Value of the Weather*. Methuen, London, 388 pp.

Maunder, W. J. (1971). The value and use of weather information. In *Transactions of the Electric Supply Authority Engineers' Institute of New Zealand Inc.*, vol. **41**, pp. 10–20.

Maunder, W. J. (1973). Weekly weather and economic activities on a national scale: an example using United States retail trade data. *Weather*, **28**, 2–18.

References

Maunder, W. J. (1974). The prediction of monthly dairy production in New Zealand through the use of weighted indices of water deficit. *New Zealand Meteorological Service Technical Note*, No. 227, 34 pp.

Maunder, W. J. (1975). The value of the weather: national and international implications. *Informationen und Beitrage zur Klimaforschung*. Geographisches Institut der Universitaet Bern, Switzerland, Autumn.

Maunder, W. J., Johnson, S. R. & McQuigg, J. D. (1971). Study of the effect of weather on road construction: a simulation model. *Monthly Weather Review*, **99**, 939–45.

Maurer, H. (1938). Die Regenveränderlichkeit in Deutsch-Ostafrika. *Annalen der Hydrographie*, p. 220.

Maury, M. F. (1855). *The Physical Geography of the Sea*. The quotation is from the 1963 edition, ed. J. Leighly. Belknap, Cambridge, Massachusetts, pp. 394–5.

Meadows, D. H., Meadows, D. L., Randers, J. & Behrens, W. W., III (1971). *The Limits to Growth*. Potomac Associates, London, 206 pp.

Meinardus, W. (1923). Meteorologische Ergebnisse der Deutschen Südpolar Expedition 1901–23. *Deutsche Südpolar Expedition III*, Meteorologie vol. 1 (1), Berlin.

Mercer, J. H. (1972). Cainozoic temperature trends in the southern hemisphere: Antarctic and Andean glacial evidence. *Palaeoecology of Africa and of the Surrounding Islands and Antarctica* **8**, 86–114.

Mercer, J. H., Fleck, R. J., Mankinen, E. A. & Sander, W. (1975). Southern Patagonia: Glacial events between 4 m.y. and 1 m.y. ago. *Royal Society of New Zealand, Bulletin* No. 13, *Quaternary Studies*, ed. R. P. Suggate & M. M. Cresswell, pp. 223–30.

Mesarovic, M. & Pestel, E. (1974). *Mankind at the Turning Point*. Hutchinson, London, 210 pp.

Meszaros, A. & Vissy, K. (1974). Concentration, size distribution and chemical nature of atmospheric aerosol particles in remote oceanic areas. *Aerosol Science*, **5**, 101–9.

Milankovitch, M. (1938). Die chronologie des Pleistozäns. *Bulletin Academy of Natural Sciences and Mathematics, Belgrade*, **4**, p. 49.

Miles, M. K. & Folland, C. K. (1974). Changes in the latitude of the climatic zones of the northern hemisphere. *Nature*, 252, p. 616 only.

Miller, D. B. & Feddes, R. G. (1971). *Global Atlas of Relative Cloud Cover 1967–1970*. National Environmental Satellite Service and Air Weather Service. Available from Director, National Climatic Center, Ashville, NC, 237 pp.

Miller, G. H. (1973). Late Quaternary glacial and climatic history of northern Cumberland Peninsula, Baffin Island, NWT, Canada. *Quaternary Research*, **3**, 561–83.

Miller, J. M. (1975) (ed.). Geophysical monitoring for climatic change. *NOAA Environmental Research Laboratories Summary Report*, No. 3, 1974, 107 pp.

408

Mitchell, J. M., Jr (1953). On the causes of instrumentally observed secular temperature trends. *Journal of Meteorology*, **10**, 244–61.

Mitchell, J. M., Jr (1961). Recent secular changes of global temperature. *Annals New York Academy of Science*, **95**, 235–50.

Mitchell, J. M., Jr (1963). On the world-wide pattern of secular temperature change. In *Changes of Climate, Arid Zone Research*, vol. 20. UNESCO, Paris, pp. 161–81.

Mitchell, J. M., Jr. (1964). A critical appraisal of periodicities in climate. *Proceedings of the Seminar on Weather and our Food Supply*. Center for Agricultural and Economic development CAED Report 20, Iowa State University, pp. 189–227.

Mitchell, J. M., Jr. (1966). Stochastic models of air–sea interaction and climatic fluctuation. *Proceedings Symposium on Arctic heat budget and atmospheric circulation*, Lake Arrowhead. California RAND Corporation, Memorandum, RM-5233-NSF, pp. 45–74.

Mitchell, J. M. Jr (1970). A preliminary evaluation of atmospheric pollution as a cause of the global temperature fluctuations of the past century. In *Global effects of environmental pollution*, ed. F. S. Singer. Springer–Verlag, New York. Reidel, Holland, pp. 138–55.

Mitchell, J. M. Jr (1972). The natural breakdown of the present interglacial and its possible intervention by human activities. *Quaternary Research*, **2**, 436–45.

Molina, M. J. & Rowland, F. S. (1974). Stratospheric sink for chlorofluoromethanes: chlorine atom-catalysed destruction of ozone. *Nature*, **249**, 810–12.

Möller, F. (1963). On the influence of changes in the CO_2 concentration in air on the radiation balance of the earth's surface and on the climate. *Journal of Geophysical Research*, **68**, 3877–86.

Moncrief, L. W. (1970). The cultural basis for our environmental crisis. *Science*, **170**, 508–12.

Monin, A. S. & Vulis, I. L. (1971). On the spectra of long-period oscillations of geophysical parameters. *Tellus*, **23**, 337–45.

Moore, R. M. & Perry, R. A. (1970). Vegetation. In *Australian Grasslands*, ed. R. M. Moore, Australian National University Press, Canberra, pp. 59–73.

Mörner, N. A. (1969). The late Quaternary history of the Kattegatt Sea and the Swedish west coast. *Sveriges Geologiska Undersökning*, series C, no. 640, 487 pp.

Mörner, N. A. (1973). Climatic changes during the last 35,000 years as indicated by land, sea and air data. *Boreas*, **2**, 33–53.

Muffatti, A. H. J. (1963). Aspects of the subtropical jet stream over Australia. *Proceedings of Symposium Tropical Meteorology*, Rotorua (New Zealand Meteorological Service, Wellington), pp. 72–88.

Müller, H. (1974). Pollenanalytische Untersuchungen und Jahresschichtenzahlungen an der holsteinzeitlichen Kieselgur von Munster-Breloh. *Geologische Jahrbuch*, **A21**, 107–40.

References

Mulvaney, D. J. & Golson, J. (eds.) (1971). *Aboriginal Man and Environment in Australia*. Australian National University Press, Canberra, 389 pp.

Muncey, R. W., Spencer, J. W. & Gupta, C. L. (1970). Methods for thermal calculations using total building response factors. 1st Symposium on use of computers for environmental engineering related to buildings, NBS Gaithersburg (USA). (Paper available, CSIRO, Division of Building Research, Highett Victoria, 3190).

Namias, J. (1970 a). Variations in sea surface temperature in the North Pacific. *Journal of Geophysical Research*, **75**, 565–82.

Namias, J. (1970 b). Climatic anomaly over the United States during the 1960s. *Science*, **170**, 741–3.

Namias, J. (1972). Large-scale and long-term fluctuations in some atmospheric and oceanic variables. In *The Changing Chemistry of the Oceans*, eds. D. Dyrssen and D. Jagner. Nobel Symposium, vol. 20. Wiley, New York, pp. 27–48.

Namias, J. (1973 a). Response of the equatorial counter-current to the subtropical atmosphere. *Science*, **181**, 1244–5.

Namias, J. (1973 b). Thermal communication between the sea surface and the lower troposphere. *Journal of Physical Oceanography*, **3**, 373–8.

National Oceanic and Atmospheric Administration (1973). The influence of weather and climate on United States grain fields: bumper crops or droughts. *A Report to the Administrator, NOAA*, Washington, DC.

Newell, R. E. (1964). Circulation of the upper atmosphere. *Scientific American*, **210** (3), 62–74.

Newell, R. E. (1974 a). Changes in the poleward energy flux by the atmosphere and ocean as a possible cause for ice ages. *Quaternary Research*, **4**, 117–27.

Newell, R. E. (1974 b). An estimate of the interhemispheric transfer of carbon monoxide from tropical general circulation data. *Tellus*, **26**, 103–7.

Newell, R. E., Kidson, J. W., Vincent, D. G. & Boer, G. J. *The General Circulation of the Tropical Atmosphere and Interactions with Extratropical Latitudes*, volume 1, 1972; volume 2, 1974. MIT Press, Cambridge, Massachusetts, and London, England.

Newman, J. E. & Pickett, R. C. (1974). World climates and food supply variations. *Science*, **186**, 877–81.

Newton, C. W. (1972). Southern hemisphere general circulation in relation to global energy and momentum balance requirements. In *Meteorology of the Southern Hemisphere, Meteorological Monographs*, vol. 13 (35). American Meteorological Society, Boston, Massachusetts, pp. 215–46.

Nix, H. A. & Kalma, J. D. (1972). Climate as a dominant control in the biogeography of northern Australia and New Guinea. In *Bridge and Barrier: the Natural and Cultural History of Torres Strait*, ed. D. Walker, *Department of Biogeography and Geomorphology, Publication BG/3*, Australian National University, Canberra, pp. 61–91.

O'Brien, J. J. & Hurlburt, H. E. (1974). Environmental jet in the Indian Ocean: theory. *Science*, **184**, 1075–7.

410

O'Brien, L. F. (1970). Heating degree days for some Australian cities. *Australian Refrigeration, Air Conditioning and Heating Journal*, **24**, 36–7.

Oliver, J. & Kington, J. A. (1970). The usefulness of ship's log books in the synoptic analysis of past climates. *Weather*, **25**, 520–7.

O'Mahony, G. (1961). Time series analysis of some Australian rainfall data. *Bureau of Meteorology Australia, Meteorological Study* 14, 65 pp.

Oort, A. H. & Rasmusson, E. M. (1971). Atmospheric circulation statistics, *NOAA Professional Paper* No. 5. National Oceanic and Atmospheric Administration, Rockville, Maryland, 323 pp.

Öpik, E. J. (1965). Climatic change in cosmic perspective. *Icarus*, **4**, 289–307.

Orheim, O. (1972). A 200-year record of glacier mass balance at Deception Island, Southwest Atlantic Ocean, and its bearing on models of global climatic change. *Institute of Polar Studies*, Report No. 42. Columbus, Ohio, 118 pp.

Osborne, P. J. (1974). An insect assemblage of early Flandrian age from Lea Marston, Warwickshire, and its bearing on the contemporary climate and ecology. *Quaternary Research*, **4**, 471–86.

Ostapoff, F. (1965). Antarctic oceanography. *Biogeography and Ecology in Antarctica*. Junk, The Hague, pp. 97–126.

Oswald, G. K. A. & Robin, G. de Q. (1973). Lakes beneath the Antarctic ice sheet. *Nature*, **245**, 251–4.

Otterman, J. (1974). Baring high-albedo soils by overgrazing: a hypothesized desertification mechanism. *Science*, **186**, 531–3.

Paddock, W. & Paddock, P. (1967). *Famine–1975!* Little Brown, Boston, 276 pp.

Painting, D. J. (1977). A study of some aspects of the climate of the northern hemisphere in recent years. *Meteorological Office Scientific Paper* No. 35 (Bracknell), 25 pp.

Palmén, E. & Newton, C. W. (1969). *Atmospheric Circulation Systems*. Academic Press, New York, 603 pp.

Paltridge, G. W. (1974). Global cloud cover and earth surface temperature. *Journal of the Atmospheric Sciences*, **31**, 1571–6.

Paltridge, G. W. (1975). Global dynamics and climate – a system of minimum entropy exchange. Part I. *Quarterly Journal of the Royal Meteorological Society*, **101**, 475–84.

Paltridge, G. W. (1976). Global dynamics and climate – a system of minimum entropy exchange, Part II. Unpublished manuscript.

Patzelt, G. (1973). Die neuzeitlichen Gletscherschwankungen in der Venedigergruppe (Hohe Tauern, Ostalpen). *Zeitschrift für Gletscherkunde und Glazialgedologie*, **9**, 5–57.

Pearman, G. I. (1975). A correction for the effect of drying of air samples and its significance to the interpretation of atmospheric CO_2 measurements. *Tellus*, **27**, 311–17.

References

Pearman, G. I. (1977). Further studies of the comparability of baseline atmospheric CO_2 measurements. *Tellus*, **29**, 171–81.

Pearman, G. I., Francey, R. J. & Fraser, P. J. (1976). Climatic implications of stable carbon isotopes in tree rings. *Nature*, **260**, 771–3.

Pearman, G. I. & Garratt, J. R. (1972). Global aspects of carbon dioxide. *Search*, **3**, 67–73.

Pearman, G. I. & Garratt, J. R. (1975). Errors in atmospheric CO_2 concentration measurements arising from the use of reference gas mixtures different in composition to the sample air. *Tellus*, **27**, 62–6.

Perry, A. H. & Barry, R. G. (1973). Recent temperature changes due to changes in the frequency and average temperature of weather types over the British Isles. *Meteorological Magazine*, **102**, 73–82.

Peterson, J. A. (1968). Cirque morphology and Pleistocene ice formation conditions in southeastern Australia. *Australian Geographical Studies*, **6**, 67–83.

Peyinghaus, W. (1974). Eine numerische Berechnung der Strahlungsbilanz und der Strahlungsernwärmung im Meridional-Vertikalschnitt: *Bonner Meteorologische, Abhandlungen*, **22**, 1–64.

Pierrehumbert, C. L. (1975). Unpublished study, Australian Bureau of Meteorology.

Pittock, A. B. (1971). Rainfall and the general circulation. In *Proceedings of the International Conference on Weather Modification*, Canberra. American Meteorological Society, pp. 330–8.

Pittock, A. B. (1973 a). Global meridional interactions in stratosphere and troposphere. *Quarterly Journal of the Royal Meteorological Society*, **99**, 424–37.

Pittock, A. B. (1973 b). How important are climatic changes? *Weather*, **27**, 262–71.

Pittock, A. B. (1974 a). Global interactions in stratosphere and troposphere. In *Proceedings of the International Conference on Structure, Composition and General Circulation of the Upper and Lower Atmospheres and Possible Anthropogenic Perturbations*, Melbourne, pp. 716–26.

Pittock, A. B. (1974 b). Ozone climatology, trends and the monitoring problem. In *Proceedings of the International Conference on Structure, Composition and General Circulation of the Upper and Lower Atmospheres and Possible Anthropogenic Perturbations*, Melbourne, pp. 455– 66.

Pittock, A. B. (1975). Climatic change and the patterns of variation in Australian rainfall. *Search*, **6**, 498–504.

Pittock, A. B. (1977). On the causes of local climatic anomalies, with special reference to precipitation in Washington State. *Journal of Applied Meteorology*, **16**, 223–30.

Plass, G. N. (1956). The carbon dioxide theory of climate change. *Tellus*, **8**, 140–53.

Plass, G. N. (1972). The influence of the combustion of fossil fuels on the climate. In *Electrochemistry of Cleaner Environments*, ed. J. O'M. Bockris. Plenum Press, New York, pp. 24–46.

Pollack, J. B., Toon, A. B., Sagan, C., Summers, A., Baldwin, B. & Van Camp, W. (1976). Volcanic explosions and climatic change: A theoretical assessment. *Journal of Geophysical Research*, **81**, 1071–83.

Pond, S. (1975). The exchanges of momentum, heat and moisture at the ocean atmosphere interface. In *Numerical Models of Ocean Circulation*. National Academy of Sciences, Washington DC, pp. 26–36.

Post, J. D. (1974). A study in trade-cycle history: the economic crisis following the Napoleonic Wars. *Journal of Economic History*, **34**, 315–49.

Priestley, C. H. B. (1949). Heat transport and zonal stress between latitudes. *Quarterly Journal of the Royal Meteorological Society*, **75**, 28–40.

Priestley, C. H. B. (1963). Some associations in Australian monthly rainfalls. *Australian Meteorological Magazine*, **41**, 12–21.

Priestley, C. H. B. (1966). The limitation of temperature by evaporation in hot climates. *Agricultural Meteorology*, **3**, 241–6.

Privett, D. W. (1960). The exchange of energy between the atmosphere and the oceans of the Southern Hemisphere. *Geophysical Memoirs, London*, **13** (104), 61 pp.

Pruppacher, H. R. (1973). The role of natural and anthropogenic pollutants in cloud and precipitation formation. In *Chemistry of the Lower Atmosphere*, ed. S. I. Rasool. Plenum Press, New York, pp. 1–67.

Quenzel, H. (1970). Determination of size distribution of atmospheric aerosol particles from spectral solar radiation measurements. *Journal of Geophysical Research*, **75**, 2915–21.

Quinn, W. H. (1971). Late Quaternary meteorological and oceanographic developments in the Equatorial Pacific. *Nature*, **229**, 330–1.

Quinn, W. H. (1974). Monitoring and predicting El Niño invasions. *Journal of Applied Meteorology*, **13**, 825–30.

Raikes, R. (1967). *Water, Weather, and Prehistory*. Arthur Barber, London, 208 pp.

Raine, J. I. (1974). Pollen sedimentation in relation to the Quaternary vegetation history of the Snowy Mountains of New South Wales. Unpublished Ph.D. Thesis, Department of Biogeography and Geomorphology, Australian National University, Canberra.

Rakipova, L. P. & Vishnyakova, O. N. (1973). The influence of carbon dioxide concentrations on the atmospheric thermal regime. *Meteorologiya i gidrologiya*, No. 5, pp. 23–31.

Ramage, C. S. (1975). Preliminary discussion of the meteorology of the 1972–1973 El Niño. *Bulletin of the American Meteorological Society*, **56**, 234–42.

Rao, K. N. & Jagannathan, P. (1963). Climatic changes in India II: Rainfall. In *Changes of Climate*, Arid Zone Research, vol. 20. UNESCO, Paris, pp. 53–66.

References

Rapp, R. R. & Warshaw, M. (1974). Some predicted climatic effects of a simulated Sahara Lake. *RAND Corporation Report* R-1415-ARPA. Santa Monica, California, 31 pp.

Rasmussen, R. A. (1972). What do the hydrocarbons from trees contribute to air pollution. *Journal of Air Pollution Control Association,* **22,** 537–43.

Rasmussen, R. A. (1974). Emission of Biogenic Hydrogen Sulfide. *Tellus,* **26,** 254–60.

Rasmussen, R. A. & Went, F. W. (1965). Material of plant origin in the atmosphere. *Proceedings of the National Academy of Sciences, USA,* **53,** 215–20.

Rasool, S. I & Schneider, S. H. (1971). Atmospheric carbon dioxide and aerosols: effects of large increases on global climate. *Science,* **173,** 138–41.

Ratcliffe, R. A. S. (1974). The use of 500 mb anomalies in long range forecasting. *Quarterly Journal of the Royal Meteorological Society,* **100,** 234–44.

Ratcliffe, R. A. S. & Murray, R. (1970). New lag associations between N. Atlantic Sea temperature and European pressure applied to long range weather forecasting. *Quarterly Journal of the Royal Meteorological Society,* **96,** 226–46.

Raynaud, D. & Delmas, R. (1977). Composition de gaz contenus dans la glace polaire. *Symposium on Isotopes and Impurities in Snow and Ice.* International Union of Geodesy and Geophysics, Grenoble, 1975 (in press).

Raynaud, D. & Lorius, C. (1973). Climatic implications of total gas content in Ice at Camp Century. *Nature,* **243,** 283–4.

Reck, R. A. (1974). Aerosols in the atmosphere: calculation of the critical absorption/backscatter ratio. *Science,* **186,** 1034–6.

Reed, R. J. (1950). The role of vertical motions in ozone-weather relationships. *Journal of Meteorology,* **7,** 263–7.

Reed, R. J. (1965). The present status of the 26 month oscillation. *Bulletin of the American Meteorological Society,* **46,** 374–87.

Reid, G. C., Isaksen, I. S. A., Holzer, T. E. & Crutzen, P. J. (1976). Influence of ancient solar-proton events on evolution of life. *Nature,* **259,** 177–9.

Reitan, C. H. (1974 *a*). A climatic model of solar radiation and temperature change. *Quaternary Research,* **4,** 25–38.

Reitan, C. H. (1974 *b*). Frequencies of cyclones and cyclogenesis for North America, 1951–1970. *Monthly Weather Review,* **102,** 861–8.

Reiter, E. R. (1975). Stratospheric-tropospheric exchange processes. *Reviews of Geophysics and Space Physics,* **13,** 459–74.

Renner, E. D. & Becker, G. E. (1970). Production of nitric oxide and nitrous oxide during denitrification by *Cory nebacterium nephridii. Journal of Bacteriology,* **101,** 821–6.

Revelle, R. & Suess, H. E. (1957). Carbon dioxide exchange between atmosphere and ocean, and the question of an increase of atmospheric CO_2 during the past decades. *Tellus*, **9**, 18–27.

Riley, J. P. & Chester, R. (1971). *Introduction to Marine Chemistry*. Academic Press, London and New York, 465 pp.

Riordan, P. (1970). Weather extremes around the world. *Earth Sciences Lab. Technical Report* 70–45–ES, US Army, Natick, Massachusetts, 38 pp.

Robertson, G. W. (1974). World Weather Watch and wheat. *World Meteorological Organization Bulletin*, **23**, 149–54.

Robin, G. de Q. (1955). Ice movement and temperature distribution in glaciers and ice sheets. *Journal of Glaciology*, **2**, 523–32.

Rockefeller Foundation (1974). Weather and climate change, food production and interstate conflict. Unpublished report of a conference held in New York City, January 1974, 33 pp.

Rodewald, M. (1972). Einige hydroklimatische Besonderheiten des Jahrzents 1961–1970 im Nordatlantik und im Nordpolarmeer. *Deutsche Hydrographische Zeitschrift, Hamburg*, **25**, 97–117.

Rodewald, M. (1973). Der trend des Meerestemperatur in Nordatlantik. *Beilage zur Berliner Wetterkarte* des Instituts für Meteorologie der Freien Universitat Berlin, S029/73, 6 pp.

Rodgers, C. D. (1975). Modelling of atmospheric radiation for climatic studies. In *The Physical Basis of Climate and Climate Modelling*, GARP Publication Series No. 16, ICSU. World Meteorological Organization, pp. 177–80.

Rodhe, H. & Virji, H. (1976). Trends and periodicities in east African rainfall data. *Monthly Weather Review*, **104**, 307–15.

Rotty, R. M. (1974). First estimates of global flaring of natural gas. *Atmospheric Environment*, **8**, 681–6.

Rowland, F. S. & Molina, M. J. (1975). Chlorofluoromethanes in the environment. *Reviews of Geophysics and Space Physics*, **13**, 1–35.

Rowntree, P. R. (1972). The influence of tropical east Pacific Ocean temperatures on the atmosphere. *Quarterly Journal of the Royal Meteorological Society*, **98**, 290–321.

Royal Meteorological Society, 1966. *World Climate from 8000 to 0 BC, Proceedings of International Symposium*, Imperial College, Royal Meteorological Society, London, 229 pp.

Ruddiman, W. F. (1971). Pleistocene sedimentation in the equatorial Atlantic: stratigraphy and faunal paleoclimatology. *Geological Society of America Bulletin*, **82**, 283–302.

Russell, J. S. (1973). Yield trends of different crops in different areas and reflections on the sources of crop yield improvement in the Australian environment. *Journal of Australian Institute of Agricultural Science*, **39**, 156–66.

Russell, P. B. & Grams, G. W. (1975). Application of soil dust optical properties in analytical models of climate change. *Journal of Applied Meteorology*, **14**, 1037–45.

References

Russo, J. A. (1966). The economic impact of weather on the construction industry of the United States. *Bulletin of the American Meteorological Society*, **47**, 967–72.

Rutford, R. H., Craddock, C., Armstrong, R. L. & White, C. (1972). Tertiary glaciation in the Jones Mountains. In *Antarctic Geology and Geophysics*, ed. R. J. Adie. Oslo, Universitetsforlaget, pp. 239–43.

Ryther, J. H. (1966). Geographic variations in productivity. In *The Sea*, ed. M. N. Hill, vol. 2. Wiley, New York, London and Sydney, pp. 347–80.

Salinger, M. J. (1976). New Zealand temperatures since 1300 AD. *Nature*, **260**, 310–11.

Salinger, M. J. & Gunn, J. M. (1975). Recent climatic warming around New Zealand. *Nature*, **256**, 396–8.

Saltzman, B. (1962). Spectral statistics of the wind at 500 mb. *Journal of the Atmospheric Sciences*, **19**, 195–206.

Sancetta, C., Imbrie, J., Kipp, N. G., McIntyre, A. & Ruddiman, W. F. (1972). Climatic record in North Atlantic deep sea core V23-82. *Quaternary Research*, **2**, 363–7.

Sanchez, W. A. & Kutzbach, J. E. (1974). Climate of the American tropics and subtropics in the 1960s and possible comparisons with climatic variations of the last millenium. *Quaternary Research*, **4**, 128–35.

Sawyer, J. S. (1974). Geomagnetism and the tropospheric circulation. *Nature*, **252**, 368–71.

Sax, R. I., Changnon, S. A. Jr, Grant, L. O., Hitschfield, W. F., Hobbs, P. V., Kahan, A. M. & Simpson, J. (1975). Weather modification: where are we now and where should we be going? An editorial overview. *Journal of Applied Meteorology*, **14**, 652–72.

Scanes, P. S. (1974). Climatic design data for use in thermal calculations in buildings – estimated clear sky solar radiation versus measured solar radiation. *Building Science*, **9**, 219–25.

Schermerhorn, V. P. (1967). Relations between topography and annual precipitation in Western Oregon and Washington. *Water Resources Research*, **3**, 707–11.

Schlegel, H. G. (1974). Production, modification and consumption of atmospheric trace gases by microorganisms. *Tellus*, **26**, 11–20.

Schmeltekopf, A. L., Goldan, P. D., Henderson, W. R., Harrop, W. J., Thompson, T. L., Fehsenfeld, F. C., Schiff, H. I., Crutzen, P. J., Isaksen, I. S. A. & Ferguson, E. E. (1975). Measurements of stratospheric $CFCl_3$, CF_2Cl_2 and N_2O, *Geophysical Research Letters*, **2**, 393–6.

Schmidt-Ten Hoopen, K. J. & Schmidt, F. H. (1951). *Kementerian Perhubungan Djawatan Meteorologi dan Geofisik, Djakarta, Verhandelingen*, No. 41.

Schneider, S. H. (1975). On the carbon dioxide–climate confusion. *Journal of the Atmospheric Sciences*, **32**, 2060–6.

416

Schneider, S. H. & Dickinson, R. E. (1974). Climate modelling. *Reviews of Geophysics and Space Physics*, **12**, 447–93.

Schneider, S. H. & Dickinson, R. E. (1975). Climate modelling methodology. In *The Physical Basis of Climate and Climate Modelling*, GARP Publications Series, No. 16, World Meteorological Organization, Geneva, pp. 142–7.

Schneider, S. H. & Mass, C. (1975). Volcanic dust, sun-spots and temperature trends. *Science*, **190**, 741–6.

Schumm, S. A. (1965). Quaternary paleohydrology. In *The Quaternary of the United States*, eds. H. E. Wright & D. G. Frey. Princeton University Press, Princeton, New Jersey, pp. 783–94.

Schutz, C. & Gates, W. L. (1974). Global climatic data for surface, 800 mb, 400 mb: October, *Rand Corporation Report* R-1425-ARPA, March, 1974. Santa Monica, California, 192 pp.

Schutz, K., Junge, C., Beck, R. & Albrecht, B. (1970). Studies of Atmospheric N_2O. *Journal of Geophysical Research*, **75**, 2230–46.

Schwarzbach, M. (1963). *Climates of the Past: An Introduction to Palaeoclimatology*. Van Nostrand, London, 328 pp.

Schwarzbach, M. (1968). Das Klima des rheinischen Tertiars: *Zeitschrift der Deutschen Geologischen Gesellschaft*, **118**, 33–68.

Schwerdtfeger, W. (1958). Climate change in Australia. Can it be documented by temperature records? *Australian Meteorological Magazine*, **23**, 61–3.

Schwerdtfeger, W. & Kachelhoffer, S. J. (1973). The frequency of cyclonic vortices over the southern ocean in relation to the extension of the pack ice belt. *Antarctic Journal of the United States*, **8**, p. 234 only.

Sellers, W. D. (1973). A new global climatic model. *Journal of Applied Meteorology*, **12**, 241–54.

Sellers, W. D. (1974). A reassessment of the effect of CO_2 variations on a simple global climatic model. *Journal of Applied Meteorology*, **13**, 831–3.

Setlow, L. N. & Karpovich, R. P. (1972). Glacial microtextures on quartz and heavy mineral sand grains from the littoral environment. *Journal of Sedimentary Petrology*, **42**, 864–75.

Shackleton, N. J. & Kennett, J. P. (1975 *a*). Paleotemperature history of the Cenozoic and the initiation of Antarctic glaciation: Oxygen and carbon isotope analysis in DSDP sites 277, 279, and 281. In *Initial Reports of the Deep Sea Drilling Project*, vol. 29. Washington, DC, US Government Printing Office, pp. 743–55.

Shackleton, N. J. & Kennett, J. P. (1975 *b*). Late Cenozoic oxygen and carbon isotope changes at D.S.D.P. site 284: implications for glacial history of the northern hemisphere and Antarctica. In *Initial Reports of the Deep Sea Drilling Project*, vol. 29. Washington, DC, US Government Printing Office, pp. 801–7.

Shackleton, N. J. & Opdyke, N. D. (1973). Oxygen isotope and palaeo-

417

References

magnetic stratigraphy of Equatorial Pacific core V28-238: oxygen isotope temperatures and ice volumes on a 10^5 year and 10^6 year scale. *Quaternary Research*, **3**, 39–55.

Shapiro, R. & Ward, F. (1962). A neglected cycle in sunspot numbers. *Journal of the Atmospheric Sciences*, **19**, 506–8.

Shaw, N. (1928). *Manual of Meteorology*, vol. 2, *Comparative Meteorology*. Cambridge University Press, Cambridge, 445 pp.

Shulman, M. D. & Brotak, E. A. (1973). Investigation of the effects of urbanization on precipitation type, frequency, areal and temporal distribution. *Report of the Water Resources Institute*, Rutgers University, New Brunswick, October 1973, 83 pp.

Shumskii, P. A. (1964). *Principles of Structural Glaciology*. Translated from Russian by D. Kraus. Dover, New York, 497 pp.

Shutts, G. J. & Green, J. S. A. (1975). Stationary Planetary Waves and Climate. Paper presented at the Australasian Conference on Climate and Climatic Change, Monash University, Victoria, 7–12 December 1975. (Available from G. J. Shutts, Atmospheric Physics, Imperial College, London.)

Simpson, J. R. & Freney, J. R. (1974). The fate of fertilizer nitrogen under different cropping systems. In *Fertilizers and the Environment*, ed. D. R. Leece. KTR Printing, Croydon, NSW, pp. 27–33.

Simpson, R. & Downey, W. K. (1975). The effect of a warm mid-latitude sea surface temperature anomaly on a numerical simulation of the general circulation of the Southern Hemisphere. *Quarterly Journal of the Royal Meteorological Society*, **101**, 847–67.

Singh, G. (1971). The Indus Valley Culture seen in the context of post-glacial climatic and ecological studies in North-West India. *Archaeology and Physical Anthropology in Oceania*, **6**, 177–89.

Singh, G. (1975). Late Quaternary vegetation and climatic oscillations at Lake George, New South Wales. Paper presented at Australasian Conference on Climate and Climatic Change, Monash University, Melbourne, Australia. Copies available from the author, Department of Biogeography and Geomorphology, ANU, Box 4, Canberra 2600, Australia.

Singh, G., Chopra, S. K., & Singh, A. B. (1973). Pollen-rain from the vegetation of North-West India. *The New Phytologist*, **72**, 191–206.

Singh, G., Joshi, R. D., Chopra, S. K. & Singh, A. B. (1974). Late Quaternary history of vegetation and climate of the Rajasthan desert, India. *Philosophical Transactions of the Royal Society*, **B267**, 467–501.

Singh, G., Joshi, R. D. & Singh, A. B. (1972). Stratigraphic and radio-carbon evidence for the age and development of three salt lake deposits in Rajasthan, India. *Quaternary Research*, **2**, 496–505.

Slatyer, R. O. (1960). Agricultural climatology of the Yass Valley. *CSIRO, Australia, Division of Land Research and Regional Survey, Technical Paper* No. 6, 30 pp.

Slicher van Bath, B. H. (1963). *The Agrarian History of Western Europe* AD *500–800*. (Trans. O. Ordish.) E. Arnold, London, 364 pp.

418

SMIC Report (1971). Inadvertent climate modification. *Report of the Study of Man's Impact on Climate.* MIT Press, Cambridge, Massachusetts, 308 pp.

Smil, V. & Milton, D. (1974). Carbon dioxide – alternative futures. *Atmospheric Environment,* **8,** 1213–23.

Smith, E. J. (1974). Rainfall modification. In *Progress in Australian Hydrology 1965–74.* Australian National Commission for UNESCO, pp. 20–3.

Smith, I. N. (1976). Numerical modelling of glaciers and their responses to mass balance oscillations. Unpublished B.Sc. (Honours) Thesis, Meteorological Department, University of Melbourne. 40 pp.

Sneath, P. H. A. (1970). *Planets and Life.* Thames & Hudson, London, 216 pp.

Snyder, C. T. & Langbein, W. B. (1962). The Pleistocene lake in Spring Valley, Nevada, and its climatic implications. *Journal of Geophysical Research,* **67,** 2385–94.

Sorenson, C. J., Knox, J. C., Larsen, J. A. & Bryson, R. A. (1971). Paleosols and the forest border in Keewatin. *Quaternary Research,* **1,** 468–73.

Stark, L. P. (1965). Positions of monthly mean troughs and ridges in the northern hemisphere, 1949–1963. *Monthly Weather Review,* **93,** 705–20.

Starr, V. P. (1968). *Physics of Negative Viscosity Phenomena.* McGraw-Hill, New York, 256 pp.

Starr, V. P. & Oort, A. H. (1973). Five-year climatic trend for the Northern Hemisphere. *Nature,* **242,** 310–13.

Starr, V. P. & White, R. M. (1954). Balance requirements of general circulation. *Department of Meteorology Final Report,* Part I. Contract No. AF19(122)-153, Massachusetts Institute of Technology, Massachusetts, pp. 186–242.

Stewart, J. (1973). *Rainfall trends in North Queensland.* James Cook University, Geography Department, Monograph Series No. 4, Townsville, 197 pp.

Stewart, J. (1975). Some changes in tropical rainfall regimes, Australia 1870–1970. In *Proceedings of the WMO/IAMAP Symposium on Long-Term Climatic Fluctuations,* Norwich (WMO-No. 421). World Meteorological Organization, Geneva, pp. 255–64.

Stommel, H. (1965). *The Gulf Stream: A Physical and Dynamical Description,* 2nd edn. University of California Press, 248 pp.

Streten, N. A. (1973). Some characteristics of satellite-observed bands of persistent cloudiness over the Southern Hemisphere. *Monthly Weather Review,* **101,** 486–95.

Sverdrup, G. M., Whitley, K. T. & Clark, W. E. (1975). Characterization of California aerosols – II Aerosol size distribution measurements in the Mojave Desert. *Atmospheric Environment,* **9,** 483–94.

Talwani, M., Udintsev, G. *et al.* (1975). Leg 38. *Geotimes,* **20,** 24–6.

Tanaka, M., Weare, B. C., Navato, A. R. & Newell, R. E. (1975). Recent African rainfall patterns. *Nature,* **255,** 201–3.

References

Taulis, E. (1934). De la distribution des pluiès au Chili. *Matériaux pour l'étude des Calamités* (Societie Géographie, Geneva), **33**, 3–20.

Taylor, J. A. (ed.) (1974). *Climatic Resources and Economic Activity: A Symposium.* David & Charles, London, 264 pp.

Taylor, T. G. (1911). *Australia in its Physiographic and Economic Aspects.* Clarendon Press, Oxford, 256 pp.

Taylor, T. G. (1916). The control of settlement by humidity and temperature (with special reference to Australia and the Empire). *Commonwealth Bureau of Meteorology Bulletin* No. 14, Melbourne.

Ten Brink, N. W. & Weidick, A. (1974). Greenland ice sheet history since the last glaciation. *Quaternary Research*, **4**, 429–40.

Terada, K. (1972). Natural disasters in Japan. *Nature*, **240**, 197–9.

Thomas, M. K. (1975). Recent climatic fluctuations in Canada. *Climatology Studies*, No. 28, Environment Canada, Toronto, 92 pp.

Thompson, P. D. (1961). *Numerical Weather Analysis and Prediction.* MacMillan, New York, 166 pp.

Thorarinsson, S. (1967). *The Eruptions of Hekla in Historical Times, a Tephrachronological Study.* H. R. Leiftur, Reykjavik, 183 pp.

Titow, J. (1960). Evidence of weather in the account rolls of the Bishopric of Winchester 1209–1350. *Economic History Review*, **12**, 360–407.

Tolstikov, E. I. (ed.) (1966). *Atlas Antarktiki*, vol. 1. G.U.C.K., Moscow. (English translation in *Soviet Geography: Reviews and Translations*, vol. 8(5–6), American Geographical Society, New York, 1967), 225 pp.

Touchebeuf de Lussigny, P. (1969). Monographie Hydrologique du Lac Tchad. *ORSTOM Service Hydrologique.*

Trenberth, K. E. (1975 a). A quasi-biennial standing wave in the Southern Hemisphere and interrelations with sea surface temperature. *Quarterly Journal of the Royal Meteorological Society*, **101**, 55–74.

Trenberth, K. E. (1975 b). Fluctuations and trends in indices of the Southern Hemispheric circulation. In *Proceedings WMO/IAMAP Symposium on Long-Term Climatic Fluctuations*, Norwich (WMO-No. 421). World Meteorological Organization, Geneva, pp. 243–53.

Trenberth, K. E. (1976). Fluctuations and trends in indices of the Southern Hemispheric circulation. *Quarterly Journal of the Royal Meteorological Society*, **102**, 65–75.

Troughton, J. H., Cord, K. A. & Björkman, O. (1974). Temperature effects on the carbon isotope ratio of C_3, C_4 and CAM plants. *Carnegie Institution Yearbook 1973.* pp. 780–4.

Troup, A. J. (1965). The 'Southern oscillation'. *Quarterly Journal of the Royal Meteorological Society*, **91**, 490–506.

Tsuchiya, M. (1970). Equatorial circulation of the South Pacific. In *Scientific Exploration of the South Pacific*, ed. W. S. Wooster. National Academy of Sciences, Washington, DC. pp. 69–74.

Tucker, G. B. (1975). Climate: is Australia's changing? *Search* (ANZAAS, Sydney), **6**, 323–8.

Tucker, G. B. (1977). An observed relation between the macroscale

420

local eddy flux of heat in the atmosphere and the mean monthly horizontal temperature gradient. *Quarterly Journal of the Royal Meteorological Society*, **103**, 157–68.

Tudor, E. R. (1973). Hydrological interpretations of diatom assemblages in two Victorian Western District crater lakes. Unpublished M.Sc. Thesis, Botany Department, University of Melbourne.

Turekian, K. K. (1965). Some aspects of the geochemistry of marine sediments. In *Chemical Oceanography*, eds. J. P. Riley & G. Skirrow. Academic Press, New York, pp. 81–126.

Twomey, S. (1971). The composition of cloud nuclei. *Journal of the Atmospheric Sciences*, **28**, 377–81.

Twomey, S. (1974). Pollution and the planetary albedo. *Atmospheric Environment*, **8**, 1251–6.

Twomey, S. & Wojciechowski, T. A. (1969). Observations of the geographical variation of cloud nuclei. *Journal of the Atmospheric Sciences*, **26**, 684–8.

Tyson, B. J., Dement, W. A. & Mooney, H. A. (1974). Volatilisation of terpenes from *Salvia mellifera*. *Nature*, **252**, 119–20.

Tyson, P. D. (1971). Spatial variation of rainfall spectra in South Africa. *Annals, Association of American Geographers*, **61**, 711–20.

Tyson, P. D., Dyer, T. G. J. & Mametse, M. N. (1975). Secular changes in South African rainfall: 1880–1972. *Quarterly Journal of the Royal Meteorological Society*, **101**, 817–33.

Uffen, R. J. (1963). Influence of the earth's core on the origin and evolution of life. *Nature*, **198**, 143–4.

Untersteiner, N. (1975). Dynamics of sea ice and glaciers and their role in climatic variations. In *The Physical Basis of Climate and Climate Modelling*, GARP Publications Series No. 16, World Meteorological Organization, Geneva, pp. 206–24.

US Committee for GARP (1975). *Understanding Climatic Change*, United States Committee for the Global Atmospheric Research Programme, National Research Council, National Academy of Sciences, Washington, DC, 239 pp.

US Navy Hydrographic Office (1957). *Oceanographic Atlas of the Polar Seas, Part I, Antarctic*. Washington, DC.

US News and World Report (1976). Food: Potent US Weapon. Interview with Earl L. Butz, Secretary of Agriculture. *US News and World Report*, 16 February, pp. 26–8.

Utterström, G. (1955). Climatic fluctuations and population problems in early modern history. *Scandinavian Economic History Review*, **3**, 3–47.

van der Hammen, T., Wijmstra, T. A. & Zagwijn, W. H. (1971). The floral record of the late Cenozoic of Europe. In *Late Cenozoic Glacial Ages*, ed. K. K. Turekian. Yale University Press, New Haven, pp. 391–424.

Van Donk, J. & Mathieu, G. (1969). Oxygen isotope compositions of foraminifera and water samples from the Arctic Ocean. *Journal of Geophysical Research*, **74**, 3396–407.

421

References

van Geel, B. & van der Hammen, T. (1973). Upper Quaternary vegetational and climatic sequence of the Fuguene area (Eastern Cordillera, Columbia): *Palaeogeography, Palaeoclimatology and Palaeoecology,* **14,** 9–92.

van Loon, H. & Jenne, R. L. (1972). The zonal harmonic standing waves in the Southern Hemisphere. *Journal of Geophysical Research,* **77,** 992–1003.

van Loon, H., Taljaard, J. J., Sasamori, T., London, J., Hoyt, D. V., Labitzke, K. & Newton, C. W. (1972). Meteorology of the Southern Hemisphere. *Meteorological Monographs,* vol. 13, No. 35, 263 pp.

van Loon, H. & Williams, J. (1976 a). The connection between trends of mean temperature and circulation at the surface. Part I – Winter. *Monthly Weather Review,* **104,** 365–80.

van Loon, H. & Williams, J. (1976 b). The connection between trends of mean temperature and circulation at the surface. Part II – Summer. *Monthly Weather Review,* **104,** pp. 1003–11.

Van Mieghem, J. (1956). The energy available in the atmosphere for conversion into kinetic energy. *Beiträge zur Physik der Atmosphäre,* **29,** 129–42.

Van Mieghem, J. (1973). *Atmospheric Energetics.* Clarendon Press, Oxford, 306 pp.

Vasari, Y., Hyvarinen, H. & Hicks, S. (eds.) (1972). *Climatic Change in Arctic Seas during the Last Ten Thousand Years.* Acta Universitatis Ouluensis, Series A, Scientium Rerum Naturum, No. 3, 511 pp.

Veitch, L. G. (1965). The description of Australian pressure fields by principal components. *Quarterly Journal of the Royal Meteorological Society,* **91,** 184–95.

Vernekar, A. D. (1972). Long-period global variations of incoming solar radiation. *Meteorological Monographs,* vol. 12, No. 34, 20 pp.

Veryard, R. G. & Ebdon, R. A. (1961). Fluctuations in tropical stratospheric winds. *Meteorological Magazine, London,* **90,** 125–43.

Vinogradov, O. W. & Lorius, C. (1972). An estimation of the results of snow accumulation measurements on the basis of Soviet and French investigations between Mirny Observatory and Vostok Station in 1964 and 1969. *Soviet Antarctic Expedition Information Bulletin,* No. 83, pp. 237–40.

Vonder Haar, T. H. (1968). Variations of the earth's radiation budget. In *Meteorological Satellite Instrumentation and Data Processing, Final Report,* Contract NASW-65, 1958–68, Department of Meteorology, University of Wisconsin, 179 pp.

Vowinckel, E. & Orvig, S. (1974). Meteorological consequences of natural or deliberate changes in the surface environment. *Department of Meteorology, McGill University, Serial* OSP 3–0233, for Atmospheric Environment Service, Canada, 37 pp.

Vowinckel, E. & van Loon, H. (1957). Das Klima des Antarkischen Ozeans, Part 3, *Archiv für Meteorologie, Geophysik und Bioklimatologie,* Ser. B. **8,** 75–102.

Wagner, A. J. (1971). Long-period variations in seasonal sea-level pressure over the northern hemisphere. *Monthly Weather Review*, **99**, 49–66.

Wahl, E. W. (1968). A comparison of the climate of the eastern United States during the 1830s with the current normals. *Monthly Weather Review*, **96**, 73–82.

Wahl, E. W. & Bryson, R. A. (1975). Recent changes in Atlantic surface temperatures. *Nature*, **254**, 45–6.

Wahl, E. W. & Lawson, T. L. (1970). The climate of the mid-nineteenth century United States compared to the current normals. *Monthly Weather Review*, **98**, 259–65.

Walker, D. (ed.) (1972). *Bridge and Barrier: the Natural and Cultural History of Torres Strait*. Research School of Pacific Studies, Australian National University, Publication BG/3 (1972), 437 pp.

Walker, G. T. & Bliss, E. W. (1932). World Weather V. *Memoirs of the Royal Meteorological Society*, **4** (36), 53–84.

Walker, G. T. & Bliss, E. W. (1937). World Weather VI. *Memoirs of the Royal Meteorological Society*, **4** (39), 119–39.

Walsh, P. J. (1976). Energy requirements for Australian dwellings for heating and cooling. Submitted to *Journal Australian Institute of Refrigeration, Air Conditioning and Heating*.

Warner, J. (1968). Supersaturation in natural clouds. *Journal de Recherches Atmosphériques*, **3**, 233–7.

Warner, J. (1974). Rain enhancement: a review. In *Proceedings of the WMO/IAMAP Scientific Conference on Weather Modification* (WMO-No. 399). World Meteorological Organization, Geneva, pp. 43–50.

Weare, B. C., Temkin, R. L. & Snell, F. M. (1974). Aerosol and climate: some further considerations. *Science*, **186**, 827–8.

Weaver, F. M. (1973). Pliocene paleoclimatic and paleoglacial history of East Antarctica recorded in deep sea piston cores. *Florida State University, Sedimentology Research Laboratory, Report* 36, 142 pp.

Webb, P. N. (1972). Wright Fjord, Pliocene marine invasion of an Antarctic dry valley. *Antarctic Journal of the US*, **7**, 225–32.

Webb, T. III & Bryson, R. A. (1972). The Late- and Post-Glacial sequence of climatic events in Wisconsin and East-Central Minnesota; quantitative estimates derived from fossil pollen spectra by multivariate statistical analysis. *Quaternary Research*, **2**, 70–115.

Webber, P. J. & Andrews, J. T. (eds.) (1973). Lichenometry: dedicated to the memory of the late Roland E. Beschel. *Arctic and Alpine Research*, **5**, 293–424.

Webster, P. J. & Curtin, D. G. (1975). Interpretations of the EOLE Experiment II. Spatial variations of transient and stationary modes. *Journal of the Atmospheric Sciences*, **32**, 1848–63.

Webster, P. J. & Keller, J. L. (1974). A strong long period tropospheric and stratospheric rhythm in the Southern hemisphere. *Nature*, **248**, 212–13.

References

Weertman, J. (1957). On the sliding of glaciers. *Journal of Glaciology*, 3, 33–8.

Weertman, J. (1961). Stability of ice-age ice sheets. *Journal of Geophysical Research*, 66, 3783–92.

Weinberg, A. M. (1972). Science and trans-science. *Minerva*, 10, 209–22.

Wendland, W. M. & Bryson, R. A. (1974). Dating climatic episodes of the Holocene. *Quaternary Research*, 4, 9–24.

Wen-Hsiung, S. (1974). Changes in China's climate. *Bulletin of American Meteorological Society*, 55, 1342–51.

Went, F. W. (1964). The nature of Aitken condensation nuclei in the atmosphere. *Proceedings of the National Academy of Sciences of the USA*, 51, 1259–67.

Went, F. W. (1966). On the nature of Aitken condensation nuclei. *Tellus*, 18, 549–56.

Wetherald, R. T. & Manabe, S. (1972). Response of the joint ocean–atmosphere model to the seasonal variation of solar radiation. *Monthly Weather Review*, 100, 42–59.

Wetherald, R. T. & Manabe, S. (1975). The effects of changing of the solar constant on the climate of a general circulation model. *Journal of the Atmospheric Sciences*, 32, 2044–59.

Weyl, P. K. (1968). The role of the oceans in climatic change: a theory of the ice ages. *Meteorological Monographs*, vol. 8, No. 30, 37–62.

Whitaker, G. L. (1971). Changes in the elevation of Great Salt Lake caused by man's activities in the drainage basin. *United States Geological Survey Professional Paper*, 750D (D187–9).

Wiesnet, D. R. & Matson, M. (1975). Monthly winter snowline variation in the northern hemisphere from satellite records, 1966–1975. *NOAA Technical Memoirs*, NESS 74. Washington, DC, 21 pp.

Wijmistra, T. A. (1975). Palynology and paleoclimatology of the last 100,000 years. In *Proceedings of the WMO/IAMAP Symposium on Long-Term Climatic Fluctuations*, Norwich (WMO-No. 421). World Meteorological Organization, Geneva, pp. 5–20.

Willett, H. C. (1974). Recent statistical evidence in support of the predictive significance of solar-climatic cycles. *Monthly Weather Review*, 102, 679–86.

Williams, G. E. (1975). A possible relation between periodic glaciation and the flexure of the galaxy. *Earth and Planetary Science Letters*, 26, 361–9.

Williams, J. (1975). The influence of snow cover on the atmospheric circulation and its role in climatic change: an analysis based on results from the NCAR Global Circulation Model. *Journal of Applied Meteorology*, 14, 137–52.

Williams, L. D. (1974). Computer simulation of glacier mass balance throughout an ablation season. *Proceedings of the Western Snow Conference*, Anchorage, Alaska, pp. 23–8.

Williams, L. D. (1975). The variation of corrie elevation and equilibrium line altitude with aspect in eastern Baffin Island, NWT, Canada. *Arctic and Alpine Research*, 7, 169–81.

Williams, L. D. (in press). The late Neoglacial (Little Ice Age) glaciation level on Baffin Island and the significance for ice sheet inception. *Paleogeography, Paleoclimatology, Palaeoecology* (in press).

Wilson, A. T. (1964). Origin of ice ages: an ice shelf theory for Pleistocene glaciation. *Nature*, **201**, 147–9.

Wilson, A. T. (1969). The climatic effects of large-scale surges of ice sheets. *Canadian Journal of Earth Science*, **6**, 911–18.

Wilson, J. W. & Atwater, M. A. (1972). Storm variability over Connecticut. *Journal of Geophysical Research*, **77**, 3950–6.

Winkler, P. (1974). Die Relative Zusammensetzung des Atmosphärischen Aerosols in Stoffgruppen. *Meteorologische Rundschau*, **27**, 129–36.

Winstanley, D. (1973 a). Recent rainfall trends in Africa, the Middle East and India. *Nature*, **243**, 464–5.

Winstanley, D. (1973 b). Rainfall patterns and the general atmospheric circulation. *Nature*, **245**, 190–4.

Winstanley, D., Emmett, B. & Winstanley, G. (1974). Climatic changes and the world food supply. Paper presented to American Society of Civil Engineers Conference, Biloxi, Mississippi. Typescript 38 pp.

Woldstedt, P. (1969). *Handbuch der Stratisgraphischen Geologie*, vol. 2. Stuttgart, 263 pp.

Wolfe, J. A. & Hopkins, D. M. (1967). Climatic changes recorded by Tertiary land floras in northwestern North America. In *Tertiary Correlations and Climatic Changes in the Pacific*. Pacific Science Congress, 11th, Tokyo, pp. 67–76.

Wollin, G., Ericson, D. B., Ryan, W. B. J. & Foster, J. H. (1971). Magnetism of the earth and climatic changes. *Earth Planetary Science Letters*, **12**, 175–83.

Wollin, G., Ericson, D. B. & Wollin, J. (1974). Geomagnetic variations and climatic changes 2,000,000 BC–1970 AD. *Colloques Internationaux du CNRS*, No. 219 – Les Méthodes Quantitatives d'Etude des Variations du Climat au Cours du Pléistocène, pp. 273–88.

Woodcock, A. H. (1953). Salt nuclei in marine air as a function of altitude and wind force. *Journal of Meteorology*, **10**, 362–71.

Wooldridge, D. D. (1971). Revised objectives for design of urban stormwater systems in the Puget Sound Region. *Report for US Department of Interior, OWRR Project* B-035-WASH, University of Washington Institute of Forest Products, Seattle, September 1971.

Wooster, W. S. & Guillen, O. (1974). Characteristics of El Niño in 1972. *Journal of Marine Research*, **32**, 387–404.

World Meteorological Organization (1970). Building climatology, *WMO Technical Note* No. 109, Geneva, 260 pp.

World Meteorological Organization (1974 a). WMO Operations Manual for sampling and analysis techniques for chemical constituents in air and precipitation. *World Meteorological Organization* No. 299, Geneva, 56 pp.

World Meteorological Organization (1974 b). Applications of meteorology to economic and social development. *WMO Technical Note* No. 132, Geneva, 130 pp.

World Meteorological Organization (1975 a). *Proceedings of the WMO/*

References

IAMAP Symposium on Long-Term climatic fluctuations, Norwich (WMO-No. 421). World Meteorological Organization, Geneva, 503 pp.

World Meteorological Organization (1975 *b*). Drought and agriculture. *WMO Technical Note* No. 138, Geneva, 127 pp.

World Meteorological Organization (1976). Statement on climate change. Adopted by World Meteorological Organization Executive Committee, June.

Wright, H. E. (1968). Climatic change in Mycenaean Greece. *Antiquity*, **42**, 123–7.

Wright, P. B. (1971). Spatial and temporal variations in seasonal rainfall in South-Western Australia. *Miscellaneous Publication 71/1, Institute of Agriculture*, University of Western Australia, 89 pp.

Wright, P. B. (1974 *a*). Seasonal rainfall in South-Western Australia and the general circulation. *Monthly Weather Review*, **102**, 219–32.

Wright, P. B. (1974 *b*). Temporal variations in seasonal rainfalls in South-Western Australia. *Monthly Weather Review*, **102**, 233–43.

Wright, P. B. (1975). An index of the Southern Oscillation. *Climatic Research Unit Report*, CRU RP4, Norwich.

Wyrtki, K. (1973 *a*). An equatorial jet in the Indian Ocean. *Science*, **181**, 262–4.

Wyrtki, K. (1973 *b*). Teleconnections in the equatorial Pacific Ocean. *Science*, **180**, 66–8.

Yamamoto, T. (1970). On the climatic change in XV and XVI centuries in Japan. *Japanese Progress in Climatology*, pp. 14–28.

Yamamoto, T. (1972). On the nature of the climatic change in Japan since the 'Little Ice Age' around 1800 AD. *Japanese Progress in Climatology*, pp. 97–110.

Yevjevich, V. (1971). Stochasticity in geophysical and hydrological time series. *Nordic Hydrology*, **2**, 217–42.

Young, S. B. & Schofield, E. K. (1973). Pollen evidence for late Quaternary climate changes on Kerguelen Islands. *Nature*, **245**, 311–13.

Zillman, J. W. (1970). Sea surface temperature gradients south of Australia. *Australian Meteorological Magazine*, **18**, 22–30.

Zillman, J. W. (1972 *a*). A study of some aspects of the radiation and heat budgets of the southern hemisphere oceans. *Meteorological Study* No. 26, Australian Government Publishing Service, Canberra, 562 pp.

Zillman, J. W. (1972 *b*). Solar radiation and sea–air interaction south of Australia. American Geophysical Union, *Antarctic Research Series*, **19**, 11–40.

Zotikov, I. A. (1961). Thermal regime of the ice sheet in Central Antarctica. *Antarktika (Moscow) Commission Reports*, Israel Program for Scientific Translations.

Zotikov, I. A. (1963). Temperatures inside Antarctic glaciers. *Antarktika (Moscow) Commission Reports*, Israel Program for Scientific Translations, pp. 64–110.

426

Author index

Author index

428

Author index

431

Author index

432

Subject index

Individual glaciers, lakes and mountains are listed under these headings

434

Subject index

baseline monitoring stations, 284
Bassiana, 112, 122–4
Bass Strait
  CO_2 concentrations, 285, 286–9
  effect of sea-level changes on, 112, 123
Bellingshausen Sea, 51, 59
Benguela Current, 35, 36, 190
Bering Strait, 7, 127, 295
biomass, 13, 15
  and CO_2, 273–4, 290–1
biosphere, 246
  and climate, 269–82, 288
  evolution of, 206, 211
  response to increasing CO_2, 291
bioturbation, 76
blocking, 237, 243
bogs, 124
Bölling interstadial, 125
Bombay, 182
bottom water formation, 37, 51
boundary conditions
  for climate modelling, 245, 250, 355–6
  ice-age, 70–1, 108
boundary layer, 236, 247
Brazil
  Current, 35, 36
  rainfall, 185–90
  sea-level record, 202
Brisbane, Queensland, 149, 196, 197
British Columbia, 172
British Isles, 159, 165
British Meteorological Service, short-term
  climatic forecasting, 339–46
Brúarjökull glacier, Iceland, 228–30
Brunhes palaeomagnetic epoch, 62, 64, 68,
  69, 216
buildings, thermal performance of, 301–8
Byrd Station, Antarctica, 78, 79, 80, 125

California
  historical data, 152
  palaeoclimate, 67
  White Mountains, tree rings in, 157
Cambrian geological period, 206
Cameron Valley, New Zealand, 158
Camooweal, Queensland, 198
Camp Century, Greenland, 79, 80, 81, 125,
  127, 155, 162, 163, 212–15, 217
Camperdown, Victoria, rainfall, 102–3
Canada, temperature fluctuations, 159, 182
  tree line, Arctic, 321
Canadian archipelago, 125
  see also Baffin Island, Ellesmere Island
Canary Current, 189, 190
Cape Town, South Africa, 182
capital works and equipment, changes in,
  309

carbonate
  abundance, 61, 62, 63, 64, 65, 68, 69
  and alkalinity of ocean, 274
  compensation depth, 58
  dissolution, 65
  sedimentation, 58, 59
carbon dioxide, atmospheric, 246, 263,
  273–6, 282–93
  annual cycle, 286–8
  and biomass, 273–4, 288
  climatic effects of, 247–8, 275–6, 282–3,
  291–3
  concentration increase, 274–5, 284–5
  and fossil fuel combustion, 275, 285,
  289–91
  fluxes of, 286–9
  and ocean, 274–6, 288
  predicted increase, 275–6, 289–91
  in primeval atmosphere, 204
  withdrawal in upper cretaceous, 207–8
carboniferous geological period, 206–7
Caribbean
  deep-sea cores, 68–9, 212–15
  region, 185–90
catastrophic climatic change, 124–34, 202,
  217, 228, 320, 334–8, 351–2, 354
  see also Antarctic ice-surges
causes of climatic change, 211–25, 256–68,
  368–70
  of abrupt change, 129–34
  see also theories of climatic change
cave
  deposits, 117
  speleothems, 212–13, 214, 217
Cenozoic, 53–68, 208–11
  geological time units of, 54
  palaeotemperature record, 56
Central America, 185–90
Central England temperatures
  spectrum, 161–2
  variation of, 156–7
  and volcanism, 256–62
Center for Climatic and Environmental
  Assessment, 332
centres of action, 164
Chad, Africa, 186
  see also Lake Chad, 97, 99
chalk, 208
Challenger Plateau, 58, 59, 62
characteristic patterns, 169, 171–4, 175, 176,
  178
Charlotte Pass, New South Wales, 148
Chile
  dendroclimatology, 158, 164
  La Serena, 161
  precipitation in, 158, 160–1
  Santiago, 158, 164, 182

436

Subject index

451